THREE PHASE PARTITIONING

Applications in Separation and Purification of Biological Molecules and Natural Products

THREE PHASE PARTITIONING

Applications in Separation and Purification of Biological Molecules and Natural Products

Edited by

MUNISHWAR NATH GUPTA

Former Emeritus Professor, Department of Biochemical Engineering and Biotechnology, Indian Institute of Technology, Hauz Khas, New Delhi, India

IPSITA ROY

Department of Biotechnology, National Institute of Pharmaceutical Education and Research (NIPER), S.A.S. Nagar, Punjab, India

ELSEVIER

Elsevier
Radarweg 29, PO Box 211, 1000 AE Amsterdam, Netherlands
The Boulevard, Langford Lane, Kidlington, Oxford OX5 1GB, United Kingdom
50 Hampshire Street, 5th Floor, Cambridge, MA 02139, United States

British Library Cataloguing-in-Publication Data
A catalogue record for this book is available from the British Library

Library of Congress Cataloging-in-Publication Data
A catalog record for this book is available from the Library of Congress

ISBN: 978-0-12-824418-0

For Information on all Elsevier publications visit our website at
https://www.elsevier.com/books-and-journals

Publisher: Susan Dennis
Acquisitions Editor: Kathryn Eryilmaz
Editorial Project Manager: Allison Hill
Production Project Manager: Bharatwaj Varatharajan
Cover Designer: Greg Harris

Typeset by Aptara, New Delhi, India

The figure in the central panel on the cover is reproduced from Chapter 4 with permission from the author.

Contents

Contributors

Rintu Banerjee
Microbial Biotechnology and Downstream Processing Laboratory, Agricultural and Food Engineering Department, Indian Institute of Technology, Kharagpur, West Bengal, India

R. Borbás
Laboratory of Interfaces and Nanostructures, Institute of Chemsitry, Eötvös Loránd University, Budapest, Hungary

Theresa Helen Taillefer Coetzer
Biochemistry, School of Life Sciences, University of KwaZulu-Natal (Pietermaritzburg campus), Scottsville, South Africa

Clive Dennison
Biochemistry Department, University of KZN, Pietermaritzburg, South Africa

Yonca Duman
Faculty of Arts and Sciences, Department of Chemistry, Kocaeli University, Umuttepe Campus, İzmit-Kocaeli, Turkey

Lauren E-A Eyssen
Biochemistry, School of Life Sciences, University of KwaZulu-Natal (Pietermaritzburg campus), Scottsville, South Africa

Mohammed Gagaoua
Food Quality and Sensory Science Department, Teagasc Ashtown Food Research Centre, Ashtown, Dublin, Ireland

James Philip Dean Goldring
Biochemistry, School of Life Sciences, University of KwaZulu-Natal (Pietermaritzburg campus), Scottsville, South Africa

Lohit Kumar Srinivas Gujjala
Advanced Technology Development Centre, Indian Institute of Technology, Kharagpur, West Bengal, India

Munishwar Nath Gupta
Former Emeritus Professor, Department of Biochemical Engineering and Biotechnology, Indian Institute of Technology, Hauz Khas, New Delhi, India

Vijay Kumar Garlapati
Department of Biotechnology and Bioinformatics, Jaypee University of Information Technology, Waknaghat, Himachal Pradesh

É. Kiss
Laboratory of Interfaces and Nanostructures, Institute of Chemsitry, Eötvös Loránd University, Budapest, Hungary.

S. P. Jeevan Kumar
Seed Biotechnology Laboratory, ICAR-Indian Institute of Seed Science, Mau Uttar Pradesh, India; ICAR-Directorate of Floricultural Research, Pune, Maharashtra, India

John H.T. Luong
School of Chemistry, University College Cork, Cork, Ireland

Sandesh J. Marathe
Department of Food Engineering and Technology, Institute of Chemical Technology, Mumbai, India

Sujata S. Patil
Department of Chemical Engineering, Institute of Chemical Technology, Mumbai, India

Virendra K. Rathod
Department of Chemical Engineering, Institute of Chemical Technology, Mumbai, India

Ipsita Roy
Department of Biotechnology, National Institute of Pharmaceutical Education and Research (NIPER), S.A.S. Nagar, Punjab, India

Nirali N. Shah
Centre for Technology Alternatives for Rural Areas, IIT Bombay, Mumbai, India

Rekha S. Singhal
Department of Food Engineering and Technology, Institute of Chemical Technology, Mumbai, India

Jing-Kun Yan
School of Food & Biological Engineering, Institute of Food Physical Processing, Jiangsu University, Zhenjiang, China

Preface

In the era of high tech, three-phase partitioning (TPP) is an outlier among separation techniques in chemical and biochemical sciences.

The first two chapters of this book recall how this simple technique found its applications in isolation and purification of different kinds of molecules from diverse natural sources. We tend to overlook that our understanding of the well-known interaction of sulphate ion with proteins is still not complete. Add the synergy of t-butanol with sulphate ion to this interaction and the situation becomes quite complicated.

Chapters 2–4 discuss what we know so far about the "mechanism" of this process. Other chapters though have also alluded to the ideas discussed in these chapters.

The applications of TPP to both upstream and downstream processes in protein production continue to dominate the range of applications (Chapters 5–10). Polysaccharides (Chapter 11) and oils (Chapters 12 and 13) form two other major classes of molecules for which TPP has been described as a useful technique.

While the word "partitioning" is a part of its name, TPP is a precipitation technique as well. Combining it with the use of enzymes, ultrasonics, microwaves, etc., has given rise to many powerful variants. This has been pointed out in many chapters (1, 2, 4, 10) but Chapter 14 discusses it in a somewhat holistic fashion.

All the chapters include a fair comparison of TPP with other alternative techniques available for the respective applications. In case of proteins and oils, separate chapters (2, 12 and 11) provide overviews of these other options.

We thank all the authors who even in this unusual year (of COVID-19 pandemic) contributed to the book; it has been a very cooperative venture! We will also like to thank two members of our former group (at IIT Delhi), Dr. Aparna Sharma and Dr. Kalyani Mondal, who explored the applications of TPP in the areas of polysaccharides and oils. That is how we became fans of this technique and that fascination has endured over the years to result in this book.

We thank Dr. Eryilmaz (Acquisition Editor) who us encouraged to work on this book and was always willing to help us patiently. Apart from rest of the team (including Ms. Hill, Project Manager) at Elsevier, we will especially like to mention Mr. Bharatwaj Varatharajan (Production Manager) who painstakingly worked on the final draft; always a pleasure to work with a true professional like him!

We hope that this book will encourage others to try this magic of salt and a solvent coming together which creates this versatile technique called three phase partitioning. Maybe we will see its applications in many more areas.

Editors

The journey in understanding interactions of salts and solvents with proteins continues!

"One of the writers of this text asked his laboratory instructor in biochemistry at a well-known medical school; how does ammonium sulphate salt out the proteins? The instructor smiled knowledgeably and said, I would be glad to explain it to you, but I don't have time just now. The question remains unanswered, in spite of the extensive research, calculation, and speculation that has been carried out by many workers in the past century……..Hofmeister noted in the last century that there are large differences in the effects of different salts on the solubility of proteins…"

From: RH Abeles, PA Fry and WP Jencks (1992) *Biochemistry.* Jones and Bartlett Publishers, Boston, MA, p. 197.

"Thirteen waters were found to be present in the first solvation shell, and the residence time of these waters has been calculated to be 23 ps"

From: WR Cannon, BM Pettitt and JA McCammon (1994) Sulphate anion in water: model structural, thermodynamic, and dynamic properties. *J Phys Chem* 98, 6225-6230.

Hofmeister series continues to be a matter of investigation!

EE Bruce, HI Okur, S Stegmaier, CI Drexler, BA Rogers, NFA van der Vegt, S Roke and PS Cremer (2020) Molecular mechanism for the interactions of Hofmeister cations with macromolecules in aqueous solution. *J Am Chem Soc* 142, 19094-19100.

C Ota, Y Fukuda, S Tanaka and K Takano (2020). Spectroscopic evidence of the salt-induced conformational change around the localized electric charges on the protein surface of Fibronectin Type III. *Langmuir* 36, 14243-14254.

..And our journey in understanding complex interactions of salts and solvents with proteins continues…..

CHAPTER 1

Three phase partitioning: some reminiscences, some science

Clive Dennison
Biochemistry Department, University of KZN, Pietermaritzburg, South Africa

Chapter outline

1.1 The origin of the TPP method

Three phase partitioning is a protein concentration and separation method, discovered by Rex Lovrien of the University of Minnesota. Lovrien had the idea that, in vivo, enzymes must operate in a mixture of salts and organic compounds, so they embarked on a systematic study of the effects of these variables (Tan and Lovrien, 1972). With the particular combination of ammonium sulfate and tertiary butanol (2-methylpropan-2-ol) (t-BuOH), the solution split into three phases, with protein precipitated in a phase between the lower aqueous and upper t-BuOH phases. Lovrien's group later developed this as a general protein isolation method (Lovrien et al. 1987), which they called three phase partitioning or TPP.

In December 1984, I went on sabbatical leave, from the Biochemistry Department of the University of Natal, in Pietermaritzburg, South Africa, to the laboratory of Irv Liener, in the Department of Biochemistry at the University of Minnesota in St Paul, USA. Rex Lovrien occupied an adjoining lab. and he extended a hand of friendship to me, in the form of an invitation to go cross-country skiing - something unheard of in South Africa.

At the end of my sabbatical period, Rex approached me with a sheaf of papers, saying, "You might be interested in this new method we are working on". When I got home, I examined the papers and found that they described the method that we now know as three phase partitioning, and how it had been discovered. Back home, we immediately tested the method on the proteinases that we were working on and were amazed at the results. For example, we were able to isolate cathepsin D in a single day (Jacobs et al. 1989) and in some cases more than 100 percent recoveries of activity were obtained. In

Three Phase Partitioning
DOI: https://doi.org/10.1016/B978-0-12-824418-0.00016-3

an attempt to better understand the method, I returned to Minnesota for further discussions with Lovrien, which led to our joint publication (Dennison and Lovrien 1997).

Rex Eugene Lovrien, was born on a farm at Eagle Grove, Iowa, on 25 January 1928 and attended a one-room country school. In 1939 the family moved to Sac City, Iowa, where he graduated from high school. At 17, he enlisted in the U.S. Army, just before V-J Day, and served with occupying forces in Asia. He subsequently graduated from the University of Minnesota, received a Ph.D. from the University of Iowa in 1958, and was a post-doctoral fellow at Yale University, the University of Indiana, and the University of Minnesota. In 1965, he became a professor in the Biochemistry Department at the University of Minnesota, St Paul campus, where he worked until his death on 24 March 2003, of congestive heart failure.

I found Rex an interesting person to talk to. It emerged that, despite coming from opposite ends of the earth, we had much in common, as we were both brought up as "farm boys" and we both had experience of operating John Deere Model D tractors.

Rex was a 'loner' and thought his own very original thoughts. For example, he didn't leap on bandwagons and compete for research grants: instead he built micro-calorimeters in a corner of his lab and used the proceeds from the sale of these to fund his research. This gave him complete freedom to pursue anything that intrigued him. His knowledge of physical chemistry, exemplified by the micro-calorimeters, was greater than mine, and many of my questions were aimed at getting me up to speed and understanding what he was saying.

1.2 How does TPP work?

Since our early publication on the use of TPP for protein concentration and purification, many researchers have applied the method to their own problems and have been surprised by the method's simplicity and effectiveness. Ward (2009), for example, who worked on green fluorescent protein, commented that, "Going back to earlier days, it often took us six months to get to the same stage of purity that we can now achieve in just an hour or two with TPP".

Many researchers were also surprised at the fact that it apparently flies in the face of conventional wisdom concerning protein isolations, for example, that organic solvents denature proteins at room temperature and that, in any step, less than 100 percent recovery of activity is to be expected. In fact, t-BuOH does not denature proteins at room temperature (so TPP can be done at room temperature) and more than 100 percent recovery of activity is commonly observed in TPP.

Surprise at the simplicity and effectiveness of TPP leads to a curiosity as to how it achieves what it does, and several authors have addressed this question (Chew, et al. 2018; Dennison. and Lovrien 1997; Dennison, 2011; Kiss et al. 1998, Ward 2009). Dennison and Lovrien (1997) explained the mechanism of TPP in terms of the kosmotropic effects of both t-BuOH and ammonium sulfate. A common property of agents that stabilize proteins (kosmotropes) is that they are excluded from the protein interior (Timasheff, et al. 1982). Ammonium sulfate is well-known as a protein stabilizer and this is attributed to the fact that the large, hydrated, sulfate ion cannot penetrate the interior of proteins. The hydrated sulfate ion is relatively large, having 13 water molecules in its first hydration layer and possibly more in a second layer (Cannon et al. 1994). t-BuOH is described as a "differentiating" cosolvent (one that emphasizes the difference between proteins) as opposed to C1 and C2 alcohols, which are described as "leveling" cosolvents (which tend to obscure the differences between proteins). Uniquely among common organic solvents, t-BuOH stabilizes proteins rather than denaturing them. In the case of t-BuOH, this may be attributed to its "bushy" structure, that prevents it from penetrating. Ammonium sulfate and t-BuOH thus both stabilize protein structure and may have synergistic effects in this regard.

It is envisaged that protein, suspended in a water/t-BuOH mixture, will equilibrate with the ambient proportions of the two solvents. If ammonium sulfate is added to the solution, the sulfate ion in particular will become hydrated, growing larger in the process and effectively sequestering water, making less available to the protein and t-BuOH. As less water becomes available, the protein will equilibrate with the new proportions of water and t-BuOH. Eventually, the stage will be reached when there is insufficient water available to keep the protein in solution and it will begin to precipitate. However, at this point it will be equilibrated with a preponderance of t-BuOH so, when it comes out of solution, it will float on the aqueous layer. Addition of more ammonium sulfate will result in precipitation of more, and perhaps different, protein and this forms the basis of a fractionation of proteins by TPP.

An advantage of TPP over conventional salting out with ammonium sulfate, is that in conventional salting out as the amount of salt added is increased, the density of the salt solution increases until a stage is reached where there is little density difference between the protein being salted out and the salt solution. At this stage it is not possible to collect the precipitate protein by centrifugation. In TPP, however, as the aqueous solution density increases with the addition of salt, the precipitated protein will float more and more easily.

1.3 Inhibition of enzyme activity by t-BuOH

In a water/t-BuOH mixture, enzyme activity generally declines with an increasing proportion of t-BuOH, reaching a maximal reduction (nearly all enzyme activity inhibited) at about 30 percent-BuOH (Dennison et al., 2000). This inhibition is reversible, upon removal of the t-BuOH by TPP. Our hypothesis is that t-BuOH increasingly makes the enzyme structure rigid and thus unable to function. The fact that the enzymes tested were all completely inhibited at 30 percent t-BuOH, suggests that 30 percent, or higher, might be an appropriate concentration of t-BuOH to use in TPP

Homogenization of liver tissue (and consequent breakdown of compartments) in a buffer, without t-BuOH, enables the lysosomal enzyme, cathepsin L, to interact with its cytoplasmic inhibitor, stefin B, forming a proteolytically active, though artefactual, complex (Pike et al. 1992). When liver tissue was homogenized in the presence of 30 percent t-BuOH, no evidence of this complex was found, supporting the contention that it may be an isolation artefact and that its formation is inhibited by 30 percent t-BuOH (Dennison, et al. 2000). This may be a model of other possible interactions that might take place in homogenization, as a result of the breakdown of compartmentalization, and suggests that routine homogenization in ≥ 30 percent t-BuOH may be advantageous in obviating the formation of homogenization artefacts.

1.4 Enhanced activity

Usually, in a protein isolation, there is a loss of activity during each step and 100 percent recovery of activity is not expected. However, in TPP, a recovery of greater than 100 percent activity is commonly found (Roy and Gupta, 2004; Gagaoua and Hafid, 2016). Singh et al. (2001) investigated the origin of the enhanced activity of TPP modified proteinase K (a 210 percent increase, in this case). The overall structure of the TPP modified enzyme was similar to the original and the hydrogen bonding of the catalytic triad was intact. The greatest difference was that the TPP-modified enzyme had a temperature factor (B-factor) more than twice that of the original enzyme, suggesting a much more flexible structure (Sun, et al. 2019).

TPP appears to be able to refold reversibly denatured proteins and to extract active proteins from inclusion bodies (Raghava et al. 2008; Roy et al. 2004; Sardar et al. 2007). This suggests that at least some of the enhanced activity observed in TPP could be due to refolding of partially denatured proteins.

Kumar et al. (2015) report that TPP increases the stability and decreases the activity of a lipase from *Thermomyces lanuginosa*. However, the activity of the TPP-treated enzyme was increased above that of the untreated one, if it was assayed in the presence of a detergent.

1.5 Other molecules

Some other molecules purified by TPP are listed in Table 1.1. TPP generally works well with proteinases and is increasingly being applied to other types of macromolecules.

1.6 TPP doesn't work in all cases

TPP doesn't work with all proteins and these exceptions might be instructive.

For reasons that have not yet been elucidated, TPP denatures hemoglobin. However, this can be useful for isolating enzymes from sources in which there is a preponderance of hemoglobin, e.g. cathepsin D from spleen (Jacobs, et al. 1989), and carbonic anhydrase I and II, catalase and superoxide dismutase from human erythrocytes (Pol et al. 1990).

Table 1.1 Some proteins and other molecules recently isolated by TPP.

Molecule isolated	Reference
Proteolytic enzymes	Gagaoua and Hafid (2016)
Ficin from *Ficus carica*	Gagaoua, M. (2014)
Zingipain from *Zingeber officinale*	Gagaoua et al. (2015)
Alkaline proteases from fish viscera	Ketnawa, et al. (2014).
Papain from *Carica papaya*	Kahina, et al. (2020)
Peroxidase from orange peels (*Citrus sinenses*)	Vetal and Rathod (2015)
Catalase from *Solanum tuberosum*	Duman and Kaya (2013)
Polyphenol oxidase from borage (*Trachystemon orientalis)*	Alici and Arabaci (2016)
Polysaccharides from Inonotus obliquus	Liu et al. (2019)
Oil from *Spirogyra sp.*	Reddy and Majumder (2014)
Oil	Pandare and Rathod (2017)
Oil from *Crotalaria juncea*	Dutta, Sarkar and Mukherjee (2015)
Bioactive Exopolysaccharides of *Phellinus baumii.*	Wang et al. (2019)
Serratiopeptidase from Serratia marcescens NRRL B 23,112 using ultrasound assisted three phase partitioning.	Pakhale and Bhagwat (2016)
Microwave assisted three phase partitioning for purification of laccase from Trametes hirsula.	Patil and Yadav (2018)

These erythrocyte enzymes all have quaternary structures, so the denaturation of hemoglobin is not simply a consequence of its quarternary structure.

Other circumstances where TPP does not work is with serum albumin, which is dimerized by TPP, and TPP is not suitable for the isolation of IgG antibodies. Although TPP commonly works well with proteinases, for bromelain it is less successful than cold acetone precipitation (unpublished observations from our laboratory).

References

Alici, E.H., Arabaci, G., 2016. Purification of polyphenol oxidase from borage (Trachystemon orientalis L.) by using three-phase partitioning and investigation of kinetic properties. Int. J. Biol. Macromol. 93, 1051–1056.

Cannon, W.R., Pettitt, B.M., McCammon, J.A., 1994. Sulfate anion in water: model structural, thermodynamic and dynamic properties. J. Phys. Chem. 98, 6225–6230.

Dennison, C., 2011. Three-phase partitioning. In: Tschesche, H. (Ed.), Methods in Protein Biochemistry. Walter de Gruyter, Berlin, Germany, pp. 1–5.

Dennison, C., Lovrien, R., 1997. Three phase partitioning: concentration and purification of proteins. Protein Expr. Purif. 11, 149–161.

Dennison, C., Moolman, L., Pillay, C.S., Meinesz, R.E., 2000. Use of 2-Methylpropan-2-ol to inhibit proteolysis and other protein interactions in a tissue homogenate: an illustrative application to the extraction of cathepsins B and L from liver tissue. Anal. Biochem. 284, 157–159.

Duman, Y.A., Kaya, E., 2013. Three-phase partitioning as a rapid and easy method for the purification and recovery of catalase from sweet potato tubers (Solanum tuberosum). Appl Biochem Biotechnol (2013) 170, 1119–1126. doi:10.1007/s12010-013-0260-9.

Dutta, R., Sarkar, U., Mukherjee, A., 2015. Process optimization for the extraction of oil from Crotalaria juncea using three phase partitioning. Ind. Crop. Prod. 71. doi:10.1016/j.indcrop.2015.03.024.

Gagaoua, M., Hafid, K., 2016. Three phase partitioning system, an emerging non-chromatographic tool for proteolytic enzymes recovery and purification. Biosens. J. 5, 1–4.

Gagaoua, M., Hoggas, N., Kahina, H., 2015. Three phase partitioning of Zingibain, a milk-clotting enzyme from Zingiber officinale Roscoe rhizomes. Int. J. Biol. Macromol. https://doi.org/10.1016/j.ijbiomac.2014.10.069.

Jacobs, G.R., Pike, R.N., Dennison, C., 1989. Isolation of cathepsin D using three-phase partitioning in t-butanol/water/ammonium sulfate. Anal. Biochem. 180, 169–171.

Kiss, É., Szamos, J., Tamása, B.T., Borbása, R., 1998. Interfacial behavior of proteins in three-phase partitioning using salt-containing water/tert-butanol systems. Colloids Surf. A. 142, 295–302.

Chew, K.W., Ling, T.C., Show, P.L., 2018. Recent developments and applications of three-phase partitioning for the recovery of proteins. Separation & Purification Reviews 48 (1), 52–64. doi:10.1080/15422119.2018.1427596.

Kahina, H., John, J., Savah, T.M., Dominguez, R., 2020. One-step recovery of latex papain from *Carica papaya* using Three Phase Partitioning and its use as milk-clotting and_meat-tenderizing_agent. Int. J. Biol. Macromol. 146, 798–810.

Ketnawa, S., Benjakul, S., Martínez-Alvarez, O., Rawdkuen, S., 2014. Three-phase partitioning and proteins hydrolysis patterns of alkaline proteases derived from fish viscera. Sep. Purif. Technol. 132, 174–181.

Kumar, M., Mukherjee, J., Sinha, M., Kaur, P., Sharma, S., Gupta, M.N., Singh, T.P., 2015. Enhancement of stability of a lipase by subjecting to three phase partitioning (TPP): structures of native and TPP-treated lipase from Thermomyces lanuginosa. Sustain Chem Process 3 article 14.

Liu, Z., Yu, D., Li, L., Liu, X., Zhang, H., Sun, W., Lin, C-C., Chen, J., Chen, Z., Wang, W., Jia, W., 2019. Three-Phase Partitioning for the Extraction and Purification of Polysaccharides from the Immunomodulatory Medicinal Mushroom Inonotus obliquus. Molecules 24, 403.

Lovrien, R., Goldensoph, C., Anderson, P.C., Odegaard, B., 1987. Three phase partitioning (TPP) via t-butanol: enzymes separations from crudes. In: Burgess, R. (Ed.), Protein Purification: Micro to Macro. A.R. Liss, New York, pp. 131–148.

Pakhale, S.V., Bhagwat, S.S., 2016. Purification of serratiopeptidase from Serratia marcescens NRRL B 23112 using ultrasound assisted three phase partitioning. Ultrason. Sonochem. 31, 532–538.

Pandare, D.C., Rathod, V.K., 2017. Three phase partitioning for extraction of oil: a review. Trends Fd Sci Technol. 68, 145–151.

Patil, P.D., Yadav, G.D., 2018. Application of microwave assisted three phase partitioning method for purification of laccase from Trametes hirsuta. Process Biochem. 65, 220–227.

Pike, R.N., Coetzer, T.H., Dennison, C., 1992. Proteolytically active complexes of cathepsin L and a cysteine proteinase inhibitor: purification and demonstration of their formation in vitro. Arch. Bichem. Biophys. 294, 623–629.

Pol, M.C., Deutsch, H.F., Visser, L., 1990. Purification of soluble enzymes from erythrocyte hemolysates by three phase partitioning. Int. J. Biochem. 22, 179–185.

Raghava, S., Barua, B., Singh, P.K., Das, M., Madan, L., Bhattacharyya, S., Bajaj, K., Gopal, B., Varadarajan, R., Gupta, M.N., 2008. Refolding and simultaneous purification by three-phase partitioning of recombinant proteins from inclusion bodies. Protein Sci. 17, 1987–1997.

Reddy, A., Majumder, A.B., 2014. Use of a combined technology of ultrasonication, three-phase partitioning, and aqueous enzymatic oil extraction for the extraction of oil from spirogyra sp. J. Engineering. doi:10.1155/2014/740631.

Roy, I., Gupta, M.N., 2004. α-Chymotrypsin shows higher activity in water as well as organic solvents after three phase partitioning. Biocatalysis Biotransformation 22, 261–268.

Roy, I., Sharma, A., Gupta, M.N., 2004. Three phase partitioning for simultaneous re-naturation and partial purification of Aspergillus niger xylanase. Biochim. Biophys. Acta Proteins Proteomics. 1698, 107–110.

Sardar, M., Sharma, A., Gupta, M.N., 2007. Refolding of denatured α-chymotrypsin and its smart bioconjugate by three-phase partitioning. Biocatalysis Biotransformation 25, 92–97.
Singh, R.K., Gsourinath, S., Sharma, S., Roy, I., Gupta, M.N., Betzel, C.H., Srinivasan, A., Singh, T.P., 2001. Enhancement of enzyme activity through three-phase partitioning: crystal structure of a modified serine proteinase at 1.5 Å resolution. Protein. Eng. 14, 307–313.
Sun, Z., Liu, Q., Qu, G., Feng, Y., Reetz, M., 2019. Utility of B-Factors in protein science: interpreting rigidity, flexibility, and internal motion and engineering thermostability. Chem. Rev. 119, 1626–1665.
Tan, K.H., Lovrien, R.E., 1972. Enzymology in aqueous-organic cosolvent mixtures. J. Biol. Chem. 247, 3278–3285.
Timasheff, S.N., Arakawa, T., Inoue, H., Gekko, K., Gorbunoff, M.J., Lee, J.C., Na, G.C., Pittz, E.P., Prakash, V., 1982. The role of solvation in protein structure stabilization and unfolding. In: Franks, F., Mathias, S.F. (Eds.), The biophysics of water. Wiley, New York, pp. 48–50.
Vetal, M.D., Rathod, V.K., 2015. Three phase partitioning a novel technique for purification of peroxidase from orange peels (Citrus sinenses). Food Bioprod. Process. 94. doi:10.1016/j.fbp.2014.03.007.

CHAPTER 2

How and why we happen to use three phase partitioning in areas other than protein purification

Munishwar Nath Gupta
Former Emeritus Professor, Department of Biochemical Engineering and Biotechnology, Indian Institute of Technology, Hauz Khas, New Delhi, India

Chapter outline

This chapter is not just about reminiscences but also has the following objectives in being placed here in the book. Sometimes science happens because of the lack of good facilities. That kind of situation forces one to think harder, provided giving it all up is not an option. That is the first message to those who have the passion but not the means to pursue research; I am sure that there are still people out there like that in developing nations. The second intent is to convey that if you stick to one area/one technique, you are more likely to get "recognition" faster. Apart from fame being ephemeral, after you gain recognition, you are asked to sit on committees and fill forms for claiming travel expenses, etc. As one who dislikes sitting on such committees and passing judgment over work of others, I did not take that route or seek that exalted level. In the bargain, developing these various applications of TPP was lots of fun. Sticking to just protein purification could have led to more papers and some "recognition", but there is no regret. The third impression I hope to convey is asking questions to oneself can be very productive. "What if we do it ..?". Below is a description of lots of answers to such questions we asked over those years while working on three phase partitioning (TPP). The last message is that there are still lots of unanswered questions and unexploited applications of TPP.

Three phase partitioning (TPP) was originally discovered/described by Lovrein and Dennison at University of Minnesota. Dennison has already described a personal account of this discovery (please see Chapter 1). Typical of the way many scientists who wrote in that era, perusal of the papers from that group shows how its importance was

Three Phase Partitioning
DOI: https://doi.org/10.1016/B978-0-12-824418-0.00010-2

understated. In fact, the claim was that it can be used for an upstream step for removal of colored material and concentration. Partial purification, if any, was just bonus like with any other upstream technique.

While naming it TPP was apt in view of the formation of the three phases at the end of the protocol, it sort of hid precipitation as the cause behind concentration. It also unfortunately juxtaposed it along with two-phase extraction methods. For example, Przybycien's group pointed out that TPP is not as mature as two-phase extraction techniques (Przybycien et al., 2004). TPP, in fact, is a combination of precipitation and extraction steps. It is this feature which partly makes it a highly versatile technique.

Sometime around late 1990s, my small research group at IIT Delhi was coming to an end of our collaboration on use of smart polymers for affinity precipitation with a Swedish group. We had published dozens of papers on the applications of affinity precipitation for protein purification. I was getting a little bored as our research was becoming repetitive. In view of our primitive infra-structure, I was looking at some work which can be carried out in test tubes. Thanks to the Indo-Swedish funding, we had a spectrophotometer! I have described this in a little more detail elsewhere (Gupta and Mukherjee, 2015). At that juncture, those early papers on TPP caught my eyes. In one of those ironical coincidences, I sort of knew Prof. Lovrein as I had done my first post-doc in 1978–79 at University of Minnesota in the same department where he was on the faculty. At that time, I did not have any idea of what he was working on. I was trained to work on protein structure, and separation techniques per se would not have held any fascination for me at that point in time. We tentatively started exploring TPP for protein purification. We were, of course, looking at it as a precipitation technique and revisiting the same systems which we had tried with affinity precipitation. We had lots of rejections when we tried to publish our early papers related to TPP. An account of that may be of interest, at least to the younger readers of this chapter. Prior to our collaboration with the Swedish group, I had spent some time in Klibanov's lab at MIT, Cambridge. That visit, again, was in search of research themes (in the broad area of proteins/enzymes). Upon my return, we started some exploratory work on behavior of enzymes in organic solvents. It was rather at a low key level since (a) we were not an organic synthesis team whereas most of the applications of non-aqueous enzymology being described in that area at that time were related to applications of enzymes in organic synthesis, (b) we did not have GC or HPLC to track product formation, and (c) I was still not comfortable with some mechanistic aspects of low water enzymology. So, we published some work on higher activity obtained with some enzymes (whose reactions we could track without HPLC/GC) in aqueous-organic co-solvent mixtures. We had also published one paper in which we described higher activity of Proteinase K as a result of brief exposure to an organic solvent at higher temperature (Gupta et al., 2000). So, when we started observing similar increases in total amounts of activity units and specific activities in case of a few enzymes after subjecting them to the TPP protocol, we tried to publish our results but we did not succeed. A particularly nasty experience was with European Journal of Biochemistry which left me with a life-long cynicism towards the peer review system (Gupta, 2013). Prof. Perham (Parham, 2013) was the editor-in-chief and he rejected one of our papers enclosing a comment which said it is not science but Harry Potter's magic. I was young and as a person trained in protein chemistry, I was in awe of people like Prof. Perham. So, I wrote to him pointing out that we were using a pure preparation of chymotrypsin and increase in total number of activity units cannot arise due to purification in any case. The reply was that it is not offensive to call somebody's science Harry Potter's magic. He did not think it necessary to address the scientific arguments. I tried to convince somebody whom I knew in an adjoining institute to carry out the X-ray diffraction of our TPP-treated chymotrypsin sample. Reluctantly, he agreed to carry out the X-ray diffraction study of Proteinase K if we could show similar changes in the enzyme activity. The results clearly showed why the enzyme activity increased as it showed very high B-factor (which I was told can be taken as increase in overall flexibility) (Singh et al., 2001). After this, we had no trouble publishing our protein purification work using TPP. Recently, I have highlighted the role of cognitive biases in science; our peer review process is definitely a victim of several of those, especially confirmation bias (Gupta and Thelma, 2019).

2.1 TPP for purification of proteins/enzymes

A few general observations first. One observation which we made in my laboratory was that estimation of proteins by most of the popular methods gave erroneous results when proteins partitioned into various phases after TPP were quantified. The numbers were little more reliable after extensive dialysis but any sulphate anion still bound to the proteins (or may be even t-butanol although none of the enzyme crystals examined by X-ray diffraction for assessing the structural changes as a result of TPP showed any bound t-butanol molecule) presumably influenced the outcome of these estimations. So, the data on any changes in total number of units of activity as a result of TPP should be considered more reliable than any changes in specific activities. It is interesting to note that increase in total number of activity units has been reported even in the presence of low amounts of water-soluble cosolvents when enzyme activities were measured in such water-cosolvent mixtures (Batra and Gupta, 1994). Similarly, many preparations of cross-linked enzyme crystals (CLEAs) obtained via precipitation with ammonium sulphate /water-miscible organic solvents have shown much higher activities than the starting enzyme preparations (Schoevart et al., 2004).

TPP exposes proteins to ammonium sulphate and (generally) t-butanol. While the effect of methanol/ethanol and polyhydric alcohols is better known, the effect of t-butanol on protein structures as such needs more extensive studies (Gray, 1993). Many organic solvents are known to promote/disrupt secondary structures such as alpha-helix and beta-sheet. We often tend to forget that enzymes did not evolve to become perfect enzymes in most of the cases. The so-called native structure (which in any case is at least equilibria between many conformational states) is not necessarily the one which corresponds to the highest possible activity. So, conformational changes can lead to higher biological activity which, of course, can be at the cost of other properties like stability, etc. The effect of ammonium sulphate is even more complicated than generally appreciated. Most of the discussion on this has focused on the Hofmeister series. Again, it is not often appreciated that we still continue to understand Hofmeister effects on protein structure (Tadeo et al., 2009; Salis and Ninham, 2014). So, it is fair to say that from a mechanistic point of view, TPP is not a closed chapter. The mechanisms discussed in this book, hence, are based upon what we understand so far.

In this regard, the results of Rather and Gupta (2013a) on the effect of various concentrations of ammonium sulphate and t-butanol at various stages of precipitations with a variety of proteins deserve more attention. While the structural changes were not drastic, they varied from protein to protein. In this context, ovalbumin presented a somewhat unique behavior (Rather and Gupta, 2013b). The precipitate obtained with ovalbumin after TPP did not dissolve in any normal aqueous buffer; about 20 percent of the protein could be dissolved with great difficulty. The physicochemical characterization of this soluble component showed unfolded structures devoid of any significant secondary structure and prone to form soluble aggregates. We, at least, are not aware of this kind of result with any other protein while carrying out TPP. As, by that time, we had found that TPP constitutes a useful process to refold many proteins from their inclusion bodies (see discussion later in this Chapter and Chapter 11), we carried out TPP of that precipitate after dissolving it in urea solution. The precipitate so obtained was soluble. In fact, many structural variants could be obtained under different conditions of TPP. One such variant showed trypsin inhibitory activity and had "marginally higher beta-sheet content and higher surface hydrophobicity" (Rather and Gupta, 2013b). This is probably the only report wherein the generation of a promiscuous activity as a result of TPP has been described. Perhaps, the promiscuous activities of TPP-treated enzymes need a further look. This is also indicated by imprinting effects we have reported during even simple precipitation (Majumdar and Gupta, 2011; Mukherjee and Gupta, 2015a, b, 2016). Catalytic promiscuity has not only been exploited in organic synthesis rather extensively, but is also implicated in evolution of new catalytic activities in nature (Gupta et al., 2011; Kapoor and Gupta, 2012; Khersonsky and Tawfik, 2010; Majumdar and Gupta, 2011). In order to see whether chymotrypsin shows similar structural changes as Proteinase K (Singh et al., 2001), X-ray diffraction work yielded a very interesting result (Singh et al., 2005) (Fig. 2.1). It was found that during crystallization, an autolytic fragment of 14 residues (Ile16–Trp29) was generated that bound to the active site of the enzyme with high affinity, rendering it inactive. Crystallization occurred only after autolysis took place for all enzyme molecules and a homogenous population existed. It cannot be ruled out that this may not be a fortuitous situation; it may have a role in controlling chymotrypsin activity *in vivo* where this autolysis may be triggered by some physiological stimulus. In as much as peptides constitute an important class of drugs via inhibiting enzymes (and proteases being one important example of that), this result at least certainly has important implications in drug design for controlling activities of serine proteases.

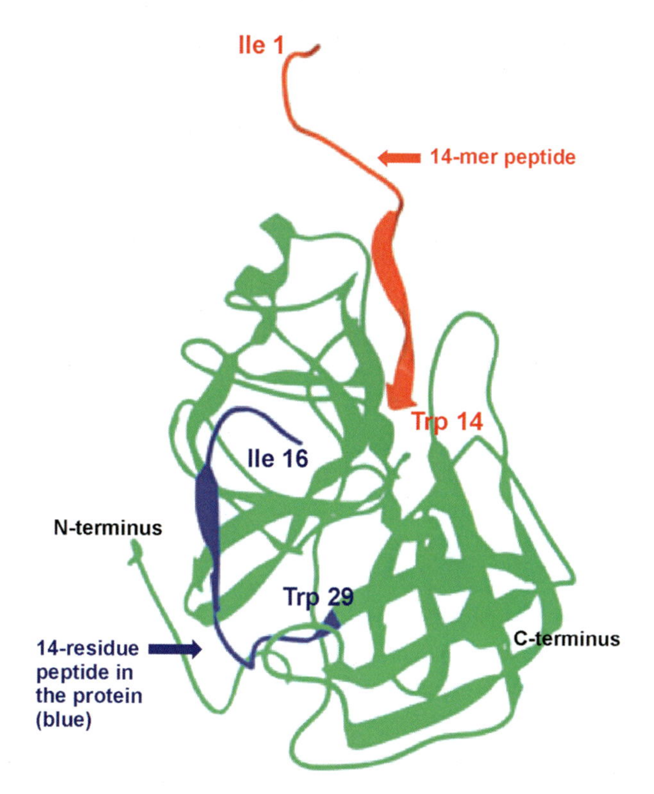

Fig. 2.1 ***Overall folding of α-chymotrypsin.*** The location of peptide in the protein structure Ile16–Trp29 is indicated in blue. The autolyzed peptide Ile1–Trp14 is shown in red at the binding site of the enzyme. The figure. was drawn with program SWISS PDB VIEWER (Guex and Peitsch, 1997). PDB: 1OXG. *Reproduced from Singh et al., 2005, with permission from the publisher.*

Before I list proteins/enzymes which we purified in my lab, I will also like to mention the following which really gave us a lot of encouragement while working with TPP when everybody, even in our country, was publishing work in more exciting areas like molecular biology, bioinformatics, etc.! Much is preached about the virtues of multidisciplinary work. Our experience was that it tends to be a rather lonely journey. We were never part of the mainstream of either biochemistry or biochemical engineering. Most of the fellow Indian scientists I happen to run into felt embarrassed as they invariably had to ask - eh, um, so what is your area exactly. Sometimes, I mischievously replied, you know IIT Delhi is actually in Delhi which is part of national capital area! The review by Przybycien et al. (2004) covered our work extensively and very fairly. It signaled that what we were doing with TPP was being read! Alkaline phosphatase from *E. coli* or calf intestine is used widely to amplify signal during ELISA based diagnostic tests. This enzyme has also found use in dephosphorylation of phosphoproteins. Chicken intestine was more accessible to us and we used TPP to purify the enzyme from that source (Sharma et al., 2000). Now, TPP is a single plate operation and one should not expect single step purification based upon just this process. In this case, following up TPP with hydrophobic interaction chromatography yielded a preparation which showed up a single band on SDS-PAGE (Sharma et al., 2000). As none of our purification targets were pharmaceutical proteins, in almost all our work, we accepted a single band on SDS-PAGE (with coomassie blue for staining) as sufficient proof of purity. Another outcome was establishing that simple ammonium sulphate precipitation gave somewhat higher yield of the enzyme but fold purification was only 7-fold; TPP using much lower concentration of ammonium sulphate led to 23-fold purification.

Phospholipase D used to be the least important among phospholipases till its role in signal transduction emerged. The enzymes from cabbage and carrot were the two originally described in the literature. Sharma and Gupta (2001b) describe a simple purification procedure from carrots using TPP. As we were in the middle of this (much to the consternation of the graduate student), the carrot season was over in Delhi; so the project cost went up as we had to get some from Bengaluru! Pectinase was another industrially important enzyme which could be obtained by using TPP; the source was tomatoes (Sharma and Gupta, 2001a). We also explored integrating metal affinity interactions with TPP, an approach which perhaps needs a relook (Roy and Gupta, 2002a).

Around this time, we had succeeded in buying a fermenter and had people with skills in protein expression in the group. This largely happened as we were into a collaborative project on protein refolding methods. GFP expressed in *E. coli* was purified using TPP (Jain et al., 2004). MLFTPP as such, of course, can also be used for purification of proteins/enzymes (Mondal et al., 2003a, b; Sharma and Gupta, 2002a, 2003]. Its development and applications are described in Chapter 10 of this book. Both TPP and MLFTPP have been successfully used in reactivating denatured enzymes and in refolding proteins from their inclusion bodies (Gautam et al., 2012; Mondal et al., 2007; Raghava et al., 2008; Roy et al., 2004a, 2005; Sardar et al., 2007; Sharma et al., 2004). That application forms Chapter 11 of this book.

As has been briefly mentioned, an intriguing result has been the higher activity shown by several enzymes after subjecting them to TPP. Examples of such results can be found in several papers published over the years (Mondal et al., 2006; Roy and Gupta, 2005; Singh et al., 2001). One of the major challenges in recent years has been to improve the catalytic activity of enzymes in low water media (Gupta, 2000). Many enzyme preparations with high purity were found to have better catalytic activity in low water containing organic solvents after being subjected to TPP (Kumar et al., 2015; Roy et al., 2004b; Roy and Gupta, 2004, 2005). An interesting behavior was displayed by chymotrypsin which formed aggregates with higher activity in low water containing organic solvents after TPP (Rather et al., 2012) (Fig. 2.2). Another outlier example is that of aldolase catalytic antibody 38C2 which gave more than one product after TPP treatment (Mondal et al., 2006). That is also an example wherein it was shown that TPP can be carried out successfully with microliter amounts of proteins.

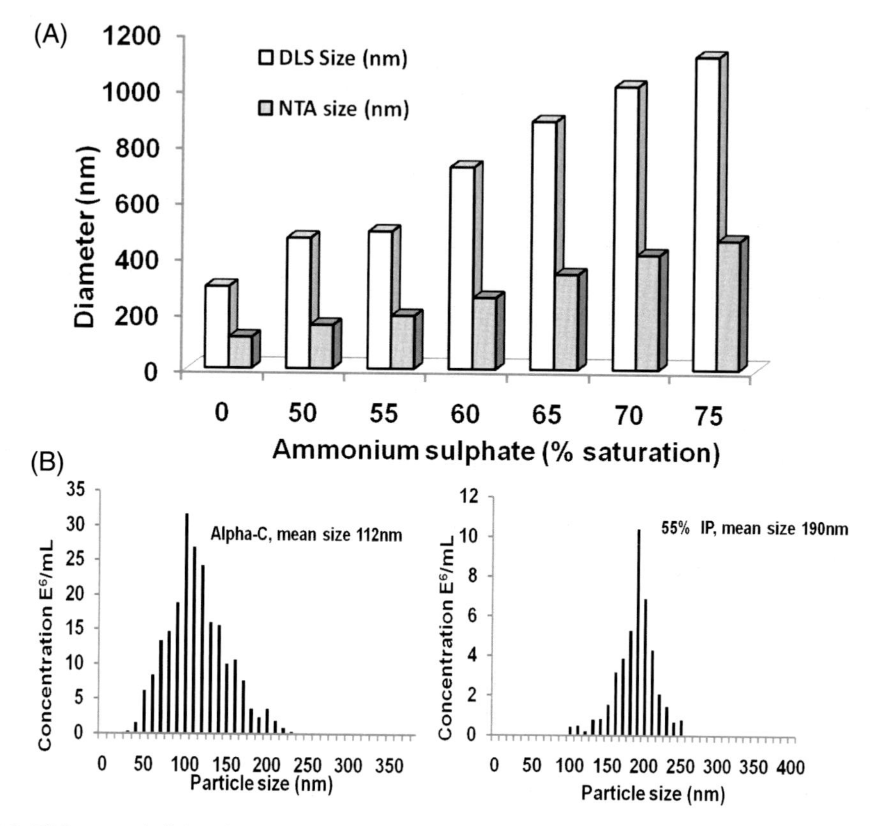

Fig. 2.2 ***TPP treated alpha chymotrypsin preparations.*** (A) Comparison of sizes by Nanoparticle tracking analysis and DLS (B) Size distribution by Nanoparticle tracking analysis. The sizes are number mean diameters (NTA data) and intensity mean diameters (DLS data). Where a suspension is not monodisperse, the intensity mean will always be larger than number mean. *Reproduced from Rather et al., 2012, under Creative Commons Attribution License, without any change.*

2.2 TPP used for edible oil extraction

One day, a Director from a national laboratory phoned and in an unusual step (at least from my perspective; as an ordinary scientist, I was not used to such graciousness from such high priests in the highly hierarchical Indian science) offered to come and see me. It transpired that he was under pressure to start exploring the use of enzymes for edible oil extraction. My reply was that I knew nothing about edible oil seeds. He said that his Institute will collaborate and take care of those aspects. He also said that generous funding will be available. Now, my laboratory, in those days, was only a slight improvement over how Ostwald described Curie's lab after visiting it. He said it was "halfway between a stable and a potato cellar" (https://digital.library.unt.edu/ark:/67531/metadc111245/m2/1/high_res_d/metadc111245.pdf, accessed on June 28, 2020). I looked at the torn curtains on the windows and decided to work on oil extraction.

Again, the mention of non-polar materials partitioning into the upper t-butanol layer after TPP in early publications from Lovrein's lab made me think of partitioning of oil into the t-butanol layer. The oil from soybean was one of the early targets and reasonable success was achieved (Sharma et al., 2002). For decades, the oil industry has struggled to find an alternative to hexane as the extracting solvent for edible oils unsuccessfully. t-Butanol is a lot more harmless solvent. The success of TPP in extraction of edible oils should be viewed against this background. Around this time, we had started looking at ultrasonication to disrupt structures like oil bodies in plant seeds and other materials. We combined ultrasonication and TPP to obtain oil from apricot, almonds and rice bran (Sharma and Gupta, 2004). Proteases could also be used to dissociate oil from proteinaceous material and thus enzyme-assisted TPP was used to obtain oil from soybean, rice bran and mango kernels (Gaur et al., 2007). Around the same period, I received a phone call from the Department of Biotechnology (Government of India). They were launching a national mission on biodiesel and I was asked to submit a proposal. Whole of my research career, I have had an aversion to do "fashionable research", so I was reluctant. However, when I looked up biodiesel on the internet, I found that it was a mixture of alkyl esters of fatty acids. I realized that biodiesel could be an attractive target for synthesis by using enzymes in low water media. So, we submitted a proposal along those lines and were given a modest amount of money. We were actually quite a misfit there. In every review meeting, most of the people would show very good looking photographs of Jatropha plants, talk about girth of the plants, etc. All we had to show was mostly a number: so much percentage conversion of the oil which was actually obtained from another Institute. It was clear that the mission project was stuck at obtaining enough Jatropha oil. In one of those setbacks which turn out to be opportunities, the Institute/group supplying us oil got fed up and sent us seeds asking us to extract oil ourselves. Application of enzyme assisted TPP gave us excellent results (Shah et al., 2004a). Apart from preparation of biodiesel, it also enabled us to initiate many other directions related to the use of enzymes under low water conditions for organic synthesis/ kinetic resolution of enantiomers (Gupta, 2000; Shah et al., 2004b; Shah and Gupta, 2007]. It is good to note (please see other chapters in this book) that the use of TPP for obtaining oils/other lipids has been pursued by many other groups since our early work.

After removing oil in t-butanol layer by TPP, the protein that precipitated in the interfacial layer in many cases is also a valuable byproduct for human nutrition or animal feed. In case of soybean flour, an integrated process was developed to recover both oil and the protein component (Kansal et al., 2006). It should be mentioned that in general, we found that the protein component recovered by this approach showed improved susceptibility to proteolytic digestion by enzymes such as trypsin and chymotrypsin *in vitro* (unpublished results). In case of the protein from soybean flour, its quality could be improved further by removing soybean trypsin inhibitor (STI) and soybean agglutinin (SBA) which are well known anti-nutritional factors known to be present in soybean meal. This was possible in a single step on an IMAC column. While from nutritional point of view itself, this was a valuable result, this approach enabled us to recover both STI and SBA in purified forms. Both proteins find diverse applications in biochemistry; an important one is that both are excellent affinity ligands. Integration of TPP and IMAC steps recovered STI and SBA in 90 percent and 93 percent yields with 26- and 15-fold purification, respectively (Kansal et al., 2006).

2.3 TPP of polysaccharides

How we happen to discover/design MLFTPP process has been described in a later chapter in this book along with the applications of this process. Briefly, our thinking was that if proteins precipitate at the interface during TPP, is it a consequence of their being macromolecular amphiphiles which have binding sites for both sulphate ion and t-butanol? We had already worked with smart polymers during our affinity precipitation work (Roy and Gupta, 2002b). Smart polymers derive their smartness from having both hydrophilic and hydrophobic segments. That led us to carry out TPP of a synthetic polymer, Eudragit™, which is commercially available as an enteric polymer. As Eudragit™ binds to xylanase, the enzyme precipitates along with Eudragit™ on the interface during MLFTPP. Our work with affinity precipitation had led us to discover unusual affinity of many polysaccharides towards some industrially relevant enzymes (Teotia et al., 2001).

Polysaccharides are still not considered exciting molecules to work with by biochemists. While monosaccharides and disaccharides have attracted a lot of attention because of their being involved in central metabolic pathways and association with important diseases (such as diabetes or lactose intolerance), oligosaccharides as glycans being parts of glycoproteins have been paid due attention in recent decades. A typical biochemist associates the word polysaccharides mostly with starch and glycogen. For biotechnologists, particularly those working in the area of biomass conversion, complex polysaccharides like lignin, pectin and chitin are very important molecules ((Bisaria and Kondo, 2014); Glazer and Nikaido, 1995; Shuddhodana et al., 2018; Sutherland, 1990; (Wood and Kellogg, 1988)). Apart from being involved in biofilms, microbial exopolysaccharides found outside the microbial surfaces also have tremendous value in biotechnological applications. Many of these are water soluble (Sutherland, 1990). So are many of the polysaccharides (or their derivatives) from marine sources as exemplified by alginates and chitosan (deacetylated chitin). Our work on affinity precipitation had exploited the smartness inherent in chitosan (a pH sensitive polymer), alginates (Ca^{2+} ion sensitive polymers) and κ-carrageenan; all water soluble polymers of marine origin (Roy and Gupta, 2002b, 2003). All these considerations led us to explore TPP of some polysaccharides.

What was even more interesting was to look at the structural changes which occurred in polysaccharides like chitosan and starch. A large number of books and reviews have been devoted to applications of chitosan. The structural changes in chitosan after its precipitation as a result of its forming an interfacial layer during TPP were tracked by UV absorption spectra, FT-IR spectra and its solubility in aqueous buffers

(Sharma et al., 2003). Notably, after being subjected to TPP, it showed decreased vulnerability to chitinase. While that may be bad news as far as exploitation of chitin as biomass is concerned, its use as a biomaterial in biomedical applications should profit from such a change (Sutherland, 1990). To the best of my knowledge, TPP-treated polysaccharides do not seem to have been evaluated/exploited so far as biomaterials. It is also noteworthy that subjecting starch to TPP also induced structural changes in the polysaccharide which reduced its hydrolysis by an alpha-amylase (Mondal et al., 2004). Not only soluble starches from potato and tapioca but cationic and oxidized forms of starches as well as an amylopectin were found to precipitate (to >85 percent of the starting amount) as a result of TPP. In view of these results, it is not surprising that TPP of carbohydrates has been pursued further as described in some subsequent chapters of this book.

2.4 TPP of microbial cells

By 2005 or so, we had succeeded in establishing basic fermentation facilities for growing microorganisms. We wondered what t-butanol would do to the cell membranes of microorganisms. We showed that permeabilization of cells as a result of TPP could be controlled and used to create holes of increasing sizes. This resulted in proteins of smaller sizes leaking first. The high molecular weight proteins like alcohol dehydrogenase, etc. were left inside the damaged cells and could be isolated in almost pure forms simply by disruption of the damaged cells (Raghava and Gupta, 2008) (Table 2.1). This effectively created a step similar to gel filtration right at the up-stream stage in production of the proteins/enzymes Table 2.2. Sadly, this innovative application of TPP has not become popular. One obvious unexplored application would be to isolate macromolecular complexes and aggregates like inclusion bodies (Mukherjee and Gupta, 2015c, 2017).

Table 2.1 Dependence of retention of proteins on their molecular mass and preincubation time in TPP permeabilized cells. Reproduced from Raghava and Gupta, 2009, with permission from the publisher.

Target	Molecular weight (kDa)	Preincubation time (min)	Retained protein (percent)	Released protein (percent)
Recombinant *E. coli* GFP_{uv}	29	15	0	100
T. thermophilus lipase	34	15	0	100
E. coli PGA	85	30	0	100
S. cerevisiae ADH	150	60	96	4
T. thermophilus ADH	170	60	100	0

PGA: Penicillin G acylase;.
ADH: Alcohol dehydrogenase.

Table 2.2 Purification of penicillin G acylase (PGA) released from permeabilized cells that were exposed to t-butanol for 30 min. Reproduced from Raghava and Gupta, 2009, with permission from the publisher.

Step	Amount of protein (mg)	Activity (U)	Specific activity (U/mg)	Yield (percent)[a]	Purification (fold)[b]
First TPP aqueous phase	48.2	82.0	1.7	96	3.4
Second TPP precipitate	7.2	74.0	10.3	87	20.6

[a]The activity released by ultrasonication was taken as 100 percent.
[b]The preparation obtained by ultrasonication (with a specific activity of 0.5 U/mg) was given the value of 1.

2.5 Isolation and purification of low molecular weight compounds

The partition component of TPP obviously allows separation of both water-soluble and lipophilic molecules which have low molecular weights. This was shown to be possible in the early work on removal of pigments, etc. from the proteins. As one of our last forays into developing various applications of TPP, we have described isolation of a natural product (Yadav et al., 2017) from a halophile. As cell debris either remains at the bottom of the vessel or forms part of the interfacial precipitate, TPP combines several upstream and downstream steps and is valuable in designing integrated processes. The expanded bed process has been described as a unit process which combines several unit processes into one (Roy et al., 2007). TPP/MLFTPP are similar processes. These, along with expanded bed chromatography or ATPS, are unique as they all have an inbuilt solid-liquid separation step which obviates the need for either centrifugation or membrane separation step at the upstream stage while isolating molecules from biological systems.

2.6 Conclusion

Various chapters in this book show that TPP has become popular, at least in some parts of the world. It has been adopted extensively (in terms of various systems) in areas like protein isolation/purification, extraction of oil and carbohydrates. Our limited work shows that it works well at very small scale (like in the case of catalytic antibody) as well. Surprisingly, scale-up to dealing with very large volumes does not seem to have been published. Also, workers in areas like protein refolding still have not exploited it in spite of our considerable successes with various systems with this application. That is still waiting for a "tipping point" to be reached for it to catch on. Using it in combination with ultrasonics and microwave irradiation should find more applications in future. Similarly, MLFTPP remains underexploited presumably because scientists still work in silos and smart polymers is a little different area. What is intriguing is its effect on protein flexibility. TPP of Proteinase K, chymotrypsin and lipase resulted in very different consequences. It is increasingly becoming clear that protein flexibility and intrinsic disorder are parts of a continuum as far as protein structure is concerned (Blundell et al., 2020; Gupta et al., 2020; Gupta and Mukherjee, 2013). Maybe this book will usher in more interest in applying TPP in these diverse and some underexploited areas. This chapter, part reminiscences (aimed at telling that our delving into these different areas was not a straightforward journey) and part highlighting the common thread of TPP which connected our forays into unchartered regions, hopefully will push others into trying this deceptively simple technique for newer applications.

References

Batra, R., Gupta, M.N., 1994. Enhancement of enzyme activity in aqueous-organic solvent mixtures. Biotechnol. Lett. 16, 1059–1064.

Blundell, T.L., Gupta, M.N., Hasnain, S.E., 2020. Intrinsic disorder in proteins: relevance to protein assemblies, drug design and host- pathogen interactions. Prog. Biophys. Mol. Biol. doi:10.1016/j.pbiomolbio.2020.06.004 (in press).

Bisaria, V.S., Kondo, A., 2014. Bioprocessing of Renewable Resources to Commodity Bioproducts. John Wiley & Sons, Inc.

Gaur, R., Sharma, A., Khare, S.K., Gupta, M.N., 2007. A novel process for extraction of edible oils: enzyme assisted three phase partitioning (EATPP). Bioresour. Technol 98, 696–699.

Gautam, S., Dubey, P., Singh, P., Varadarajan, R., Gupta, M.N., 2012. Simultaneous refolding and purification of recombinant proteins by macro-(affinity ligand) facilitated three-phase partitioning. Anal. Biochem. 430, 56–64.

Glazer, A.N., Nikaido, H., 1995. Microbial Biotechnology. W. H. Freeman, New York, pp. 325–391.

Gray, C.J., 1993. Stabilisation of enzymes with soluble additives. In: Gupta, MN (Ed.), Thermostability of Enzymes. Springer-Verlag, Heidelberg, pp. 124–143.

Guex, N., Peitsch, M.C., 1997. SWISS-MODEL and the Swiss-PDB Viewer: an environment for comparative protein modeling. Electrophoresis 18, 2714–2723.

Gupta, M.N., 2000. Methods in Non-Aqueous Enzymology. Birkhauser, Basel.

Gupta, M.N., 2013. Peer review: past, present and future. Curr. Sci. 105, 159–161.

Gupta, M.N., Alam, A., Hasnain, S.E., 2020a. Protein promiscuity in drug, discovery, drug repurposing and antibiotic resistance. Biochimie 175, 50–57.

Gupta, M.N., Kapoor, M., Majumdar, A.B., Singh, V., 2011. Isoenzymes, moonlighting proteins and promiscuous enzymes. Curr. Sci. 100, 1152–1162.

Gupta, M.N., Mukherjee, J., 2013. Enzymology: some paradigm shifts over the years. Curr. Sci. 104, 1178–1186.

Gupta, M.N., Mukherjee, J., 2015. Skills in biocatalysis are essential for establishing a sound biotechnological base in India. Proc. Indian Natn. Sci. Acad. 81, 1113–1132.

Gupta, M.N., Perwez, M., Sardar, M., 2020b. Protein crosslinking: uses in chemistry, biology and biotechnology. Biocat. Biotransform. 38, 178–201.

Gupta, M.N., Thelma, B.K., 2019. Ethics in measurement practice. In: Muralidhar, K., Ghosh, A., Singhvi, A.K. (Eds.), Ethics in Science Education, Research and Governance. Indian National Science Academy, New Delhi, pp. 45–63.

Gupta, M.N., Tyagi, R., Sharma, S., Karthikyan, K., Singh, T.P., 2000. Enhancement of catalytic efficiency of enzymes through exposure to organic solvent at 70 degree C: three-dimensional structure of a treated serine protease at 2.2 Å resolution. Proteins 39, 226–234.

Jain, S., Singh, R., Gupta, M.N., 2004. Purification of recombinant green fluorescent protein by three phase partitioning. J. Chromatogr. A 1035, 83–86.

Kansal, S., Sharma, A., Gupta, M.N., 2006. An integrated process for obtaining oil, protease inhibitors and lectin from soybean flour. Food Res. Int. 39, 499–502.

Kapoor, M., Gupta, M.N., 2012. Lipase promiscuity and its biochemical applications. Process Biochem. 47, 555–569.

Khersonsky, O., Tawfik, D.S., 2010. Enzyme promiscuity: a mechanistic and evolutionary perspective. Annu. Rev. Biochem. 79, 471–505.

Kumar, M., Mukherjee, J., Sinha, M., Kaur, P., Sharma, S., Gupta, M.N., Singh, T.P., 2015. Enhancement of stability of a lipase by subjecting to three phase partitioning (TPP): structures of native and TPP-treated lipase from Thermomyces lanuginosa. Sust. Chem. Proc. 3, 14.
Majumdar, A.B., Gupta, M.N., 2011. Increasing the catalytic efficiency of candida rugosa lipase for the synthesis of tert-alkyl butyrates in low water media. Biocat. Biotransform. 29, 238–245.
Mondal, K., Ramesh, N.G., Roy, I., Gupta, M.N., 2006. Enhancing the synthetic utility of aldolase antibody 38C2. Bioorg. Med. Chem. Lett. 16, 807–810.
Mondal, K., Raghava, S., Barua, B., Varadarajan, R., Gupta, M.N., 2007. Role of stimuli-sensitive polymers in protein refolding: α-Amylase and CcdB (controller of cell division or death B) as model proteins. Langmuir 23, 70–75.
Mondal, K., Sharma, A., Gupta, M.N., 2003a. Macroaffinity ligand -facilitated three phase partitioning of glucoamylase and pullulanase using alginate. Protein Exp. Purif. 28, 190–195.
Mondal, K., Sharma, A., Gupta, M.N., 2004. Three phase partitioning of starch and its structural consequences. Carbohydr. Polym. 56, 355–359.
Mondal, K., Sharma, S., Lata, L., Gupta, M.N., 2003b. Macroaffinity ligand -facilitated three phase partitioning [MLFTPP] of alpha amylase using a modified alginate. Biotechnol. Prog. 19, 493–494.
Mukherjee, J., Gupta, M.N., 2015a. Enhancing the catalytic efficiency of subtilisin for transesterification by dual bioimprinting. Tetrahedron Lett. 56, 4397–4401.
Mukherjee, J., Gupta, M.N., 2015b. Molecular bioimprinting of lipases with surfactants and its functional consequences in low water media. Int. J. Biol. Macromol. 81, 544–551.
Mukherjee, J., Gupta, M.N., 2015c. Paradigm shifts in our views on inclusion bodies. Curr. Biochem. Engg. 2, 1–9.
Mukherjee, J., Gupta, M.N., 2016. Dual imprinting of T. lanuginosus lipase for synthesis of biodiesel. Biotechnol. Rep. 10, 38–43.
Mukherjee, J., Gupta, M.N., 2017. Protein aggregates: forms, functions and applications. Int. J. Biol. Macromol. 97, 778–789.
Parham, P., 2013. R.N. Perham at the helm: 1998-2013. FEBS J. 280, 6279–6279.
Przybycien, T.M., Puzar, N.S., Steele, L.M., 2004. Alternative bioseparation operations: life beyond packed-bed chromatography. Curr. Opin. Biotechnol. 15, 469–478.
Raghava, S., Barua, B., Singh, P.K., et al., 2008. Refolding and simultaneous purification by three-phase partitioning of recombinant proteins from inclusion bodies. Protein Sci. 17, 1987–1997.
Raghava, S., Gupta, M.N., 2008. Tuning permeabilisation of microbial cells by three phase partitioning. Anal. Biochem. 385, 20–25.
Rather, G., Gupta, M.N., 2013a. Three phase partitioning leads to subtle structural changes in proteins. Int. J. Biol. Macromol. 60, 134–140.
Rather, G., Gupta, M.N., 2013b. Refolding of urea denatured ovalbumin with three phase partitioning generates many conformational variants. Int. J. Biol. Macromol. 60, 301–308.
Rather, G.M., Mukherjee, J., Halling, P.J., Gupta, M.N., 2012. Activation of alpha chymotrypsin by three phase partitioning is accompanied by aggregation. PLoS One 7, e49241.
Roy, I., Gupta, M.N., 2002a. Three phase partitioning of proteins. Anal. Biochem. 300, 11–14.
Roy, I., Gupta, M.N., 2002. Macro-affinity ligands in bioseparation. In: Gupta, M.N. (Ed.), Methods for Affinity-Based Separation of Enzymes and Proteins. Birkhauser, Basel, pp. 130–147.
Roy, I., Gupta, M.N., 2003. Smart polymeric materials: emerging biochemical applications. Chem. Biol. 10, 1161–1171.
Roy, I., Gupta, M.N., 2004. α-Chymotrypsin shows higher activity in water as well as organic solvents after three phase partitioning. Biocatal. Biotransf. 22, 261–268.
Roy, I., Gupta, M.N., 2005. Enhancing reaction rate for transesterification reaction catalyzed by Chromobacterium lipase. Enzym. Microb. Technol. 36, 896–899.
Roy, I., Mondal, K., Gupta, M.N., 2007. Leveraging protein purification strategies in proteomics. J. Chromatogr. B 849, 32–42.
Roy, I., Sharma, A., Gupta, M.N., 2004a. Three-phase partitioning for simultaneous renaturation and partial purification of Aspergillus niger xylanase. Biochim. Biophys. Acta 1698, 107–110.
Roy, I., Sharma, S., Gupta, M.N., 2004b. Obtaining higher transesterification rates with subtilisin Carlsberg in nonaqueous media. Bioorg. Med. Chem. Lett. 14, 887–889.

Roy, I., Sharma, A., Gupta, M.N., 2005. Recovery of biological activity in reversibly inactivated proteins by three phase partitioning. Enzym. Microb. Technol. 37, 113–120.
Salis, A., Ninhan, B.W., 2014. Models and mechanisms of hofmeister effects in electrolyte solutions, and colloid and protein systems revisited. Chem. Soc. Rev. 43, 7358–7377.
Sardar, M., Sharma, A., Gupta, M.N., 2007. Refolding of a denatured α-chymotrypsin and its smart bioconjugate by three-phase partitioning. Biocatal. Biotransf. 25, 92–97.
Schoevaart, R., Wolbers, M.W., Golubovic, M., Ottens, M., Kieboom, A.P.G., van Rantwijk, F., van der Wielen, L.A.M., Sheldon, R.A, 2004. Preparation, optimization, and structures of cross-linked enzyme aggregates (CLEAs). Biotechnol. Bioeng. 87, 754–762.
Shah, S., Sharma, A., Gupta, M.N., 2004a. Extraction of oil from Jatropha curcas L. seed kernels by enzyme assisted three phase partitioning. Ind. Crop Prod. 20, 275–279.
Shah, S., Sharma, S., Gupta, M.N., 2004b. Biodiesel preparation by lipase-catalyzed transesterification of Jatropha oil. Energy Fuels 18, 154–159.
Sharma, A., Gupta, M.N., 2001a. Purification of pectinase from tomato using three phase partitioning. Biotechnol. Lett. 23, 1625–1627.
Sharma, S., Gupta, M.N., 2001b. Purification of phospholipase D from Dacus carota by three phase partitioning and its characterization. Protein Expr. Purif. 21, 310–316.
Sharma, A., Gupta, M.N., 2002a. Macro[affinity ligand] facilitated three phase partitioning [MLFTPP] for purification of xylanase. Biotechnol. Bioeng. 80, 228–232.
Sharma, A., Gupta, M.N., 2004. Oil extraction from almond, apricot and rice bran by three-phase partitioning after ultrasonication. Eur. J. Lipid Sci. Technol. 106, 183–186.
Shah, S., Gupta, M.N., 2007. Kinetic resolution of [+/-]-1-phenylethanol in [Bmim][PF_6] using high activity preparations of lipases. Bioorg. Med. Chem. Lett. 17, 921–924.
Sharma, A., Khare, S.K., Gupta, M.N., 2002. Three phase partitioning for extraction of oil from soybean. Bioresour. Technol. 85, 327–329.
Sharma, A., Mondal, K., Gupta, M.N., 2003. Some studies on characterization of three phase partitioned chitosan. Carbohyd. Polym. 52, 433–438.
Sharma, A., Roy, I., Gupta, M.N., 2004. Affinity precipitation and macroaffinity ligand facilitated three phase partitioning for refolding and simultaneous purification of urea-denatured pectinase. Biotechnol. Prog. 20, 1255–1258.
Sharma, A., Sharma, S., Gupta, M.N., 2000. Purification of alkaline phosphatase from chicken intestine by three phase partitioning and use of phenyl-sepharose 6B in the batch mode. Bioseparation 9, 155–161.
Shuddhodana, Gupta, M.N., Bisaria, V.S., 2018. Stable cellulolytic enzymes and their application in hydrolysis of lignocellulosic biomass. Biotechnol. J. 13, 1700633.
Singh, R.K., Gourinath, S., Sharma, S., Roy, I., Gupta, M.N., Betzel, C., Srinivasan, A., Singh, T.P., 2001. Enhancement of enzyme activity through three-phase partitioning: crystal structure of a modified serine proteinase at 1.5 Å resolution. Protein. Eng. 14, 307–313.
Singh, N., Jabeen, T., Sharma, S., Roy, I., Gupta, M.N., Bilgrami, S., Somvanshi, R.K., Dey, S., Perbandt, M., Betzel, C., Srinivasan, A., Singh, T.P., 2005. Detection of native peptides as potent inhibitors of enzymes: crystal structure of the complex formed between treated bovine α-chymotrypsin and an autocatalytically produced fragment, IIe-Val-Asn-Gly-Glu-Glu-Ala-Val-Pro-Gly-Ser-Trp-Pro-Trp, at 2.2 angstroms resolution. FEBS J. 272, 562–572.
Sutherland, I.W., 1990. Biotechnology of Microbial Exopolysaccharides. Cambridge University Press, Cambridge.
Tadeo, X., Lopez- Mendez, B., Castano, D., Trigueros, T., Millet, O., 2009. Protein stabilisation and the hofmeister effect: the role of hydrophobic solvation. Bipophys. J. 97, 2595–2603.
Teotia, S., Khare, S.K., Gupta, M.N., 2001. An efficient purification process for sweet potato beta-amylase by affinity precipitation with alginate. Enzym. Microb. Technol. 28, 792–795.
Wood, W.A., Kellogg, S.T., 1988. Biomass, Part B: Lignin, Pectin, and Chitin, Methods in Enzymology, 161. Academic Press, San Diego.
Yadav, N., Gupta, M.N., Khare, S.K., 2017. Three phase partitioning and spectroscopic characterization of bioactive constituent from halophilic Bacillus subtilis EMB M15. Bioresour. Technol. 242, 283–286.

CHAPTER 3

Fundamental aspects of protein isolation and purification

John H.T. Luong
School of Chemistry, University College Cork, Cork, Ireland

Chapter outline

3.1 Introduction

Proteins come from different sources such as plants, animals, and microorganisms. Proteins can also be obtained from mammalian cell cultures, tissues, body fluids, or their overexpression in bacteria, yeast, or mammalian cells. Hybridoma cells are a source of monoclonal antibodies. Plant-based protein ingredients are produced from various crops for diversified applications. Consumption of soy protein has been increasing due to its health benefits and low cost. Soybean protein is used for manufacturing dry strength additives, which are essential for maximizing the mechanical integrity of papermaking products. Soybean protein is inexpensive compared to carboxymethyl cellulose (CMC),

Three Phase Partitioning
DOI: https://doi.org/10.1016/B978-0-12-824418-0.00013-8

chitosan, polyacrylamides, etc. New industrial applications are expected to drive the supply for plant proteins in the coming year. The demand for animal protein ingredients is likely to increase significantly considering their health benefits. As an example, whey proteins are proven to enhance immunity and glutathione (GSH) levels in cancer patients undergoing chemotherapy. Several keys roles of GSH in a multitude of cellular processes encompass cell differentiation, proliferation, and apoptosis (Traverso et al., 2013). Microalgae are another potential source of renewable proteins, carbohydrates, lipids, and pigments (Vanthoor-Koopmans et al., 2013). These intracellular products are in the cytoplasm, internal organelles, or bind to cell membranes. Several yeasts, notably *K.marxianus* (*K.lactis*), *Candidapseudotropicalis*, *Candida kefyr*, and *Torulopsis cremoris*, are potentially good sources of single cell protein (SCP). The issue of ethanol as a concomitant during their growth in cheese whey can be circumvented using Crabtree-negative mutants or mixed cultures. Currently, most SCP is used for animal feed. Proteins are increasingly produced by cloning in various host expression systems (Hartley, 2006; Tripathi and Shrivastava, 2019). Recombinant bacteria, yeasts, or filamentous fungi can be genetically created for producing microbial proteins; however, the *E. coli* expression system is still predominant. The use of yeasts is noteworthy due to their fast growth with high cell densities together with an expression level, which is regulated by simple medium manipulation.

Amino acid sequence, size, shape, charge, and aqueous solubility of each protein are unique, governing its physicochemical characteristics. As the number of proteins in a cell can be up to 10,000, separation to isolate one specific protein is always a daunting task. A purified protein might be subject to unfolding, aggregation, and degradation during purification. Other important factors include process scalability, cost-effectiveness, and the purpose of purification. As discussed later, a fusion tag inserted in recombinant proteins is designed to reduce the number of purification steps. Techniques for protein extraction and isolation vary depending on the starting material quantity, the protein location within the cell, and the subsequent purification steps. For labile proteins, purification is often performed at 4 °C to preserve protein structural integrity and minimize proteolysis (in case of contaminating proteases). A purification scheme should aim at both high fold-purification and high recovery of activities. Pure proteins allow the probing of amino acid sequences or the subsequent preparation of protein crystals, the first step toward the understanding of their tertiary structure. For academic research, the quantity of a required protein is often small in milligram ranges, but its purity must be as high as possible, at least over 95 percent. In contrast, for applied purposes, the quantity is much higher with the purity level varying according to intended applications. Different chromatographic techniques are often used in combination to purify proteins, based on their physicochemical properties such as mass, charge, hydrophobicity, and affinity to a specific ligand. During their flow crossing the surface, molecules will move fast if they exhibit infrequent interactions with the surface

Table 3.1 Common chromatographic techniques used in protein purification.

Chromatographic technique	Basis of separation
Ion-exchange chromatography (IEC)	Charge differences at a given pH
Hydrophobic interaction chromatography (HIC)	Surface hydrophobic differences
Gel permeation or gel filtration chromatography (GFC)	Differences in mass and shape
Affinity chromatography	Specific binding to a ligand

and vice versa. Spherical beads with functional groups are tightly packed in a column to achieve maximum ligand loadings (Table 3.1).

Chromatofocusing, a technique based on differences in the isoelectric point (pI) is not discussed in this chapter due to its low capacity and this method should ideally be used for partially pure samples. Hydroxyapatite chromatography is not fully understood and not discussed here albeit it has been used to purify proteins and DNA after the pioneering work of Tiselius et al. (1956). The separation principle is based on complex interactions between proteins and calcium phosphate-based media. Analytical aspects of proteins by mass spectroscopy, x-ray crystallography, NMR spectroscopy, etc. for protein characterization are available in the literature (Franks, 1988; Kaur et al., 2019). Of considerable interest is the probing of atomic-level characterization of protein-protein association (Pan et al., 2019).

3.2 Fusion tags and protein solubility

Recombinant protein production systems aim to maximize the accumulation of a soluble protein in the bacterial cell, e.g., *E. coli*, one of the predominant hosts. However, over-expression in *E. coli* generally leads to the formation of inclusion bodies. (Sevastsyanovich et al., 2010). The solubility of the native protein is at the mercy of nature, however, advanced recombinant technology (genetic and protein engineering) offers a real potential to enhance the protein solubility, enabling a single-step purification step with high purity (Rosano and Ceccarelli, 2014). Different strategies have been advocated to address this issue such as lower expression temperature, engineered host strains, cultivation conditions, and co-production of molecular chaperones and folding modulators (Costa et al. 2014). Gene fusion technology enables the host cell to produce a target protein linked with a peptide or protein molecules to mediate its solubility and purification. Some tags, e.g., His_6 is mainly designed so the target protein can be purified by metal chelate chromatography as discussed later. This subject is very broad, and many review papers are available in the literature. This section is limited to the performances of those fusion tags that enhance solubility as well as enable purification by an affinity-based separation technique. Other solubility enhancer tags are discussed in a review paper of Costa et al. (2014). Despite several approaches with different tags have been investigated, sometimes one has to try few to suit the intended use of the purified protein. Of course, other tags will emerge in the future, a suitable tag must fulfill the following requirements: (i) high level of soluble protein, (ii) ease of removal of the tag, and (iii) no adverse effect on the protein activity.

-MBP (maltose binding protein, 43 kDa). It promotes the solubility of a target protein by showing chaperone intrinsic activity (Fox et al., 2001). This tag works best when placed at the N-terminus of recombinant proteins (Sachdev and Chirgwin, 2000).

-TrxA (Trx). This intracellular thermostable protein of *E. coli* (12-kDa) offers enhanced solubility of target proteins, comparable to MBP. It is more effective at the N-terminal of target proteins (Young et al. 2012).

-GST (Glutathione-S-transferase, 26 kDa). Its performance is more efficient when it is sited at the N-terminal (Malhotra, 2009). GST might also promote the production of soluble protein in *E. coli* and. However, the fusion protein more often ends in inclusion bodies.

-The Strep II tag with only 8 amino acids binds specifically to streptavidin (Schmidt et al., 1996). This tag can be placed within the protein sequence or at the N- or C-terminal ends. Strep II-fused proteins bind to the streptavidin -column from which these can be eluted easily under simply by incorporating biotin derivatives in the elution buffer (Li, 2010; Terpe, 2003).

The rationale behind the use of fusion tags has been discussed by Butt et al. (2005) and Nallamsetty and Waugh (2007).

3.3 Cell lysis

The first step in protein separation and purification is the use of an appropriate buffer at a physiological pH during protein extraction by cell lysis. The buffer concentration must be above 25 mM to ensure buffering capacity and should not interfere with the analytical method of protein analysis. Four most common buffers are summarized in Table 3.2 with detailed information on their advantages and disadvantages.

It is advisable to add some protease inhibitors in the lysis buffer to inhibit endogenous proteases. Ethylenediaminetetraacetic acid (EDTA) and ethylene glycol-bis (β-aminoethyl ether)-N,N,N′,N′-tetraacetic acid (EGTA) are two common metal chelators, which often added to the storage buffer. Their binding to Mg^{2+} prevents cleavage of the purified protein by contaminating metalloproteases. Several proteins produced by microorganisms or animal cell cultures are released into the media in very dilute form. Initial product recovery only involves the separation of cells from the liquid medium by filtration or centrifugation, depending on the quantity. For intracellular proteins, cell lysis or cellular disruption is a process to disrupt parts of the cell wall for extracting and separating the organelles, proteins, and other biomolecules of microorganisms (bacteria, algae, and fungi), mammalian tissues, and plant cells. This is a primary step in protein

Table 3.2 Four common buffers used in protein separation

Buffer, pH range	Characteristics, advantages, and disadvantages
Tris, 7.5–9.0	Inexpensive, but might interfere with some enzymes\ activities. Its pH is dependent on temperature and concentration. Transparent in the pH range and it is important for protein analysis by UV detection
Phosphate, 5.8–8.0	Inexpensive, but cannot be used with divalent cations. Its pH is independent of temperature. The precipitation of proteins with Na-phosphate buffer or K-phosphate buffer is often performed.
MOPS, 6.5–7.9	The pH depends slightly on temperature. It cannot be autoclaved but has a high buffering capacity at physiological pH. MOPS form no complex with most metal ions and serves as a non-coordinating buffer in solutions with metal ions
HEPES, 6.8–8.2	pH is slightly dependent on temperature. It cannot be autoclaved and forms radicals under certain conditions. HEPES exhibits negligible metal ion binding.

MOPS: 3-(N-morpholino) propanesulfonic acid; HEPES: 4-(2-hydroxyethyl)−1-piperazineethanesulfonic acid).

isolation and purification from the hosts, e.g., *E. coli* and *Saccharomyces cerevisiae*, which do not excrete products. In this context, some fundamental aspects of cell membrane deserve a brief discussion here. The cell has an outer cell membrane, which regulates the transport of various species across it. Mammalian cells have only one boundary called the cytoplasmic membrane, whereas bacteria have several layers., The plasma membrane of Gram-positive bacteria is covered by the cell wall or the peptidoglycan layer. This layer makes up 50–80 percent of the envelope and its 10 percent is associated with teichoic acid. The outermost membrane is made of lipopolysaccharides, consisting of polysaccharides, lipids, and proteins. The cytoplasmic membrane or plasma membrane is a phospholipid bilayer with a thickness of 4 nm (Madigan et al., 2008), comprising highly hydrophobic (fatty acid) and hydrophilic (glycerol) moieties. Eukaryotic cells have sterols, rigid, and planar molecules in their membranes whereas prokaryotic cells have hopanoids, which are comparable to sterols their membranes. Different techniques are available for cell disruption and they can be classified as mechanical and non-mechanical methods (Fig. 3.1).

3.3.1 Mechanical procedures

French press. As one of the earliest techniques, the French pressure cell press (or French press) was pioneered by Charles Stacy French in 1950 (Fork and David, 2006). It is used to disrupt cell walls and cell membranes of small samples. It is equipped with a hydraulic pump that drives a piston to force the liquid sample through a tiny discharge valve under high pressure, about 2000 bar where the pressure drops to atmospheric. The

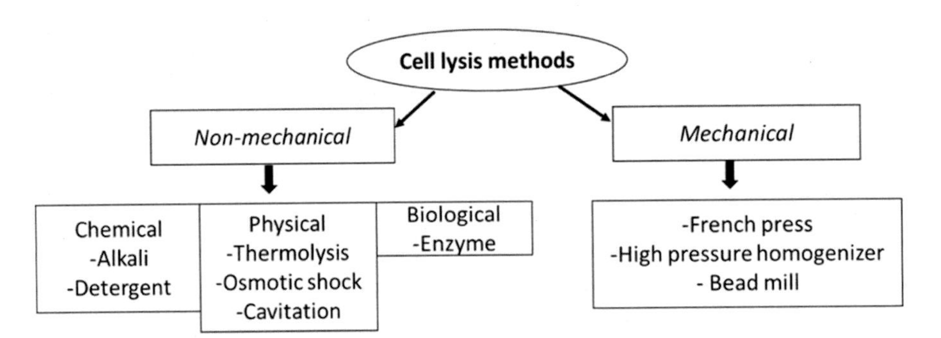

Fig. 3.1 ***Techniques used for cell lysis comprising different mechanical and non-mechanical methods.***

cells are subject to shear stress, leading to cellular disruption. The cells also encounter decompression and subsequently expand and rupture. This method is mainly used for lysing microbial cells, and other bacteria for the isolation of proteins and cellular components. The technique involves no chemicals that might denature the target proteins or must be removed in subsequent purification steps, depending upon the intended applications. The shear force can be regulated by setting the piston pressure and delicate biological structures are less damaging with a single pass. However, this method has several shortcomings: (i) the sample size of below 40–50 mL, especially with viscous samples, frequently logs of the tiny valve. The setup occupies considerable space and the operation is less user-friendly and the equipment is expensive relative to the volume of processed sample. Frequent cleaning is required, and the operation is awkward to manipulate. The sample must also be free of large cell clumps, i.e., a pre-homogenization step is required, so the overall operation becomes less attractive. A modified French press, known as the Hughes press, is equipped with a piston and a pressure-resistant cylinder to process concentrated frozen samples with a maximum pressure of 5000 bar The suspension is forced through a narrow orifice at the cylinder. The disruption is attributed to high shear stress and the abrasiveness of ice crystals. Another modification of this technique, known as the X-press allows the frozen sample press back and forth between two chambers. In general, all soluble proteins are expected to be released together with the target protein, a major drawback of this procedure. In some cases, the target protein is overwhelmed with contaminants, i.e., the purification step might be more involved with reducing protein yield.

High pressure homogenizers. Perhaps, high pressure homogenization (20–100 MPa) and ultra-high pressure homogenization (300–400 MPa) are the most efficient fluid processing equipment for cell lysis for the laboratory to large scale production facilities. A typical homogenizer is equipped with a high pressure pump and a homogenizing valve. A cell suspension enters the valve before a stream at high velocity falls on an impact ring where the cell disruption takes place. Among several valve designs, tapered cell-disruption, and fat globule dispersion (Middelberg, 1995) are used for cell disruption

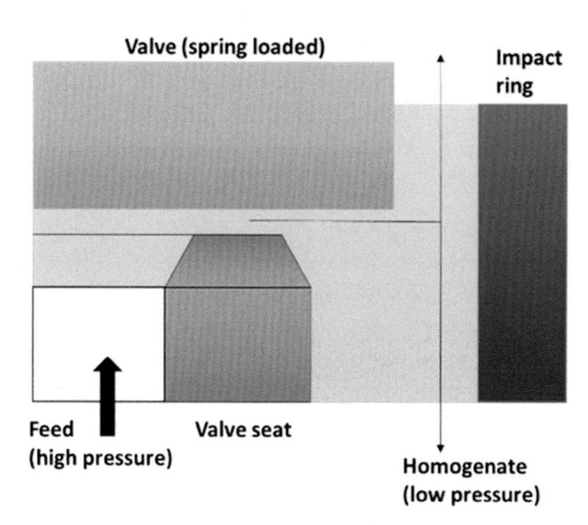

Fig. 3.2 ***Typical cross-section of a high-pressure homogenizer valve.***

(Fig. 3.2). The operation involves three major forces for efficient cell lysis: cavitation, shear, and impact (Kleinig and Middelberg, 1998). High-velocity jets are achieved by maintaining small valve gaps with short impact-ring diameters. The membrane is disrupted by high shear force when the cell enters the orifice under compression and expansion upon discharge. Commercial homogenizers are easy to operate and produce a high yield cell lysis in a shorter time. Working with filamentous fungi, the valves of the plunger pumps might not be closed completely to prevent the pressure from building up. Nevertheless, they are still most suitable for processing Gram-negative bacteria. Satisfactory disruption often requires multiple passes, i.e., a single pass is not sufficiently effective. In the disruption of *E. coli* from 5 to 150 g/L wet weight, homogenization efficiency decreases slightly with increasing cell concentrations. The fractional release of protein (D) is related to the applied pressure (P), the cell concentration (X), and the number of passes (N) as follows (Kleinig et al., 1995):

$$\log\left(\frac{1}{1-D}\right) = (0.0149 - 2.75.10^{-5}X)N^{0.71}P^{1.165} \tag{3.1}$$

Eq. (3.1) is limited to the disruption of *E. coli* using a Gaulin (http://gaulinhomogenizer.com) 15 MR high-pressure homogenizer. The pressure is related to the distance between the valve and the seat, depending on the cell types, 5–150 MPa (Engler and Robinson, 1981; Harrison, 1991).

Older homogenizers, the French press, and Manton-Gaulin homogenizer have an operating pressure of 6000–10,000 psi. Thus, 2–3 passes are needed to achieve adequate lysis. Modern homogenizers equipped with temperature control are often continuous and operated at higher pressures, 15,000 psi (100 MPa). Under this high pressure, Gram-negative bacteria, e.g., *E. coli* can be lysed in one passage (Rossi and Capponi, 2005).

Ultrasonication: In ultrasonication, waves are converted into mechanical oscillations by a titanium probe, immersed into the cell suspension (> 1 cm). A sound wave propagated in a medium carries a certain amount of energy. Applied ultrasound effectively compresses the liquid, followed by expansion (or rarefaction). Small, oscillating bubbles of gaseous substances are then released as a result of the sudden pressure drop. These bubbles (~ tens of micrometers) keep expanding after each cycle of ultrasonication until they reach an unstable size. At this critical threshold, they collide and/or violently collapse, resulting in a violent shock wave that passes through the medium to disrupt the cell membrane. Cells suspended in the liquid are also accelerated with high velocities by shock waves to trigger cell-to-cell collisions, causing cell damage and disruption. The frequencies between 20 kcycle/s (20 kHz) and 1000 kcycle/s (1 MHz) are commonly used in ultrasonic chemical processes. This specification is derived from the following simple relationship:

$$c = \lambda v = \sqrt{\frac{1}{\beta \rho}} \tag{3.2}$$

where c the velocity of ultrasound in water is ~ 1500 m/s, and the wavelength (λ) is in the range of 0.15 to 7.6 cm.

The shortcomings of sonication include excess heat release, yield variation, and the generation of free- radicals from water that can react with proteins. Therefore, sonication is best performed in multiple short bursts (30 s on and off for 90 s) while the sample is immersed in an ice bath. This procedure might be repeated 3–5 times to lyse most of the cells. Both time and power are cell-dependent and must be optimized from cell to cell. Bacterial cells can be disrupted within 1–2 min but up to 10 min is required for yeast. Small-sized particles in cell debris are problematic in later steps. For small samples, centrifugation at g-force of 12,000–25,000 g at 4 °C) for 20 min is sufficient to yield clear supernatants (Ames 1994). Sonication using 10–100 W acoustic power requires no chemicals and remains a simple method for cell disruption in the laboratory. Probes with small diameters are commercially available for processing a small sample volume (1 mL). However, for cell quantities larger than 50 g, sonication becomes less effective because of the difficulty in maintaining low temperatures and the long sonication times required to reach adequate lysis. Sonication can also lyse mammalian and insect cells.

Hydrodynamic cavitation generated in a venturi tube or an orifice plate (1 mm in diameter), results in high flow velocity (Balasundaram et al., 2009; Lee et al., 2015). This is a process of vaporization, bubble generation, collapse, and implosion in the liquid to release tremendous localized energy in the form of shock waves. Some reported results also indicate that the hydrodynamic cavitation is more energy efficient than acoustic cavitation (sonication) for cell disruption (Capocellia et al., 2014; Lee et al., 2015).

Bead mill (bead beating method). Commercial beads are prepared from glass, steel, or ceramic with different sizes. Under agitation at high speeds, cells are broken due to their collision with beads to release proteins and other intracellular components by shear force. For the best result, the process must be optimized for bead diameter and density, cell concentration, and agitation speed. Beads of 0.25–0.5 mm in diameters are effective for lysis of bacteria and yeast with high efficiency (Chisti and Moo-Young, 1986; Taskova et al., 2006). As cells are broken into cell debris, all intracellular components are released along with the target protein and thereby separation and purification of this mixture become more complicated. Heat generation due to the collision between beads and cells might also degrade proteins and other biomolecules of interest.

3.4 Non-mechanical procedures

3.4.1 Physical methods

(a) Thermolysis: Heat treatment of microorganisms above 60 °C for 60–90 min in a stirring tank can be effective to disrupt the bacterial cell walls and release their intracellular proteins and other biomolecules. This process often denatures proteins, therefore, it is limited for the isolation of thermostable recombinant enzymes such as esterase (Ren et al., 2007), inorganic pyrophosphatase (Hoe et al., 2001), catalase (Andrews and Martin, 1979), Fe-superoxide dismutase (He et al., 2007), etc. Like chemical permeabilization, the cells are not destroyed by thermolysis. Proteases are likely deactivated, resulting in improved protein recovery. Short cycles of freezing and thawing can also be used for breaking the cells. For recombinant proteins from *E. coli*, 2 min freezing in dry ice/ethanol bath, and 8 min thawing on ice water bath as a cycle carried out 3 times worked well (Johnson and Hecht, 1994). The isolation of the desired protein by thermo-precipitation is limited to the recovery of thermophilic enzymes/proteins. These proteins are still stable at high temperatures even after they have been expressed in *E. coli*.

(b) Osmotic shock: Equilibrating cells with high salt solution [typically IM] leads to a build-up of osmotic pressure inside the cells. For *E. coli*, this estimated pressure is ~ 2 atm (Grayson et al., 2006). When the cell is exposed to lower salt concentrations, its membrane is permeable to a large amount of water. This phenomenon, known as osmosis, elicits cell swelling, and eventual bursting. In general, this technique is more applicable for disrupting the membrane of mammalian cells due to its fragile structure. This method only recovered 60 percent of recombinant creatinase from *E. coli*, however, the efficiency can be up to 75 percent if the cells are treated with Ca^{2+} or Mg^{2+}(Chen et al., 2004).

3.4.2 Chemical methods

(a) Detergents: Mild surfactants (surface acting agents) have been used for the disruption of cell membranes to release proteins and other intracellular materials. Biological membranes, the most obvious hydrophobic/hydrophilic interfaces, are disrupted by surfactants, known as amphiphilic molecules with both hydrophilic and hydrophobic regions. Sodium dodecyl sulfate (SDS) is very effective in solubilizing biological membranes, however, it often unfolds (denatures) cytosolic proteins and partition membrane proteins into small droplets, known as micelles. Anionic sodium deoxycholate is derived from bile salts and used in cell lysis. Its cationic counterpart, cetyltrimethylammonium bromide (CTAB), another well-known detergent, is also widely used to isolate DNA from plants. Zwitterionic detergents have no charge because they possess both anionic and cationic groups. The isolation of membrane proteins is also performed by a popular zwitterionic derivative of cholic acid, known as CHAPS. Non-ionizing detergents with an uncharged polar head, e.g., a glycoside (sugar) or polyethylene chain, are mild but still capable of dissociating proteins, which are loosely bound to membranes.

In aqueous environments, detergents will spontaneously form small particles or micelles at a critical micelle concentration (CMC) due to the orientation of their hydrophilic heads and the hydrophobic tails. In the narrow concentration range wherein the formation of micelles starts, the surfactant solutions show a drastic change of specific conductivity, surface tension, light scattering, etc. Thus, different methods apply to the determination of the CMC. As an example, the CMC of an ionic surfactant can be related to the specific conductivity (κ) as $\kappa = f(C)$. The slope change or discontinuity of κ/C dependence gives the CMC as follows:

$$\frac{d\kappa}{dC} = A_o + \frac{A_{1-}A_2}{1+e^{\frac{(C-C_o\}}{\Delta C}}} \tag{3.3}$$

where ΔC = the width of the transition width, C_o (CMC) = the center of the sigmoidal, A_1 = large values of surfactant concentrations and A_2 = small values of surfactant concentrations (horizontal asymptote). The second derivative of specific conductivity ($d^2\kappa/dC^2$) versus surfactant concentration can also be used for the determination of CMC:

$$\frac{d^2\kappa}{dC^2} = \left(\frac{d^2\kappa}{dC^2}\right)_{C=0} = \frac{A}{\omega\sqrt{\frac{\pi}{2}}} exp^{\frac{-2(C-C_o)^2}{\omega^2}} \tag{3.4}$$

where A = the total area under the curve and ω = the width of the peak at half-height. Detailed information for the determination of CMC by equations 1–2 can be found elsewhere (Goronja et al., 2016). Micelles, ranging from 1,200 to 80,000 Daltons, disrupt cell membranes, and pick up proteins dissociated from cellular membranes. Detergents and their use are application-specific and their effects on biological systems will vary greatly.

(b) Alkali treatment: In alkaline lysis, all kinds of cells can be subjected to a lysis buffer consisting of NaOH and SDS, pH = 11.5–12.5 (Harrison, 1991). The OH^- ion hydrolyzes the fatty acid-glycerol ester bonds of the cell membrane, which becomes more susceptible to SDS solubilization. The process is very simple and uses two inexpensive chemicals, however, it might take 6 to 12 h. The isolation of bacterial plasmid DNA is commonly conducted by this convenient method (Feliciello and Chinali, 1993). As most of the proteins denature under these conditions, this method is not used for isolating proteins in their functional forms.

(c) Enzyme lysis: Several enzymes such as protease zymolase, glycanase, lysozyme, lysostaphin, and cellulase can be used for small scale or large-scale lysis. The method is specific because enzymes are effective against specific classes of cells: chitinase for yeast cells, lysozyme for bacterial cells. Lysozyme breaks the glycosidic bond of peptidoglycan; therefore, it is mainly used to lyse Gram-positive bacteria. For Gram-negative bacteria, a detergent is needed to remove the outer membrane, followed by the attack of lysozyme. Tris is often used as a buffer in lysis methods to permeabilize outer membranes very effectively. The addition of EDTA (1 mM) is also helpful as EDTA chelates the magnesium ions that stabilize membranes. During cell lysis from some specific samples, significant DNA is liberated, therefore, it is necessary to add DNase to reduce the solution viscosity. A typical lysis buffer consists of 50 mM Tris–HCl pH 7.5, 50–200 mM NaCl, 5 percent glycerol (v/v), 1 mM DTT (dithiothreitol is also widely used for disruption of protein disulfide bonds), and 1 mM PMSF (phenylmethylsulfonyl fluoride). Enzymatic cell lysis can be carried out on any scale, but the lysozyme and DNase become costly for large-scale preparations (Islam et al. 2017).

The protocols for enzyme lysis from various commercial enzyme suppliers are available and well documented. The cell lysis by the enzymes is also discussed by Andrews and Asenjo (1987) and Salazar and Asenjo (2007). A simple protocol for cell lysis is described in Box 3.1:

Different methods can be used for cell disruption and have their advantages and disadvantages. Chemical methods can be performed in any wet chemistry lab but complete cell disruption may not be achieved. Enzyme lysis is cell-dependent and costly but it is very specific and does not alter the protein functionality. Mechanical procedures

Box 1 A typical procedure for cell lysis using lysozyme

1. Resuspend the cells in chilled lysis buffer: 1 g cell wet weight/ 1 mL buffer
2. Add the PMSF* (10 µL PMSF @100 mM) per mL of cell suspension).
3. Add lysosyme at 300 µg/mL and incubate the cell suspension at 4 °C for 4 h
4. Add 5 µL $MgCl_2$ (1M) and 1 µL DNase (1 mg/mL)/ per mL of cell suspension.
5. Incubation at 4 °C for 30 min.
6. Remove cell debris by ultracentrifugation at 4 °C for 30 min at 20,000 – 30,000 g.

*PMSF (phenylmethylsulfonyl fluoride) is unstable in water and should only be added to lysis buffers with a final concentration of 1 mM just before use to protect protease.

require some capital investment for equipment and a high cost for maintenance. Ideally, non-mechanical and mechanical methods can be combined to increase the efficiency of lysis and avoid protein denaturation. Of course, the ultimate choice will be dictated by the value of the target protein and the intended use. A well-known product, yeast autolysate or yeast extract of *S. cerevisiae*, consists of the "water-soluble components of the yeast cell such as amino acids, peptides, carbohydrates, salts, etc. As one of the important component of several bacterial culture media, its autolysis encompasses the flowing steps: (i) autoclaved the yeast suspension for 10 min at 115 °C and cooled on ice, (ii) removal of the cell debris by filtration or centrifugation, (iii) the supernatant is subject to autoclave, cooling and spray drying to obtain the powder (Zarei et al. 2016). However, physical, chemical, and enzymatic methods are combined to disrupt the yeast cell wall, followed by other separation methods to yield odorless, tasteless, and colorless product (Potman and Wesdorp, 1994) as required for its use in food or cosmetic industries.

3.5 Protein precipitation

Protein extracts of microbial cells, plant cells, or tissues contain endogenous organic compounds and their removal is a prerequisite. Among such protein sources, plant tissues contain lipids, phenolics, carbohydrates, organic acids, pigments, etc., which interfere with protein extraction and purification. As an important part of downstream processing, a target protein is first separated from endogenous contaminants by different techniques.

A given protein should have an overall charge due to the presence of its amino acids with different charges: positive, negative, or neutral. Proteins exhibit a net positive charge at pH below pI but a net negative charge at pH above their pI. The charge of proteins forms an important basis for protein separation by IEC as discussed later. Hydrophobic residues of proteins are mostly embedded in their globular core to minimize their interaction with water. However, some residues exist in patches on the surface, effecting their low solubility in an aqueous solvent. Interaction between ionic groups in the solvent and charged and polar surface residues improves the solubility of a protein.

3.5.1 Salting-in and salting out

Water molecules close to a protein exhibits different interactions, compared to bulk water molecules. The protein exhibits three main interactions with water: (i) ion hydration of charged side chains (*Asp, Glu, Lys*, etc.), hydrogen bonding between polar groups and water (*Ser, Thr, Tyr*, and the main chain of the protein), and hydrophobic hydration (*Val, Ile, Leu, Phe*). The hydration layer (0.3 to 0.4 g water per gram protein (Rupley et al., 1983), closely associated with the protein surface plays an important role in the protein solubility and folding (Fig. 3.3).

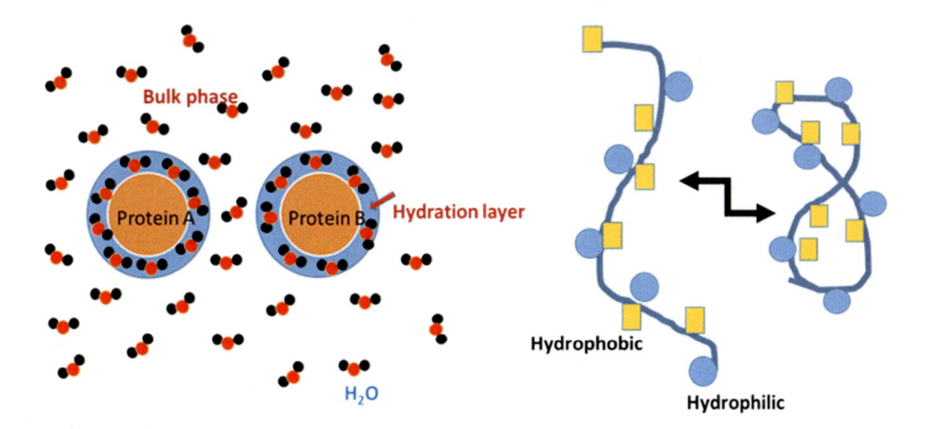

Fig. 3.3 (Left) The formation of a hydration layer between water and protein. Polar groups on the protein surface form hydrogen bonds with water, one of the key factors of protein hydration. Charged amino acid sidechains (as well as the N- and C-termini) also make electrostatic interactions with water. (Right) Hydrophobic interactions are important for the folding of proteins to reduce undesirable interactions with water.

Among different types of neutral salts, ammonium sulfate $(NH_4)_2SO_4$ exhibits high aqueous solubility (~ 70 g/100 mL) and antimicrobial property as most microorganisms cannot survive in high salinity due to cytoplasmic protein aggregation (Baker et al. 2019). The ammonium (NH_4^+) and sulfate (SO_4^{2-}) ions of this salt are attracted to the opposite charges on the protein. This attraction of opposite charges prevents the water molecules from interacting with the protein. In general, globular proteins become more soluble at a low salt concentration, < 0.15–0.5 M (Fig. 3.3). The "salting-in" behavior is also known as the Hofmeister effect (Hofmeister, 1888). Like urea and guanidine hydrochloride, $MgCl_2$ and NaSCN destabilize native proteins due to their more favorable interaction with dissociated and unfolded proteins (Arakawa and Timasheff 1984; Nozaki and Tanford, 1963, 1970; Lee and Timasheff, 1974). However, protein solubility decreases with increasing salt concentration as more hydrophobic domains of the protein are exposed and interact with one another, leading to protein precipitation or "salting out". In other words, the water surface tension increases when salt is added to the solution, resulting in increased hydrophobic interaction between protein and water. In response, the protein is folded to minimize its surface area in contact with the solvent. This eventual self-association leads to protein precipitation. A detailed discussion of these complex effects can be found elsewhere (Kita et al., 1994; Timasheff and Arakawa, 1997). Proteins differ markedly in their solubility at high ionic strength, therefore, "salting out" is a very useful procedure to assist in the purification of the desired protein. Fibrinogen, a blood-clotting protein is precipitated by 0.8 M ammonium sulfate, whereas a concentration of 2.4 M is needed to precipitate serum albumin.

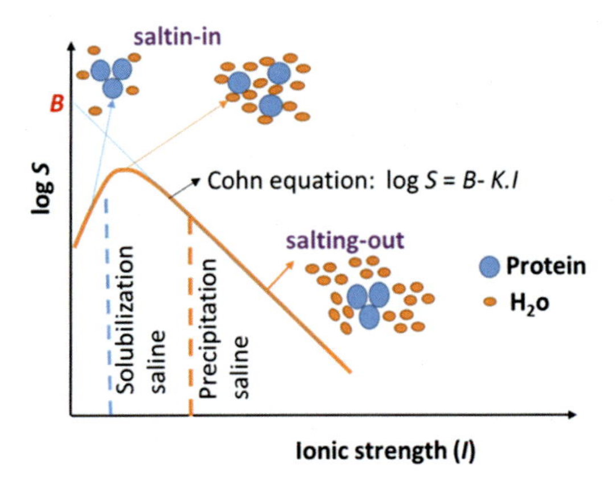

Fig. 3.4 ***A typical characteristic solubility curve of a protein*:** "salting-in" at low salt concentrations and "salting-out" at higher salt concentrations.

The solubility of a protein and increasing ionic strength of the solution is governed by the Cohn equation. (Fig. 3.4):

$$logS = B - K.I \tag{3.5}$$

where S = the protein solubility, B is idealized solubility, K is a salt-specific constant and I is the solution ionic strength.

$$I = \sum_{k=1}^{n} C_k Z_k^2 \tag{3.6}$$

where Z_k is the ion charge of the salt and C_k is the salt concentration.

The Cohn equation. with historical significance (Cohn, 1925) was originally suggested to relate the protein solubility (g/L) as a function of the molecular concentration of ammonium sulfate [M]

$$logS = -aM + b \tag{3.7}$$

α is considered constant (average 4.4 for pseudoglobulin), over a narrow range of hydrogen ion concentrations (pH range), whereas β (average 7.1) forms a non-linear relationship with pH even within this narrow range. The β constant yields the solubility of a salt-free protein solution.

Ammonium sulfate with two ions high in the Hofmeister series can also stabilize a protein structure (Scheme 3.1)

The optimal conditions including the ammonium sulfate concentration are protein dependent and for the best protein yield, this process is often performed at 4 °C to preserve the protein functionality. Various ammonium sulfate (MW = 131.14) concentrations must be precisely prepared and incubated with a protein sample overnight. Saturated ammonium sulfate is temperature-dependent: 4.1 M (541.8 g/L) at 25 °C,

← Increasing precipitation (salting-out)

Anions: PO_4^{3-} > $\mathbf{SO_4^{2-}}$ > CH_3COO^- > Cl^- > Br^- > ClO_4^- > SCN^-

Cations: $\mathbf{NH_4^+}$ > Rb^+ > K^+ > Na^+> Li^+ > Mg^{2+} > Ca^{2+} > Ba^{2+}

Increasing chaotropic effect (salting -in) →

Scheme 3.1 Salting-in and salting-out in the Hofmeister series. The term chaotropic refers to the ability of the ion to produce chaos in the water structure.

4.06 M (536.49 g/L) at 20 °C, 3.97 M (524.6 g/L) at 10 °C, 3.93 M (519.1 g/L) at 4 °C and 3.9 M (515.35 g/L) at 0 °C. The specific volume of ammonium sulfate is also dependent on the temperature, 0.54 ml/g, i.e., adding 1 g of ammonium sulfate to an aqueous solution will increase the volume by 0.54 mL. The preparation of this salt is facilitated by using a monogram/table available in several textbooks (Englard and Seifter, 1990) or an ammonium sulfate calculator (https://www.encorbio.com/protocols/AM-SO_4.htm). The basic theory of protein precipitation by the addition of ammonium sulfate is discussed elsewhere (Wingfield, 2016).

3.5.2 Trichloroacetic acid (TCA) and other procedures

Trichloroacetic acid (TCA), 10 percent w/v, or 20 percent w/v, is most widely used for protein precipitation in dilute solution. However, this acid is reported to be less effective in precipitating low amounts of protein (0.02–0.03 mg) in dilute protein solution (Manik and Aftab 1999; Wang et al. 2006; Ngo et al., 2015). TCA solution is stable for months and is commonly used for plant and animal proteins. This procedure minimizes protein degradation and inhibits the activity of endogenous proteases. The other two classical procedures for the extraction of hydrophobic proteins are based on TCA/acetone and phenol (Novak and Havlicek, 2016). The precipitated proteins are more difficult to dissolve and NaOH must be added to increase their solubilization (Nandakumar et al., 2003). There are some attempts to elucidate the mechanism of protein precipitation induced by TCA. The precipitation extent versus TCA concentration can be described as a U-shaped curve with 15 percent TCA as the optimal point, implying hydrophobic aggregation is the dominating mechanism (Novak and Havlicek, 2016). The acidic nature of TCA invokes protein precipitation by sequestering the protein-bound water (Tanford, 1964) and conformational changes (Ahmad et al., 2000). TCA-induced protein precipitation might also involve the reversible association of a stable partially structured intermediate (Rajalingam et al., 2009), a discussion topic for another chapter in this book. A nonionic surfactant; Triton-X100 and deoxycholate, can be used with TCA/acetone to extract and precipitate protein (Feist and Hummon,

2015), however, the surfactant must be removed from the purified protein and each surfactant has its own limitations.

3.5.3 Protein precipitation by organic solvents

Proteins in the native state exhibit low solubility in the presence of organic solvents at low temperatures. In contrast to salting-out salts, the use of solvents at high concentrations or high temperatures should be avoided due to their interactions with hydrophobic groups of the proteins (Arakawa et al., 1990, 2007; Inoue and Timasheff, 1972; Pittz and Timasheff 1978;). Protein precipitation by acetone requires at least a 4:1 ratio of acetone to the aqueous protein solution (Zhang et al., 2015), a major drawback for processing a large protein volume. The use of 2-methyl-2, 4-pentanediol is important as this organic solvent precipitates protein in the native state (Pittz and Bello, 1971, 1973). Ethanol (10–40 percent) is also an effective solvent that fractionates plasma proteins into various functional therapeutic products (Cohn et al., 1947; Morgenthaler, 2000). Mechanistic insights into egg-white lysozyme precipitation by ethanol unravel that the solubility of this enzyme is reduced with increasing ethanol concentration. Its native conformation is changed to α-helical rich structure in the presence of 60 percent ethanol (Yoshikawa et al., 2012). The protein conformational change is reversible and can be attributed to the favorable interaction of ethanol with non-polar groups and the unfavorable interaction with polar groups. Other organic solvents, e.g., a mixture of acetonitrile/TCA, methanol/chloroform, and phenol/ammonium acetate/methanol have also been used for protein precipitation (Simpson, 2006).

The use of butanol and pentanol for the precipitation of proteins from plants, animals, microorganisms, etc. is worth mentioning. A protein sample is treated with an organic solvent (normally t-butanol as it acts as a differentiating solvent) and then a salt solution, mainly ammonium sulfate. Tertiary (*t*-) butanol is not miscible with water but can be induced to form a separate phase by the addition of salt. If protein is present in the solution it may form a third phase, between the aqueous and t-butanol phases. After stirring or centrifugation, the resultant exhibits a three-phase mixture comprising an upper organic phase, a lower salt-rich aqueous phase, and a central protein or intermediate layer with precipitated proteins (Dennison and Lovrien, 1997). This phenomenon is known as three-phase partitioning (TPP) (Bayraktar and Onal, 2013; Li et al., 2013; Roy et al., 2004) to which the rest of this book is devoted.

The upper phase is enriched with pigments, lipids, and other hydrophobic materials in protein samples, whereas the lower phase consists of more polar compounds. Thus, TPP is a simple but powerful method for separating proteins with different hydrophobic/hydrophobic activities from other contaminants. TPP is very effective in separating the target proteins form various contaminants including some low molecular weight or hydrophobic molecules. A 10-fold purification factor with a yield of 76 percent is achieved by TPP for the recovery of pectinases from *Aspergillus niger* (Sharma and Gupta, 2001). TPP is applicable for the separation of polymers. TPP is also applied to extract and purify chitosan from shrimp shell (Sharma et al., 2003), alginates from *Dunaliella salina* (Sharma and Gupta, 2002), tapioca starch and tapioca starch derivatives (Mondal et al., 2004), levan and hydrolyzed levan from levansucrase of *Zymomonas mobilis* (de Oliveira Coimbra et al., 2010), aloe polysaccharides (Tan et al., 2015), and *Corbicula fluminea* polysaccharides (Yan et al., 2017), polysaccharide-protein complexes (PSP) from *C. fluminea* (Wang et al., 2017), polysaccharides from the immunomodulatory medicinal mushroom *Inonotus obliquus* (Liu et al., 2019). Albeit this topic is out of the scope of this chapter, it is essential to stress that protein samples prepared from plants, animals, and microorganisms contain polysaccharides that need to be separated from target proteins. Therefore, protein complexing agents, e.g., tannins, lipids, phenolics, and other enzyme inhibitors from plants and micro-organisms are extracted into the upper butanol phase, thus facilitating the purification of proteins. This method can be easily scaled up from the bench scale to pilot plant and production facilities. The formulation of a TPP mixture depends on pH relative to the pl, protein molecular weight, protein concentration, temperature, and the amounts of t-butanol and ammonium sulfate (Pike and Dennison, 1989). The method can be optimized for isolating a specific single-chain protein with a known isoelectric point. The effects of protein concentration and MW are also significant, and it is difficult to predict the behavior of an unknown system.

3.5.4 Metal-chelate protein precipitation

The metal ions form complexes with histidine, cysteine, tryptophan, and arginine residues of proteins (Porath et al., 1975). This binding event forms the basis for the birth of metal chelate affinity chromatography for protein separation. Polyhistidine tags (His-tags) can be fused to the N or C terminus of recombinant proteins to enable the purification of such proteins by metal-chelate adsorbents (Hochuli et al., 1988). Two common metals, Ni^{2+} and Cu^{2+} bind strongly to iminodiacetic acid (IDA) and nitrilotriacetic acid (NTA), which can be conjugated with the polymers. This affinity of the proteins with His-tags towards metal chelates can also be exploited in the precipitation mode. For precipitation, the polymers must be sensitive to temperature, pH, ionic strength, etc., but thermosensitive polymers are preferred because no chemicals are added to the protein sample (Galaev et al., 1997; Kumar et al., 1999). However, the conjugation of iminodiacetic acid (IDA) or nitrilotriacetic acid (NTA) to the thermosensitive polymers adversely affects the efficiency of polymer precipitation. Of interest is the complex between imidazole and Cu^{2+}albeit the association binding constant of $\log K_{ass} = 3.76$ is not sufficient metal chelation. Therefore, several imidazole ligands must be attached to the polymer chain. In this context, imidazole is incorporated into a NIPAM (N-isopropylacrylamide)-based polymer and co-polymerized with 1-vinylimidazole. The resulting polymer exhibits good complex formation with proteins (Galaev et al., 1997) and can be precipitated at 20–30 °C with the aid of 0.4 M NaCl (Kumar et al., 1999). Metal chelate affinity precipitation has been used for purification of natural metal-ion binding proteins and recombinant proteins with His-tag residues (Carter et al., 2005; Kumar et al., 2003; Nguyen and Luong, 1989; Stiborova et al., 2003). A protocol for metal-chelate affinity precipitation of proteins using responsive polymers is well described elsewhere (Mattiasson et al., 2007).

3.6 Affinity precipitation and immunoprecipitation

Affinity precipitation is based on a single step of binding between the target protein and the ligand to form a stable complex with an association binding constant of at least 10^5 M^{-1}. The ligands can be small or macromolecules and unless they are target specific antibodies, the number of protein ligands is very limited. A stable ligand-protein pair facilitates the removal of contaminants in the column by washing steps. The bound protein is eluted from the affinity ligand and this procedure is perhaps the simplest method for separating a target protein in the presence of other unwanted biomolecules from crude extracts.

Some low–molecular weight ligands are *p*-aminobenzamidine, a trypsin inhibitor (Luong et al., 1988; Galaev and Mattiasson, 1992; Chen and Jang, 1995) and triazine dyes (Guoqiang et al., 1995; Linné-Larsson and Mattiasson, 1994) used for purification of nucleotide-dependent enzymes. Of interest is the development of polymers with multiple binding sites for affinity ligands with low binding constants with their corresponding proteins, enabling the affinity precipitation procedure. A typical example is the binding between the concanavalin A-D-glucose pair with an association constant of $K_{ads} = 6.5 \pm 5.4 \times 10^3\ M^{-1}$ (Kussrow et al., 2009). Immunoprecipitation (IP) is a specific method based on a specific antibody immobilized to a solid support, e.g., magnetic beads, agarose resin in a microcentrifuge tube. A solution with a target antigen is added to the beads, followed by mixing and incubation. The beads pelleted to the tube bottom by magnet or centrifugation are isolated from the supernatant. Bioconjugation has been developed extensively and IP is widely used for purifying a small amount of proteins/biomolecules for immunoassays (Vashist and Luong, 2018), Western blotting, and other assay techniques.

3.7 Protein purification

3.7.1 Dialysis

After cell lysis and protein precipitation, samples with proteins are readily available for further purification and concentration. In general, the sample also has many unwanted compounds from the sources of protein or chemicals used in the cell lysis and protein precipitation step. On a bench scale, dialysis is commonly used for the removal of such small unwanted compounds. Commercial flat dialysis tubing with different MW cut-off (MWCO= membrane pore size in Angstrom (Å)), ranging from 1–1,000,000 kDa, are widely available and this step is straightforward. Pore sizes of 1 K to 50 K MWCO membranes range from ~ 10–100 Å The MWCO tells that the dialysis membrane is only suitable if the molecular weight of the desired protein is significantly higher than MWCO. Even then, small protein loss (up to 5–10 percent) generally accompanies dialysis A dialysis membrane is a semi-permeable film, fabricated from regenerated cellulose or cellulose ester with various sized pores. Cellulose ester-based membranes with various MWCOs are supplied as wet products. However, regenerated cellulose exhibits minimal protein adsorption, better chemical compatibility, and heat stability, compared to cellulose acetate (cellulose ester). It is also more resistant to organic solvents and diluted acids.

Small molecules selectively diffuse through a semipermeable membrane whereas proteins are retained as they are larger than the membrane pores. This process is time-consuming and requires multiple buffer exchanges. As an example, the dialysis of 1 mL of the sample will need at least 100 mL of dialysis buffer, i.e., the concentration of unwanted dialyzable substances decreases 100-fold after equilibrium. After two extra buffer changes of the same volume, the contaminant level in the sample is reduced by 10^6-fold. Many membrane-based designs for carrying out dialysis are available commercially. Stirring the outside buffer during dialysis helps in two ways. The equilibration is faster, and it breaks up the Nernst layer (about 200–300 molecules thick), the molecules of which can reenter the dialysis bag. The diffusion is very slow for molecules if their mass is comparable to that of the membrane. For practical dialysis, unwanted molecules must be at least 20–50 times smaller than the membranes MWCO rating. In this context, it is not practical to separate a mixture of protein of 10 kDa to 30 kDa by a 20K rated dialysis membrane.

A concept of "forced dialysis" deserves a brief comment here as this method has become popular recently for protein separation of solution volumes less than 100 μL and as low as 10 μL (Zumstein 2001). A microcentrifuge tube (0.5–1.5 mL) with the sample is covered with a dialysis membrane and secured by a tube cap with a predrilled hole. Gentle centrifugation is performed with the microcentrifuge tube inverted and immersed in a dialysis buffer. Indeed, such special tubes are also commercially available.

3.7.2 Size exclusion chromatography (SEC)

Proteins with different sizes are separated by gel filtration or size exclusion chromatography, based on porous beads made from polyacrylamide, dextran, or agarose. Commercial beads are typically 100 μm (0.1 mm) in diameter. Small proteins easily penetrate such porous beads and spend some time within these holes, resulting in a slow flow through a column. In contrast, large proteins move out of the column quickly owing to their minimum interactions with porous beads. A mixture of proteins with sizes above the column size range co-elute cannot enter the pores and elute together. Proteins with molecular masses below the permeation limit have full access to the pores but also co-elute eventually. In this context, the molecular weight of an eluted protein is inversely related to the total elution volume, i.e., more the elution volume is needed for a small protein and vice versa. With proteins of known mass, e.g., bovine serum albumin (66.5 kDa), protein A (42 kDa), Con A (104–112 kDa), a standard curve between the elution volume and protein mass can be correlated, i.e., the mass of a protein in a mixture can be estimated from the elution volume. This method is mainly useful for purifying native proteins with different sizes but with similar physicochemical properties such a charge, polarity, hydrophobicity, etc.

Like membrane dialysis, size exclusion chromatography can be used for protein desalting using resin beads with pores for the penetration of only salts, whereas proteins flow freely out of the column. The column performance is very sensitive to column packing and some stationary phases can exhibit nonspecific interactions between protein and resin, resulting in decreasing resolution. Only a small sample can be used for adequate resolution and this method often offers low resolution for complex protein mixtures.

Commercial resins are specifically designed to process samples with different concentrations and sample sizes. Their performances will depend on the protein molecular weight and the separation conditions as specified by the suppliers (Table 3.3). Of notice is the stationary phases based on cross-linked agarose and dextran as they offer the highest resolution and compatible with most solvents. For the reuse of the stationary phase, it must be autoclavable to prevent plausible contamination. Column packing can have a significant effect on protein resolution; therefore, this task requires considerable expertise and training. Many of the current gel-filtration media can withstand high pressure and pre-packed HPLC columns are available commercially from many vendors.

Table 3.3 Some key features of selected commercial stationary phases.

Matrix (stationary phase)	Key features
Highly cross-linked agarose (Superose)	Compatible to aqueous/organic solvents and viscous solvents Autoclavable and might display hydrophobic interactions with proteins.
Cross-linked agarose (Sepharose)	Non-autoclavable, higher resolution, and yield of molecules over a larger fractionation range.
Cross-linked agarose and dextran (Superdex)	Autoclavable and compatible with aqueous and organic solvents. High resolution and recovery.
Cross-linked dextran by epichlorohydrin (Sephadex)	Autoclavable; some types may shrink in organic media. Solvent-compatible matrices are available. It is useful for group separation and molecular weight determination with ease of buffer exchange.
Cross-linked allyl dextran and N, N - methylene bisacrylamide (Sephacryl)	Autoclavable and compatible with aqueous and organic solvents. It can separate proteins over a large molecular weight range with high recovery.

3.7.3 Ion exchange (IEC) chromatography

The separation by IEC is based on the protein charge at a given pH, thus, the isoelectric point (*pI*) of the protein of interest must be determined in advance. The surface charge of proteins is pH dependent, positive at low pH as the NH_2 is protonated and becomes negative at high pH due to the deprotonation of the carboxyl group. At its isoelectric point (pI), a protein becomes neutral and cannot interact with the surrounding medium, leading to precipitation. The surface charge of proteins can be easily estimated by several commercial instruments. A parameter, denoted as the zeta potential, indicates the degree of repulsion or attraction among neighboring charged proteins in a dispersion. Any change in this charge reflects one of the following events: surface modification, unfolding, and denaturation of protein, aggregation, and conformation changes. The zeta potential value depends on biomolecules, ionic strength, and the solution pH (Salgın et al., 2012). A solution with high salt content must be used as low salt content provides erroneous results due to decreasing electrostatic shielding. Phase analysis light scattering with high sensitivity is conveniently used for probing the motion of charged proteins. The zeta potential also serves as an indicator for predicting the stability as well as any conformational changes of the target protein in response to ligand interaction. Detailed information on this technique can be found elsewhere (Tscharnuter, 2001).

A cation exchange resin, e.g., cellulose or agarose beads with a carboxylate group binds to positively charged proteins; an anion exchange resin, e.g., a diethylaminoethyl group, binds to proteins with opposite charges. Ion exchange resins are classified as "weak" or "strong", depending on the ionization state of their functional groups with pH. Cation exchange resins with sulfonic acid functional groups are strong while resins with carboxylic acids are weak. Similarly, anionic-exchange resins with quaternary amines are strong while resins with secondary or tertiary amines are weak. A protein with a net negative charge can bind to a negatively charged cation exchanger as its surface often possesses some positively charged clusters.

A buffer solution (mobile phase) with its pH set between the pl or protein pK_a and the bead pK_a is needed to flow the protein sample through the column. Non-binding proteins (neutral or the same charge as the beads) pass through, whereas proteins with the opposite charge exhibit ionic interactions with the beads and stay in the column. To elute the bound proteins, the column is often flushed with a salt (commonly NaCl) gradient where the salt concentration is gradually increased. Molecules with the weak binding are eluted first, followed by molecules with stronger binding to the column. A simple salt like NaCl plays a dual role as the Na^+ ion competes with the positively charged bound protein for the bead negative functional group in cation exchange separation. In contrast, the Cl^- ion competes with the bound protein for the bead positive functional group in anion exchange chromatography. A low pH buffer is also often used as the elution buffer.

The more acidic conditions lower the net charge of the protein or render it more positive. The protein now acquires a positive net charge and is repelled by the like-charged resin to come out of the column pure. At pH < pI, the protein becomes positive. Conversely, the protein becomes negative overall if the pH is higher than its pI. IEX is more effective in the separation of proteins with significantly different pI. Several commercial resins are available to offer high throughput and high selectivity with excellent purity and yield. Long linear polymer chains are preferred as they are more accessible with a tighter binding of target substances. In most cases, such information is proprietary. The protein binding capacity of some commercial resins for bovine serum albumin are listed in Table 3.4 (Muller, 2005). In comparison to the resins used in affinity and hydrophobic chromatography, ion exchange resins display the highest binding capacities for proteins.

Albeit the Langmuir isotherm still has some limitations, it is often used to correlate equilibrium adsorption data for proteins to estimate the binding capacity (Langmuir, 1918):

$$q_{eq} = \frac{q_m.K.C_{eq}}{1+K.C_{eq}} \tag{3.8}$$

where q_m is the total binding capacity of the resin, K is a specific constant, q_{eq} is the amount protein binding at a specific time and C_{eq} is the corresponding equilibrium protein concentration.

As discussed earlier, ICE is useful to isolate and purify protein samples based on charge differences. Therefore, peptide tags can be inserted to a recombinant protein with a specific pI, out of the range of other contaminants to facilitate separation.

Materials used in the stationary phase of IEC deserve a brief discussion here because of the widespread use of this method in other fields. Silica-based materials are still the most important ion exchangers for use in HPLC columns. Silica particles can be coated with a polymer layer such as fluorocarbon, silicone, and polystyrene. The thin layer is subject to chemical modification to introduce functional groups. The particle sizes are 5 (Hordijk and Cappenberg, 1985) to 44 μm (Fritz et al., 1980) with a binding capacity ranges from 5 to 1000 μg/g (Buytenhuys, 1981). Compared to entire polymeric particles, the diffusion within a thin layer is much faster. Silica can also be functionalized with a chemical group, which is bonded to a silica particle. Zipax (Du Pont) is an important

Table 3.4 Selected binding capacity for anion exchange resins.

Resin	Protein Binding Capacity [mg/mL] as specified by selected suppliers
Q Sepharose® XL	> 130 (BSA)
Ceramic Hyper D	> 180 mg/mL (BSA)
Fractogel® EMD TMAE Hicap (M)	> 180 mg/mL (BSA)
Toyopearl® SuperQ-650M	> 150 mg/mL (BSA)
Q Sepharose® Fast Flow	> 120 mg/mL (HSA)
Toyopearl® QAE-550C	60 – 80 mg/mL

example consisting of spherical glass beads of 30 μm with a thin porous silica layer (in μm range of thickness) on the surface. This type of material is known as pellicular materials. Microparticulate materials are fully porous particles with functional groups distributed throughout their internal pore structure. They can be spherical with a diameter of 5 μm or irregular in shape with similar chromatographic efficiencies. Both pellicular and microparticulate materials provide comparable chromatographic efficiencies. Pellicular materials with low active surface areas are limited to process a small sample but they are more easily packed into columns, compared to microparticulate materials. Other commercial ion exchange resins are rigid polymeric beads with covalent surface chemistries. They have good physical and chemical stability with ease of handling and packing.

Silica tends to have high solubility in aqueous solutions at a very low or high pH. Therefore, silica-based IC materials cannot be operated at $pH < 2.0$ or samples of alkaline pH. This limitation might be problematic for the processing of proteins as the column must be subject to harsh conditions to prevent contamination. Metal ions or any positively charged molecules display electrostatic interactions with silica-based anion-exchange IC columns, i.e., they compete with proteins for binding to IC materials and reduce chromatographic efficiencies. The protein sample must be pretreated to remove such metal ions.

Of also interest is the ion-exchange behavior of hydrous oxides such as ($A1_20_3.nH_20$), silica ($SiO_2.nH_20$), and zirconia ($ZrO_2.nH_20$) as they display both anion and cation-exchange characteristics in a pH region. Consequently, both anions and cations in a sample mixture can be separated by a column packed with these materials (Fig. 3.5).

Technical problems related to IEX are frequently encountered such as no/low protein binding, low resolution, the protein loses activity or proteins do not come out after elution. Thus, the column must be equilibrated and washed carefully. The operating pH plays an important role in the binding and elution steps, i.e., the ionic strength of binding buffer and elution buffer must be adjusted correctly, and pH should not be far away from the protein pI. The issue of protein with low activity could stem from the loss of its co-factor during the purification.

High-resolution ion exchange columns and anion/cation exchanger are also commercially available and they are designed for high-resolution separation of proteins/biomolecules with high throughputs. Resins have a particle size of 5–10 μm and the column is operated under a maximum pressure of 500 psi. Based on the column binding capacity, high resolution is often achieved with a low sample load and/or reduced flow rate. Linear gradient elution is most effective in small scale IEX applications. In the past, IEX was based on hydrophilic gel matrices but high-resolution performance requires nonporous solid supports, which are inorganic, organic, or a hybrid. Various proteins can be purified with high-resolution by high-performance ion-exchange chromatography (Regnier, 1982). Trends in this field with some pros and cons have been discussed in the literature (Bhusnure and Mali, 2015). Except for the diameter of resins, detailed information of such commercial solid supports is propriety but based on the principle

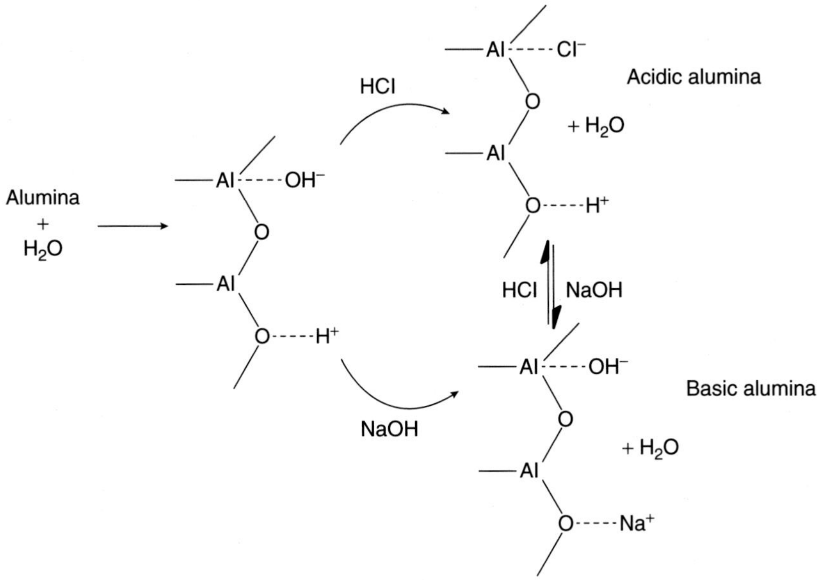

Fig. 3.5 ***Ion-exchange properties of alumina.*** Alumina can, therefore, exhibit marked anion- and cation-exchange characteristics in acidic and alkaline media, respectively. *The plot was adapted from Ref. (Churms, 1966).*

of IEX, one would expect considerable similarity among mobile phases and elution protocols used in such commercial columns. With nonporous solid and polymer-based supports, there is no limitation of molecular weights of proteins that can be resolved. At least one commercial product is prepared from 100 percent silica-based biol coated columns with a diameter of 1.7 μm (Feldthus, 2012) Fast recoveries of proteins by high performance IEC columns are always critical for processing very labile proteins.

3.7.4 Hydrophobic chromatography/ hydrophobic interaction chromatography (HIC)

Hydrophobic interactions drive many colloidal self-assembly and important biological processes. During such assembly, the hydration shells of the associating solutes are disrupted as hydrophobic–water contacts are replaced by hydrophobic–hydrophobic interactions. This technique separates proteins based on their relative degree of hydrophobicity. Based on mild operating conditions without the use of organic solvents and biocompatible matrices, this useful technique is capable of purifying proteins while maintaining their biological activity.

In HIC, interaction occurs between hydrophobic ligands of the stationary phase and hydrophobic regions of the target protein. However, such hydrophobic groups are covered or shielded by a network of hydrogen bonds from water molecules. As discussed in "sating-out", the water shield is disrupted by a chaotropic salt solution, e.g., ammonium sulfate (1–2 M) or sodium chloride (3 M) and a buffer to control pH (e.g., phosphate, $6 \le pH \le 7$). The salt ions interact with water and thus, expose the hydrophobic groups that interact with the hydrophobic patches on the surface of proteins. The retention mechanism is governed by the adsorption-desorption equilibrium. Hydrophobic proteins will bind strongly to the matrix whereas hydrophilic proteins are stripped out of the column. After equilibrium binding, the column is eluted with a low-salt solution to promote the reformation of the water shield. Bound proteins are eluted out in the order of increasing hydrophobicity using a decreasing salt gradient. For strong hydrophobic interactions, the elution step may also be beneficial by adding mild organic modifiers or detergents to the elution solution.

The HIC stationary phase often has straight-chain alkyl ligands with the hydrophobic character or aryl ligands, which show both aromatic and hydrophobic interactions (Hofstee and Otillio, 1978). Like any modes of chromatography, stationary phases for HICHPLC are based on silica bonded with ligands of butyl, phenyl, ether, amide, or propyl. Stationary phases are often nonporous but porous polymethacrylate-based particles are also available. HIC columns packed with microparticles (2.5–5 μm in diameter) are operated at 100–400 bar. Alkyl ligands, butyl or octyl group is often used as their binding capacity is related to the alkyl chain length. The functional groups bind proteins based entirely on the protein hydrophobicity.

HIC mobile phases are based on sodium or potassium phosphate with a pH ranging from 5 to 7. Thus, the effect of pH on protein binding is not significant over moderate pH ranges. Separation is normally conducted at room temperature and the result might not be reproduced in a cold room. HIC has been used extensively in analytical, environmental, and clinical chemistry. However, commercial HIC columns for the pilot plan and preparative scale are also available for proteins with relative hydrophobicity and pH stability 1–10. Some proteins do not only unfold at high temperatures, and surprisingly also at low temperatures, known as cold denaturation (Vajpai et al., 2013). Enhanced hydrophobic interactions with increasing temperature might be attributed to temperature-induced conformational changes of proteins (Wu et al., 1986). Temperature effects on dansyl derivatives of amino acids from 5- 50 °C in HIC were reported by Haidacher et al. (1996). The hydrophobic force is strongly temperature-dependent and becomes weaker at lower temperatures (van Dijk et al., 2015), therefore HIC is conducted at moderate temperatures. Some commercial stationary phases for hydrophobic interaction are based on phenyl sepharose (Diao et al., 2013; Okada et al., 2009).

The key points of the protein purification by hydrophobic chromatography are summarized in Box 3.2.

3.7.5 Affinity interaction chromatography (AIC)

This powerful purification method exploits the specific binding between a protein and its corresponding ligand via affinity interactions, e.g., avidin and biotin. Agarose

Box 2 Three main factors for protein separation by hydrophobic chromatography

1. The nature and hydrophobicity of the adsorbent as well as the protein,
2. The conditions of the adsorption,
3. The conditions of the elution of the adsorbed material.
4. pH should be close to pI of the protein (no charge) to maximize hydrophobic interaction.
5. The effect of hydrophobicity is difficult to predict.
6. The use of additive must be carefully considered as they might affect the protein structure and activity.

and polyacrylamide bead derivatives are useful in the affinity purification of proteins (Cuatrecasas, 1970). Both cross-linked dextran (Sephadex) and the beaded agarose derivatives have many of the desirable features required for the performance of affinity purification. The stationary phase must be mechanically and chemically stable and have minimal nonspecific interaction with proteins. Contaminants are easily removed through wash steps, whereas the protein of interest is tightly bound and remains in the column. The bound protein is then stripped (eluted) from the support with high purity. The binding event is very specific, hence in affinity separation proteins can be purified many folds in one step. Like other chromatographic procedures, a crude protein is passed over a solid support column. Non-target molecules with no binding to the immobilized ligand are easily washed away whereas the target molecules are retained in the column. The target protein is then eluted by using a specific buffer, resulting in a highly purified and concentrated protein. In selective purification, a ligand specific for a protein or a covalently attached tag is used. Many ligands are non-selective or class-selective, e.g., heparin for DNA-binding proteins, lectin for glycoproteins chromatography, and protein A, G, L for immunoglobulin. Some common ligands used in affinity ligands are summarized in Table 3.5 (Kimple et al., 2013). This procedure is widely used for purifying recombinant proteins, antibodies, and transcription factors.

Table 3.5 Key features of some popular affinity ligands

Affinity ligands	Key features and characteristics	Elution buffer
Heparin	DNA-binding protein	High ionic strength
Antigenic peptide	Antibody 9antigen-specific)	Free peptide
Myc-specific antibody	Myc-tagged protein	Low pH
Protein A, G, L, or protamine	Antibody-class specific	Extreme pHs
Reduced glutathione	GST-tagged protein	Free glutathione
FLAG-specific antibody	FLAG-tagged protein	FLAG peptide or low pH

3.7.6 Metal chelate chromatography

As described in metal chelate precipitation, some proteins with histidine and cysteine bind to metal ions, mainly Cu^{2+} and Ni^{2+} at around neutral pH (pH 6–8). Sepharose beads can be conjugated with iminodiacetic acid via a short spacer arm to chelate a metal ion, Ni^{2+}, which then forms a complex with two histidine molecules of proteins (Fig. 3.6). This purification method is also excellent for purifying many natural proteins

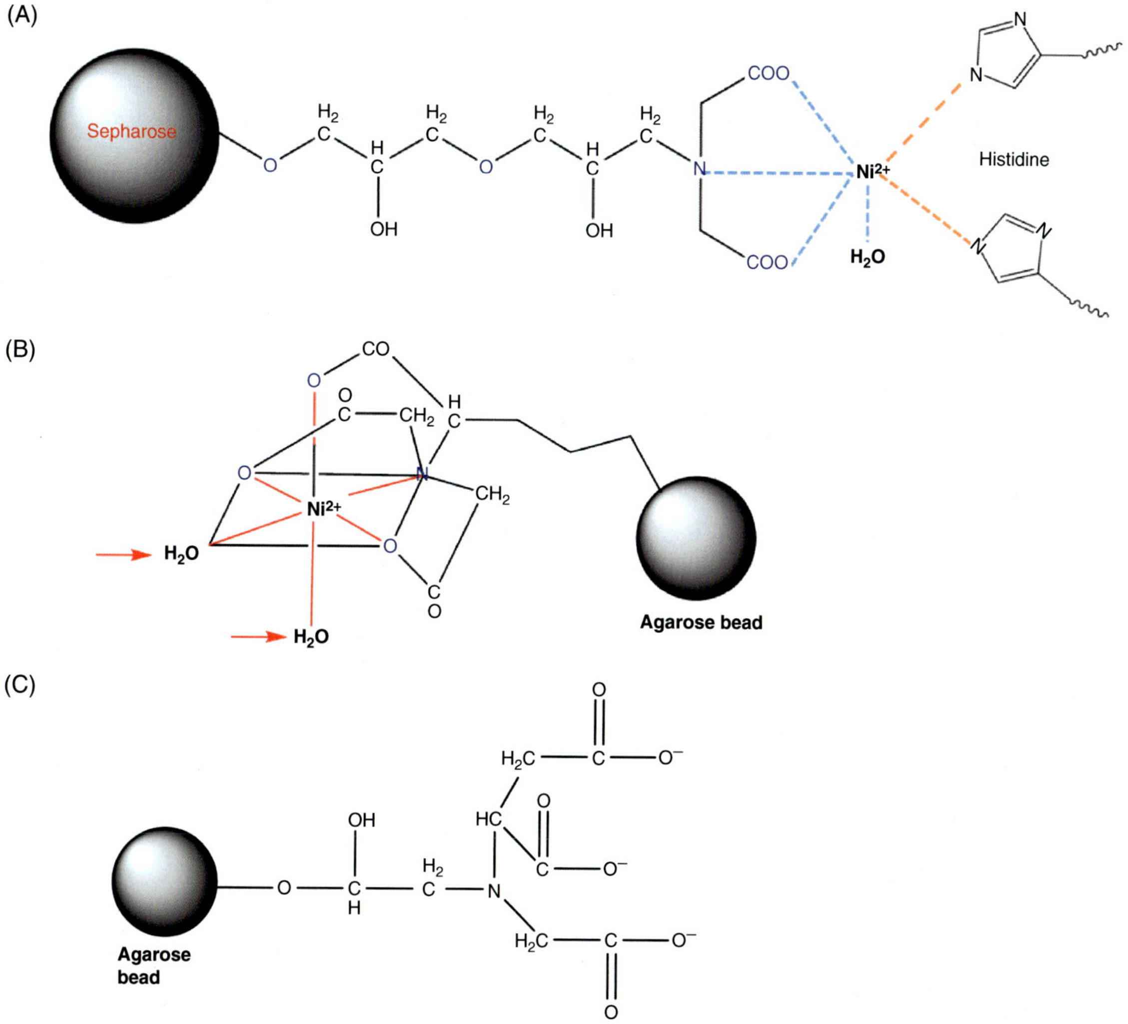

Fig. 3.6 (A) Sepharose is functionalized with iminodiaceric acid to chelate Ni^{2+}, which then forms a complex with two histidine molecules. (B) Ni-NTA (nitrilotriacetic acid) ligand is covalently attached to a cross-linked agarose matrix. Histidine residues can coordinate with the Ni^{2+} ion by replacing the bound water molecules (indicated by red arrows).

with histidine or recombinant $(His)_6$ fusion proteins. A polyhistidine-tag with at least six histidine (His) residues is positioned at the C- or N-terminus of the protein. In comparison to other tags, histidine tags are less disruptive to the properties of the host protein, i.e., their removal from the host protein after purification is optional. Functionalized Sepharose beads are packed in a column and loaded with a solution of divalent metal ions such as Ni^{2+}, Cu^{2+}, Zn^{2+}, etc. to saturate all the binding sites of functionalized Sepharose. This complex then chelates with histidine moieties of the target protein. The binding to a target protein depends on pH and after equilibrium, the bound protein is simply eluted by reducing the pH or increasing the buffer ionic strength. Another option is the addition of EDTA or imidazole in the buffer. Among several divalent cations, Cu^{2+} exhibits strong binding, and some proteins only bind to Cu^{2+} whereas Ni^{2+} is widely used for poly (His) fusion proteins. Zinc ions give a weaker binding but Zn^{2+} can be exploited to achieve selective elution of a protein mixture. It should be noted that metalloproteins should not be purified by metal chelating chromatography due to their ability to scavenge the metal ions from the stationary phase.

Commercial GE chelate stationary phases are available such as iminodiacetic acid coupled to Sepharose beads (34 μm) via an ether bond with a capacity of 23 μmoles Cu^{2+}/ml and iminodiacetic acid can also be coupled Sepharose beads (90 μm) via a spacer arm using epoxy coupling with a capacity of 22–30 μmoles Zn^{2+}/ml. Besides iminodiacetic acid, the use of nitrilotriacetic acid (NTA) (Fig. 3.6B) as metal chelating ligands is also popular (Arnau et al., 2006).

3.8 Conclusions and outlooks

Significant progress in protein chemistry, materials chemistry, and advanced instrumentation has been achieved during the last 200 years for protein purification. New matrices have been invented for ion exchange, hydrophobic, and affinity chromatography. Purified proteins can be well characterized by advanced instrumentation for their properties. Molecular cloning allows modification and heterologous overexpression of almost any protein. Bacteria, yeast, mammalian cells, and insect cells are common hosts to produce such heterologous proteins (Li et al., 2016). Ideal purifications should meet three requirements: high purity, high activity, and high yield. The protein purification remains the costliest step in protein production, particularly, recombinant proteins. Of notice is the use of fusion tags for protein solubility in the *E. coli* expression system, the dominant host for recombinant protein production. However, foreign proteins often aggregate into insoluble inclusion bodies and this is the major setback of the *E. coli* expression system (Costa et al., 2014). Among several fusion tags, a novel Fh8 Fusion System (Hitag®), Fh8 (GenBank ID: AF213970.1) with a low molecular weight of 8 kDa is one of the promising new fusion technologies to enhance protein solubility and serves as a robust purification handle (Costa et al., 2013a, 2013b).

Protein purification can be performed in any well-equipped labs with from milligrams for research using automated chromatography systems. All required lysis buffer, elution buffer, stationary phases are commercially available. Affinity chromatography is routinely used to purify affinity-tagged proteins with high purity. Of notice is one-step purification of a glutathione S-transferase (GST)-tagged protein and on-column tag removal with high purity. The GST (MW 26 kDa) tag often increases the target protein solubility and can be removed by a sequence-specific protease. Recombinant proteins can also be expressed with a histidine tag, mainly hexa-histidine (His_6-tag) (Li et al., 2016), thus they can be purified by metal chelate chromatography. Histidine tags with 6 histidine molecules are less disruptive than other tags and its removal after purification might not be a critical issue. Perhaps, this is the reason why the His_6-tag is widely used as compared to other fusion tags. One critical issue is if the protein of interest attached to such tags may undergo a significant conformational change. Of course, this problem is protein dependent, but the protein attached to GTS (glutathione S-transferase) alters its native state (Benard and Bokoch, 2002; Ren et al., 2003). Such a result is not completely unexpected as GST has 220 amino acids, which is very large compared to His_6-tag or FLAG with 8 amino acids (AspTyrLysAspAspAspAspLys). Insoluble protein aggregates are often accumulated in inclusion bodies in the high-level expression of MBP (maltose binding protein) fusion proteins in *E. coli*. Like GST, the large size of the MBP tag may alter protein function (Kimple et al., 2013). The Halo (33-kDa) protein is obtained from on a modified haloalkane dehalogenase found in bacteria and has some important applications such as in vitro protein labeling, purification, and in vivo labeling. The protein of interest can be cleaved from HaloTag by TEV (tobacco etch virus) protease (Kimple et al., 2013). With a chitin-binding domain, the intein-chitin binding domain (intein-CBD) tag is useful for purifying a native recombinant protein without the need for a protease. The CBD tag is relatively small, compared to GST and MBP, and it is unlikely to affect the properties of the protein of interest (Kimple et al., 2013). An overall review of current fusion tags for protein purification can be found in the literature (Gautam et al., 2012; Kimple et al., 2013).

Unlike tagged proteins, untagged proteins obtained by overexpression or from a natural source require several purification steps from cell lysis, protein precipitation, and at least one chromatography method. The isoelectric point of a specific protein can be estimated or quickly measured by zeta-sizer. Therefore, it is straightforward to purify it by IEC considering both strong and weak cation/anion exchange resins are commercially available. In some cases, hydrophobic chromatography is preferred because it is compatible with high-salt samples eluted from IEC columns. Hydrophobic properties are difficult to predict, so some trial runs to screen different media are necessary.

The dominant host for recombinant protein production is still *E. coli*, which offers benefits of low cost, ease of use, and scale-ups. However, the aggregation of overexpressed proteins into insoluble inclusion bodies is a major shortcoming of the recombinant *E. coli*, as discussed earlier.

Micropurification deserves a brief discussion here since many proteins are present at tiny amounts from tissues or various model organisms and their purification requires the use of small, high-resolution purification systems. As a typical example, mouse brain tissue only allows the use of < 200 μL buffer for sonication, centrifugation, and purification. Fortunately, such miniaturized equipment is also commercially available for this demanding task.

In the past, the purification of a target protein from microorganisms was very difficult and time-consuming, resulting in only a few micrograms. Process-scale protein purification is now capable of producing several grams of a protein by fast protein liquid chromatography. This is generally applied only to proteins due to a wide choice of resins and buffers. The buffer pressure used is < 5 bar with a flow rate is of 1–5 mL/min, which is considered relatively fast. In the past, many affinity separations were based on agarose or polyacrylamide with low performance. The use of advanced materials like silica and monolithic supports results in high-performance affinity chromatography.

Protein-based materials will demand significant amounts of proteins for diversified applications. Of interest is the development of protein-based materials that push for more production of specialty proteins at reasonable costs. A few proteins of interest silk are from spiders and silkworm, elastin from mammalian blood vessels, skin, etc., resilin from insects, collagen from mammalian tissues, and keratin from reptilian scales, mammalian wool, avian feathers, etc. with unique amino acids and amino acid sequences (Abascal and Regan, 2018). The use of magnetic particles (MP) for the purification of proteins is also worth mentioning. MPs are commercially available and can be prepared using well-established procedures. They are made of iron oxide nanoparticles (5–50 nm) and decorated with polymers with diameters ranging from 35 nm to 4.5 μm. Magnetic beads can be functionalized with a carboxyl group or stearic acid to facilitate bioconjugation. Thus, ligands with amino groups are covalently conjugated to the carboxyl group of the coating polymer by carbodiimide chemistry (Luong and Vashist, 2020).

The last point is the potential use of tag combinations for tandem affinity purification (TAP). The TAP tag with two affinity modules forms a complex with the target protein, which can be isolated in two consecutive purification steps. This approach is useful for expression and purification of proteins, which are receptive to proteolysis (Panagiotidis et al., 1995). Besides currently known tags, other affinity tags will emerge and can be used in combination to improve protein purity using only one buffer system. Instrumentation will be more advanced with data acquisition and analysis and columns with new materials for various chromatographic schemes will be more robust with high binding capacities to provide high-resolution and high flow rates. The most noticeable is the particle size and column length provided by various commercial suppliers. In the past, most of the chromatographic columns were based on typically 7.8 mm in diameter and 150 to 300 mm in length. The long column lengths must be sufficiently long to accommodate large particles with low mechanical strength. Many standard columns use microscale particles (<3 μm), packed in a column with a diameter of typically 4.6 mm and 5 cm in length. This trend has been supported by some key developments in materials and particle chemistries.

References

Abascal, N.C., Regan, L., 2018. The past, present and future of protein based materials. Open Biol 8, 180113.

Ahmad, A., Madhusudanan, K.P., Bhakuni, V., 2000. Trichloroacetic acid and trifluoroacetic acid-induced unfolding of cytochrome c: stabilization of a native-like folded intermediate. Biochim. Biophys. Acta 1480, 201–210.

Ames, G.F.-L., 1994. Isolation and purification of periplasmic binding proteins. Methods Enzymol. 235, 234–241.

Andrews, B.A., Asenjo, J.A., 1987. Enzymatic lysis and disruption of microbial cells. Trends Biotechnol. 5, 273–277.

Andrews, G.P., Martin, S.E., 1979. Heat inactivation of catalase from Staphylococcus aureus MF-31. Appl. Environ. Microbiol. 37, 1180–1185.

Arakawa, T., Bhat, R., Timasheff, S.N., 1990. Why preferential hydration does not always stabilize the native structure of globular proteins. Biochemistry 29, 1924–1931.

Arakawa, T., Kita, Y., Timasheff, S.N., 2007. Protein precipitation and denaturation by dimethyl sulfoxide. Biophys. Chem. 131, 62–70.

Arakawa, T., Timasheff, S.N., 1984. Mechanism of protein salting in and salting out by divalent cation salts: balance between hydration and salt binding. Biochemistry 23, 5912–5923.

Arnau, J., Lauritzen, C., Petersen, G.E., Pedersen, J., 2006. Current strategies for the use of affinity tags and tag removal for the purification of recombinant proteins. Protein Expr. Purif. 48, 1–13.

Baker, S.L., Munasinghe, A., Kaupbayeva, B., Kang, N.B., Certiat, M., Murata, H., et al., 2019. Transforming protein-polymer conjugate purification by tuning protein solubility. Nat. Commun. 10, 4718.

Balasundaram, B., Harrison, S., Bracewell, D.G., 2009. Advances in product release strategies and impact on bioprocess design. Trends Biotechnol. 27, 477–485.

Bayraktar, H., Onal, S., 2013. Concentration and purification of alpha-galactosidase from watermelon (Citrullus vulgaris) by three phase partitioning. Sep. Purif. Technol. 118, 835–841.

Benard, V., Bokoch, G.M., 2002. Assay of Cdc42, Rac, and Rho GTPase activation by affinity methods. Methods Enzymol. 345, 349–359.

Bhusnure, O.G., Mali, S.N., 2015. Recent trends in ion-exchange chromatography. Int. J. Pharma. Drugs Anal. 3, 403–416.

Buytenhuys, F.A., 1981. Ion chromatography of inorganic and organic ionic species using refractive index detection. J. Chromatogr. 218, 57–64.

Capocellia, M., Prisciandaro, M., Lancia, A., Musmarra, D., 2014. Comparison between hydrodynamic and acoustic cavitation in microbial cell disruption. Chem. Eng. 38, 13–18.

Carter, S., Rimmer, S., Sturdy, A., Webb, M., 2005. Highly branched stimuli responsive poly[(N-isopropylacrylamide)-co-(1,2-propandiol-3-methacrylate)]s with protein binding functionality. Macromol. Biosci. 5, 373–378.

Chen, J.-.P., Jang, F.-.L., 1995. Purification of trypsin by affinity precipitation combining with aqueous two-phase extraction. Biotechnol. Tech. 9, 461–466.

Chen, Y.-.C., Chen, L.-.A., Chen, S.-.J., Chang, M.-.C., Chen, T.-.L., 2004. A modified osmotic shock for periplasmic release of a recombinant creatinase from Escherichia coli. Biochem. Eng. J. 19, 211–215.

Chisti, Y., Moo-Young, M., 1986. Disruption of microbial cells for intracellular products. Enzym. Microb. Technol. 8, 194–204.

Churms, S.C., 1966. The effect of pH on the ion-exchange properties of hydrated alumina: part 1. Capacity and selectivity. S. Afr. J. Chem. 19, 98–107.

Cohn, E.J., 1925. The physical chemistry of the proteins. Physio. Rev. 5, 349–437.

Cohn, E.J., Hughes, Jr., W.L., Weare, J.H., 1947. Preparation and properties of serum and plasma proteins. XIII. Crystallization of serum albumins from ethanol-water mixtures. J. Am. Chem. Soc. 69, 1753–1761.

Costa, S.J., Almeida, A., Castro, A., Domingues, L., Besir, H., 2013a. The novel Fh8 and H fusion partners for soluble protein expression in Escherichia coli: a comparison with the traditional gene fusion technology. Appl. Microbiol. Biotechnol. 97, 6779–6791.
Costa, S.J., Coelho, E., Franco, L., Almeida, A., Castro, A., Domingues, L., 2013b. The Fh8 tag: a fusion partner for simple and cost-effective protein purification in Escherichia coli. Protein Expr. Purif. 92, 163–170.
Costa, S., Almeida, A., Castro, A., Domingues, L., 2014. Fusion tags for protein solubility, purification, and immunogenicity in Escherichia coli: the novel Fh8 system. Front. Microbiol. 5, 63.
Cuatrecasas, P., 1970. Protein purification by affinity chromatography: derivatizations of agarose and polyacrylamide beads. J. Biol. Chem. 245, 3059–3065.
De Oliveira Coimbra, C.G., Lopes, C.E., Calazans, G.M.T., 2010. Three-phase partitioning of hydrolyzed levan. Bioresour. Technol. 101, 4725–4728.
Dennison, C., Lovrien, R., 1997. Three phase partitioning: concentration and purification of proteins. Protein Expr. Purif. 11, 149–161.
Diao, J., Burré, J., Vivona, S., Cipriano, D., Sharma, M., Kyoung, M., et al., 2013. Native α-synuclein induces clustering of synaptic-vesicle mimics via binding to phospholipids and synaptobrevin-2/VAMP2. Elife 2, e00592 001.
Englard, S., Seifter, S., 1990. Precipitation techniques. Methods Enzymol. 182, 285–300.
Engler, C.R., Robinson, C.W., 1981. Disruption of Candida utilis cells in high pressure flow devices. Biotechnol. Bioeng. 23, 765–780.
Feist, P., Hummon, A.B., 2015. Proteomic challenges: sample preparation techniques for microgram-quantity protein analysis from biological samples. Int. J. Mol. Sci. 16, 3537–3563.
Franks, F., 1988. Characterization of Proteins. Humana Press, NJ, USA.
Feliciello, I., Chinali, G., 1993. A modified alkaline lysis method for the preparation of highly purified plasmid DNA from Escherichia coli. Anal. Biochem. 212, 394–401.
Feldthus, A., 2012. https://www.waters.com/webassets/cms/library/docs/local_seminar_presentations/DA_SEC_IEX_for_Bio_UPLC.pdf.
Fork, G., David, C., 2006. Charles Stacy French. National Academy of Sciences Biographical Memoirs 88, 62–89.
Fox, J.D., Kapust, R.B., Waugh, D.S., 2001. Single amino acid substitutions on the surface of Escherichia coli maltose-binding protein can have a profound impact on the solubility of fusion proteins. Protein Sci. 10, 622–630.
Fritz, J.S., Gjerde, D.T., Becker, R.M., 1980. Cation chromatography with a conductivity detector. Anal. Chem. 52, 1519–1522.
Galaev, I.Y., Kumar, A., Agarwal, R., Gupta, M.N., Mattiasson, B., 1997. Imidazole: a new ligand for metal affinity precipitation. Precipitation of Kunitz soybean trypsin inhibitor using Cu (II)-loaded copolymers of 1-vinylimidazole with N-vinylcaprolactam or N-isopropylacrylamide. Appl. Biochem. Biotechnol. 68, 121–133.
Galaev, I.Y., Mattiasson, B., 1992. Affinity thermoprecipitation of trypsin using soybean trypsin inhibitor conjugated with a thermo-reactive polymer, poly (N-vinyl caprolactam). Biotechnol. Tech. 6, 353–358.
Gautam, S., Mukherjee, J., Roy, I., Gupta, M.N., 2012. Emerging trends in designing short and efficient protein purification protocols. Am. J. Biochem. Biotechnol. 8, 230–254.
Goronja, J.M., Janošević Ležaić, A.M., Dimitrijević, B.M., Malenović, A.M., Stanisavljev, D.R., Nataša, D., Pejić, N.D., 2016. Determination of critical micelle concentration of cetyltrimethylammonium bromide: different procedures for analysis of experimental data. Hem. Ind. 70, 485–492.
Grayson, P., Evilevitch, A., Inamdar, M.M., Purohit, P.K., Gelbart, W.M., Knobler, C.M., et al., 2006. The effect of genome length on ejection forces in bacteriophage lambda. Virology 348, 430–436.
Guoqiang, D., Benhura, M.A.N., Kaul, R., Mattiasson, B., 1995. Affinity thermoprecipitation of yeast alcohol dehydrogenase through metal ion promoted binding with Eudragit-bound Cibacron Blue 3GA. Biotechnol. Prog. 11, 187–193.
Haidacher, D., Vailaya, A., Horvath, C., 1996. Temperature effects in hydrophobic interaction chromatography (thermodynamics/hydrophobic effect/molecular chromatography), Proc. Natl. Acad. Sci. USA, 93, 2290–2295.
Harrison, S.T.L., 1991. Bacterial cell disruption: a key unit operation in the recovery of intracellular products. Biotechnol. Adv. 9, 217–240.

Hartley, J.L., 2006. Cloning technologies for protein expression and purification. Current Opinion Biotechnol 17, 359–366.

He, Y.-.Z., Fan, K.-.Q., Jia, C.-.J., Wang, Z.-.J., Pan, W.-.B., Huang, L., et al., 2007. Characterization of a hyperthermostable Fe-superoxide dismutase from hot spring. Appl. Microbiol. Biotechnol. 75, 367–376.

Hochuli, E., Bannwarth, W., Döbeli, H., Gentz, R., Stüber, D., 1988. Genetic approach to facilitate purification of recombinant proteins with a novel metal chelate adsorbent. Bio/Technol 6, 1321–1325.

Hoe, H.S., Kim, H.K., Kwon, S.T., 2001. Expression in Escherichia coli of the thermostable inorganic pyrophosphatase from the Aquifex aeolicus and purification and characterization of the recombinant enzyme protein. Protein Expr. Purif. 23, 242–248.

Hofmeister, F., 1888. Zur Lehre Von Der Wirkung Der Salze (About the Science of the Effect of Salts). Naunyn-Schmiedeberg's Arch Pharmacol 25, 1–30.

Hofstee, B.H., Otillio, N.F., 1978. Non-ionic adsorption chromatography of proteins. J. Chromatogr. 159, 57–69.

Hordijk, C.A., Cappenberg, T.E.J., 1985. Sulfate analysis in pore water by radio-ion chromatography employing 5-sulfoisophthalic acid as a novel eluent. Microbiol. Methods 3, 205–214.

Inoue, H.S.N, Timasheff, S.N., 1972. Preferential and absolute interactions of solvent components with proteins in mixed solvent systems. Biopolymers 11, 737–743.

Islam, M.S., Aryasomayajula, A., Selvaganapathy, P.R., 2017. A review on macroscale and microscale cell lysis methods. Micromachines (Basel) 8, 83.

Johnson, B.H., Hecht, M.H., 1994. Recombinant proteins can be isolated from E. coli cells by repeated cycles of freezing and thawing. Bio/Technology 12, 1357–1360.

Kaur, H., Bhagwat, S.R., Sharma, T.K., Kumar, A., 2019. Analytical techniques for characterization of biological molecules – proteins and aptamers/oligonucleotides. Bioanalysis 11, 103–117.

Kimple, M.E., Brill, A.L., Pasker, R.L., 2013. Overview of affinity tags for protein purification. Curr. Protoc. Protein Sci. 73, 9.9 Unit.

Kita, Y., Arakawa, T., Lin, T.-.Y., Timasheff, S.N., 1994. Contribution of the surface free energy perturbations to protein-solvent interactions. Biochemistry 33, 1517–1589.

Kleinig, A.R., Mansell, C.J., Nguyen, Q.D., Badalyan, A., Middelberg, A.P.J., 1995. Influence of broth dilution on the disruption of Escherichia coli. Biotech. Tech. 9, 759–762.

Kleinig, A.R., Middelberg, A.P.J., 1998. On the mechanism of microbial cell disruption in high-pressure homogenisation. Chem. Eng. Sci. 53, 891–898.

Kumar, A., Galaev, I.Y., Mattiasson, B., 1999. Metal chelate affinity precipitation: a new approach to protein purification. Bioseparation 7, 185–194.

Kumar, A., Wahlund, P.-.O., Kepka, C., Galaev, I.Y., Mattiasson, B., 2003. Purification of histidine-tagged single chain Fv-antibody fragments by metal chelate affinity precipitation using thermo-responsive copolymers. Biotechnol. Bioeng. 84, 495–503.

Kussrow, A., Kaltgrad, E., Wolfenden, M.L., Cloninger, M.J., Finn, M.G., Bornhop, D.J., 2009. Measurement of mono- and polyvalent carbohydrate-lectin binding by back-scattering interferometry. Anal. Chem. 81, 4889–4897.

Lee, A.K., Lewis, D.M., Ashman, P.J., 2015. Microalgal cell disruption by hydrodynamic cavitation for the production of biofuels. J. Appl. Phycol. 27, 1881–1889.

Langmuir, I., 1918. The adsorption of gases on plane surface of glass, mica and platinum. J. Am. Chem. Soc. 40, 1361–1402.

Lee, J.C., Timasheff, S.N., 1974. Partial specific volumes and interactions with solvent components of proteins in guanidine hydrochloride. Biochemistry 13, 257–265.

Li, Y., 2010. Commonly used tag combinations for tandem affinity purification. Biotechnol. Appl. Biochem. 55, 73–83.

Li, P., Li, L., Zhao, Y., Sun, L., Zhang, Y.J., 2016. Selective binding and magnetic separation of histidine-tagged proteins using Fe3O4/Cu-apatite nanoparticles. Inorg. Biochem. 156, 49–54.

Li, Z., Jiang, F., Li, Y., Zhang, X., Tan, T., 2013. Simultaneously concentrating and pretreating of microalgae Chlorella spp. by three-phase partitioning. Bioresour. Technol. 149, 286–291.

Linné-Larsson, E., Mattiasson, B., 1994. Isolation of concanavalin A by affinity precipitation. Biotechnol. Tech. 8, 51–56.

Liu, Z., Yu, D., Li, L., Liu, X., Zhang, H., Sun, W., et al., 2019. Three-phase partitioning for the extraction and purification of polysaccharides from the immunomodulatory medicinal mushroom Inonotus obliquus. Molecules 24, 403.
Luong, J.H.T., Male, K.B., Nguyen, A.L., 1988. Synthesis and characterization of a water-soluble affinity polymer for trypsin purification. Biotechnol. Bioeng. 31, 439–446.
Luong, J.H.T., Vashist, S.K., 2020. Chemistry of biotin-streptavidin and the growing concern of an emerging biotin interference in clinical immunoassays. ACS Omega 5, 10–18.
Madigan, M.T., Martinko, J.M., Dunlap, P.V., Clark, D.P., 2008. Brock Biology of Microorganisms, 12th edition. Pearson, London.
Malhotra, A., 2009. Tagging for protein expression. In: Burgess, R.R., and Deutscher, M.P. (Eds.), Guide to Protein Purification, 2nd ed., vol. 463. Elsevier, San Diego, p.p. 239–258.
Manik, L.D., Aftab, A., 1999. Accurate Total Protein Assay by Mixing Sample with Acid, Adding Precipitating Agent, Collecting Precipitate, then Spectrophotometric Determination of Concentration; Overcomes Interference by Common Non-Protein Agents in Solution. U.S. Patent # US5900376 A.
Mattiasson, B., Kumar, A., Ivanov, A.E., Galaev, I.Y., 2007. Metal-chelate affinity precipitation of proteins using responsive polymers. Nat. Protoc. 2, 213–220.
Middelberg, A.P.J., 1995. Process-scale disruption of microorganisms. Biotech. Adv. 13, 491–551.
Mondal, K., Sharma, A., Gupta, M.N., 2004. Three phase partitioning of starch and its structural consequences. Carbohydr. Polym. 56, 355–359.
Morgenthaler, J.J., 2000. New developments in plasma fractionation and virus inactivation. Vox Sang. 78, 217–221.
Muller, E., 2005. Properties and characterization of high capacity resins for biochromatography. Chem. Eng. Technol. 28, 1295–1305.
Nallamsetty, S., Waugh, D.S., 2007. Mutations that alter the equilibrium between open and closed conformations of Escherichia coli maltose-binding protein impede its ability to enhance the solubility of passenger proteins. Biochem. Biophys. Res. Commun. 364, 639–644.
Nandakumar, M.P., Shen, J., Raman, B., Marten, M.R., 2003. Solubilization of trichloroacetic acid (TCA) precipitated microbial proteins via NaOH for two-dimensional electrophoresis. J. Proteome Res. 2, 89–93.
Ngo, A.N., Ezoulin, M.J.M., Youm, I., Youan, B.-B.C., 2015. Optimal concentration of 2,2,2-trichloroacetic acid for protein precipitation based on response surface methodology. J. Anal. Bioanal. 5 (4), 198.
Nguyen, A.L., Luong, J.H.T., 1989. Synthesis and application of water-soluble reactive polymers for purification and immobilization of biomolecules. Biotechnol. Bioeng. 34, 1186–1190.
Novák, V., Havlícek, V., 2016. Protein Extraction and Precipitation. In: Ciborowski, P., Silberring, J. (Eds.), Proteomic Profiling and Analytical ChemistrySecond edition. Elsevier, B.V., Amsterdam, Netherlands, pp. 51–62.
Nozaki, Y., Tanford, C., 1963. The solubility of amino acids and related compounds in aqueous urea solutions. J. Biol. Chem. 238, 4074–4081.
Nozaki, Y., Tanford, C., 1970. The solubility of amino acids, diglycine, and triglycine in aqueous guanidine hydrochloride solutions. J. Biol. Chem. 245, 1648–1652.
Okada, C., Yamashita, E., Lee, S., Shibata, S., Katahira, J., Nakagawa, A., et al., 2009. A high-resolution structure of the pre-microRNA nuclear export machinery. Science 326, 1275–1279.
Pan, A.C., Jacobson, D., Yatsenko, K., Sritharan, D., Weinreich, T.M., Shaw, D.E., 2019. Atomic-level characterization of protein–protein association. Proc. Natl. Acad. Sci. USA 116, 4244–4249.
Panagiotidis, C.A., Silverstein, S.J., 1995. pALEX, a dual-tag prokaryotic expression vector for the purification of full-length proteins. Gene 164, 45–47.
Pike, R.N., Dennison, C., 1989. Protein fractionation by three phase partitioning (TPP) in aqueous/t-butanol mixtures. Biotechnol. Bioeng. 33, 221–228.
Pittz, E.P., Bello, J., 1971. Studies on bovine pancreatic ribonuclease A and model compounds in aqueous 2-methyl-2, 4-pentanediol. 1. Amino acid solubility, thermal reversibility of ribonuclease A, and preferential hydration of ribonuclease-A crystals. Biochem. Biophys. 146, 513–524.
Pittz, E.P., Bello, J., 1973. Studies on bovine pancreatic ribonuclease A and model tyrosyl compounds in aqueous 2-methyl-2, 4-pentanediol: II. Spectral investigation of salvation. Biochem. Biophys. 156, 437–447.

Pittz, E.P., Timasheff, S.N., 1978. Interaction of ribonuclease A with aqueous 2-methyl-2, 4-pentanediol at pH 5.8. Biochemistry 17, 615–623.

Porath, J., Carlsson, J., Olsson, J., Belfrage, G., 1975. Metal chelate affinity chromatography: a new approach to protein fractionation. Nature 258, 598–599.

Potman, R.P., Wesdorp, J. 1994. Method for the preparation of a yeast extract, said yeast extract, its use as a food flavour, and a food composition comprising the yeast extract. U.S. Patent # 5,288,509.

Rajalingam, D., Loftis, C., Xu, J.J., Kumar, T.K., 2009. Trichloroacetic acid-induced protein precipitation involves the reversible association of a stable partially structured intermediate. Protein Sci. 18, 980–993.

Regnier, F.E., 1982. High-performance ion-exchange chromatography of proteins: the current status. Anal. Biochem. 126, 1–7.

Ren, L., Chang, E., Makky, K., Haas, A.L., Kaboord, B., Walid-Qoronfleh, M., 2003. Glutathione S-transferase pull-down assays using dehydrated immobilized glutathione resin. Anal. Biochem. 322, 164–169.

Ren, X., Yu, D., Yu, L., Gao, G., Han, S., Feng, Y., 2007. A new study of cell disruption to release recombinant thermostable enzyme from Escherichia coli by thermolysis. J. Biotechnol. 129, 668–673.

Rosano, G.L., Ceccarelli, E.A., 2014. Recombinant protein expression in Escherichia coli: advances and challenges. Front Microbiol 5, 172.

Rossi, M., Capponi, L., 2005. Process for the Purification of Bacterially Expressed Proteins PCT, WO2995/068500 A1.

Roy, I., Sharma, A., Gupta, M.N., 2004. Three-phase partitioning for simultaneous renaturation and partial purification of Aspergillus niger xylanase. Biochim. Biophys. Acta 1698, 107–110.

Rupley, J.A., Gratton, E., Careri, G., 1983. Water and globular proteins. Trends Biochem. Sci. 8, 18–22.

Sachdev, D., Chirgwin, J.M., 2000. Fusions to maltose-binding protein: control of folding and solubility in protein purification. Methods Enzymol. 326, 312–321.

Salazar, O., Asenjo, J.A., 2007. Enzymatic lysis of microbial cells. Biotechnol. Lett. 29, 985–994.

Salgin, S., Salgin, U., Bahadir, S., 2012. Zeta potentials and isoelectric points of biomolecules: the effects of ion types and ionic strengths. Int. J. Electrochem. Sci. 7, 12404–12414.

Schmidt, T.G., Koepke, J., Frank, R., Skerra, A., 1996. One-step affinity purification of bacterially produced proteins by means of the "Strep tag" and immobilized recombinant core streptavidin. J. Mol. Biol. 255, 753–766.

Sevastsyanovich, Y.R., Alfasi, S.N., Cole, J.A., 2010. Sense and nonsense from a systems biology approach to microbial recombinant protein production. Biotechnol. Appl. Biochem. 55, 9–28.

Sharma, A., Gupta, M.N., 2001. Purification of pectinases by three-phase partitioning. Biotechnol. Lett. 23, 1625–1627.

Sharma, A., Gupta, M.N., 2002. Three phase partitioning of carbohydrate polymers: separation and purification of alginates. Carbohydr. Polym. 48, 391–395.

Sharma, A., Mondal, K., Gupta, M.N., 2003. Some studies on characterization of three phase partitioned chitosan. Carbohydr. Polym. 52, 433–438.

Simpson, R.J., 2006. Precipitation of proteins by organic solvents. Cold Spring Harb. Protoc. doi:10.1101/pdb.prot4310. 2006(1):pdb.prot4310.

Stiborova, H., Kostal, J., Mulchandani, A., Chen, W., 2003. One-step metal affinity purification of histidine-tagged proteins by temperature-triggered precipitation. Biotechnol. Bioeng. 82, 605–611.

Tan, Z.J., Wang, C.Y., Yi, Y.J., Wang, H.Y., Zhou, W.L., Tan, S.Y., Li, F.F., 2015. Three phase partitioning for simultaneous purification of aloe polysaccharide and protein using a single-step extraction. Process Biochem. 50, 482–486.

Tanford, C., 1964. Cohesive forces and disruptive reagents. Brookhaven Symp. Biol. 17, 154–183.

Taskova, R.M., Zorn, H., Krings, U., Bouws, H., Berger, R.G., 2006. A comparison of cell wall disruption techniques for the isolation of intracellular metabolites from Pleurotus and Lepista sp. Z. Naturforsch. C J. Biosci. 61, 347–350.

Terpe, K., 2003. Overview of tag protein fusions: from molecular and biochemical fundamentals to commercial systems. Appl. Microbiol. Biotechnol. 60, 523–533.

Timasheff, S.N., Arakawa, T., 1997. Membrane protein purification Stabilization of protein structure by solvents. In: Creighton, T.E. (Ed.), Protein Structure: A Practical Approach2nd edition. IRL Press, Oxford University, pp. 349–364.

Tiselius, A., Hjerten, S., Levin, O., 1956. Protein chromatography on calcium phosphate columns. Arch. Biochem. Biophys. 65, 132–155.

Traverso, N., Ricciarelli, R., Nitti, M., Marengo, B., Furfaro, A.L., Pronzato, M.A., et al., 2013. Role of glutathione in cancer progression and chemoresistance. Oxid. Med. Cell Longev., 972913.

Tripathi, N.K., Shrivastava, A., 2019. Recent developments in bioprocessing of recombinant proteins: expression hosts and process development. Front. Bioeng. Biotechnol. 7, 420 Article.

Vajpai, N., Nisius, L., Wiktor, M., Grzesiek, S., 2013. High-pressure NMR reveals close similarity between cold and alcohol protein denaturation in ubiquitin. Proc. Natl. Acad. Sci. USA 110, E368–E376.

van Dijk, E., Hoogeveen, A., Abeln, S., 2015. The hydrophobic temperature dependence of amino acids directly calculated from protein structures. PLoS Comput. Biol. 11, e1004277.

Vanthoor-Koopmans, M., Wijffels, R.H., Barbosa, M.J., Eppink, M.H.M., 2013. Biorefinery of microalgae for food and fuel. Bioresour. Technol. 135, 142–149.

Vashist, S.K., Luong, J.H.T., 2018. Handbook of Immunoassay Technologies: Approaches, Performances, and Applications, 1st ed. Academic Press, London.

Wang, A., Wu, C.J., Chen, S.H., 2006. Gold nanoparticle-assisted protein enrichment and electroelution for biological samples containing low protein concentration–a prelude of gel electrophoresis. J. Proteome Res. 5, 1488–1492.

Wang, Y.Y., Qiu, W.Y., Wang, Z.B., Ma, H.L., Yan, J.K., 2017. Extraction and characterization of anti-oxidative polysaccharide-protein complexes from Corbicula fluminea through three-phase partitioning. RSC Adv. 7, 11067–11075.

Wingfield, P.T., 2016. Protein precipitation using ammonium sulfate. Curr. Protoc. Protein Sci. 84. doi:10.1002/0471140864.psa03fs84. A.3F.1-A.3F.9.

Wu, S.-.L., Figueroa, A., Karger, B.L., 1986. Protein conformational effects in hydrophobic interaction chromatography. Retention characterization and the role of mobile phase additives and stationary phase hydrophobicity. J. Chromatogr. 371, 3–27.

Yan, J.K., Wang, Y.Y., Qiu, W.Y., Shao, N., 2017. Three-phase partitioning for efficient extraction and separation of polysaccharides from Corbicula fluminea. Carbohydr. Polym. 163, 10–19.

Yoshikawa, H., Hirano, A., Arakawa, T., Shiraki, K., 2012. Mechanistic insights into protein precipitation by alcohol. Int. Biol. Macromol. 50, 865–871.

Young, C.L., Britton, Z.T., Robinson, A.S., 2012. Recombinant protein expression and purification: a comprehensive review of affinity tags and microbial applications. Biotechnol. J. 7, 620–634.

Zarei, O., Dastmalchi, S., Hamzeh-Mivehroud, M., 2016. A simple and rapid protocol for producing yeast extract from Saccharomyces cerevisiae suitable for preparing bacterial culture media. Iran J. Pharm. Res. 15, 907–913.

Zhang, Y., Bottinelli, D., Lisacek, F., Luban, J., Strambio-de-Castillia, C., Varesio, E., et al., 2015. Optimization of human dendritic cell sample preparation for mass spectrometry-based proteomic studies. Anal. Biochem. 484, 40–50.

Zumstein, L., 2001. Dialysis and Ultrafiltration. Curr. Protoc. Mol. Biol. doi:10.1002/0471142727.mba03cs41.

CHAPTER 4

The multiple facets of three-phase partitioning in the purification, concentration, yield and activity of enzymes and proteins

James Philip Dean Goldring
Biochemistry, School of Life Sciences, University of KwaZulu-Natal (Pietermaritzburg campus), Scottsville, South Africa

Chapter outline

4.1 Introduction

Three-phase partitioning (TPP) has its origins from the observation by Tan and Lovrien (1972) that several enzymes had increases in activity in the presence of *t*-butanol. Twelve years later, Lovrien's group (Odegaard et al., 1984) isolated cellulases from *Trichoderma reesei* with three-phase partitioning. Lovrien et al. (1987) went on to show that the technique could be used to isolate many different proteins. The three-phases in the TPP resulted after Lovrien et al. (1987) added ammonium sulfate followed by *t*-butanol to a soluble extract in buffer. After equilibration the solution may separate into three-phases and this can be assisted by low g-force, under 5000 x *g*, centrifugation. The solution separates into a lower buffered ammonium sulfate phase, a precipitated phase, which is often rich in protein and DNA, that floats on the lower phase and an upper butanol phase (Tan and Lovrien, 1972) thus producing the three-phases of the technique.

Three Phase Partitioning
DOI: https://doi.org/10.1016/B978-0-12-824418-0.00001-1

Pike and Dennison (Pike and Dennison, 1989) first added 30 percent v/v *t*-butanol to the buffered solution and then added the ammonium sulfate and noted that samples can be mixed vigorously, including vortexing. There was no foaming and enzyme activity did not decrease, so they concluded that the *t*-butanol preserved protein structure. Both *t*-butanol and ammonium sulfate have been shown to stabilize proteins (Dennison and Lovrien, 1997). A preference is to add *t*-butanol first, as ammonium sulfate on its own precipitates proteins. When adding the *t*-butanol first the protein is stabilized and then precipitated with ammonium sulfate. Additionally ammonium sulfate has a higher density in *t*-butanol than in water as the *t*-butanol abstracts water from the aqueous phase (Pike and Dennison, 1989). In practice many groups have added either *t*-butanol or ammonium sulfate first and obtained the desired result.

The ammonium sulfate, when added to aqueous protein solutions takes up water of hydration and thus removes water from around proteins in solution, has an exclusion crowding effect, and the salt ions interact with opposite charges on a protein (Dennison, 2012; Dennison and Lovrien, 1997). Dennison and Lovrien describe the effects of the salt in TPP in terms of dehydration, ionic strength effects, kosmotropy, cavity surface tension enhancement, and exclusion crowding which leads to proteins precipitating from solution (Dennison and Lovrien, 1997). Tertiary butanol is soluble in water (Lovrien et al., 1987). When sufficient ammonium sulfate is present with *t*-butanol, *t*-butanol ammonium sulfate mixtures separate into two phases with *t*-butanol on top. Wilson and Walker (2010) describing precipitation with ammonium sulfate suggest that at low concentrations the salt removes the easily accessible water molecules around the protein surface then water molecules that were "forced into contact with hydrophobic groups on the protein surface" are removed to solvate the sulfate anion. The exposed hydrophobic regions then interact with the exposed hydrophobic regions on neighboring proteins and the proteins precipitate out of solution (Wilson and Walker, 2010). Dennison suggests that *t*-butanol in the TPP procedure "presumably binds to surface hydrophobic patches" on the protein and the *t*-butanol lowers the density of the protein and the protein floats at the interface of the aqueous and organic phases (Dennison, 2012). One interpretation from these suggestions is that at a particular concentration the sulfate anion removes water from the surface of the protein and this exposes the hydrophobic regions of the protein which in turn bind to *t*-butanol (an interpretation supported by C. Dennison, personal communication). This is the author's interpretation and definitive experimental evidence is not yet available. There is relatively little salt in the interface protein layer (Dennison, 2003, 2012).

There are four parameters that should be optimized in turn when developing a TPP protocol. The parameters are: ratio of *t*-butanol volume to sample volume, percent ammonium sulfate, pH and temperature. Some studies start off with 30 percent ammonium sulfate and then vary the volume of *t*-butanol to obtain the optimum *t*-butanol to sample volume ratio followed by optimizing percent ammonium sulfate, then pH and lastly temperature at the established *t*-butanol and ammonium sulfate concentrations (Gagaoua et al., 2014). There are two important practical considerations. First *t*-butanol has a melting temperature of 25 °C, so needs to be warmed before use in cold environments or during winter months. Secondly it is important to give the ammonium sulfate and the *t*-butanol time to equilibrate in turn with the protein solution and most authors use 30 to 60 min for each addition. Following a step by step optimization as illustrated here has successfully led to the isolation of many of the proteins listed in Table 4.1.

The advantages of using TPP were outlined by Lovrien et al. (1987). TPP is a rapid technique which, in many instances involves a single procedure (TPP) in place of multiple precipitation or chromatographic steps. The technique requires few pieces of equipment often used in routine laboratories, namely a weighing balance, centrifuge and spectrophotometer. The technique concentrates proteins in two ways, firstly unwanted proteins are removed which concentrates the desired protein relative to unwanted proteins and secondly, the isolated protein is in a smaller volume of liquid. The technique can be performed at room temperature. TPP often increases the activity of enzymes. The technique is simple to execute and it is easy to retrieve the desired protein. An overview in the form of a graphical abstract is presented in Fig. 4.1. Each of these facets of TPP will be discussed more fully in the following text.

4.2 TPP as a rapid single step procedure to isolate and concentrate proteins

TPP is rapid, taking just over two hours to perform assuming time is given to enable both the *t*-butanol and the ammonium sulfate to equilibrate. TPP is quoted in several publications as a "single step" technique which may be misleading. Most authors using the technique begin with either a homogenization or clarification step with centrifugation or a combination of the two. Thus the entire separation requires two or three steps, but once the material is prepared, the single TPP step replaces multiple chromatographic or precipitation steps. The three-phases can form without centrifugation (Narayan et al., 2008), but centrifugation produces a more compact protein layer which is preferred and easier to work with (personal observation). The following are examples of proteins from diverse sources isolated in a "single step" TPP protocol. Invertase from yeast, *Sacchromyces cereviscae* (Akardere et al., 2010) or from tomato *Lycopersicon esculentum* (Özer et al., 2010). α-*galactosida*se from tomato *Lycopersicon esculentum* (Çalci et al., 2009) and β-galactosidase from chick peas, *Cicer arietinum* (Duman and Kaya, 2013a). Exo-polygalacturonase from the fungus *Aspergillus sojae* (Dogan and Tari, 2008) and catalase from potato, *Solanum tuberosum* (Duman and Kaya, 2013b). Several proteases have been isolated from plants and two examples are ficin (ficain) from *Ficus carica* (Gagaoua et al., 2014) and papain from *Carica papaya* (Hafid et al., 2020). Both of these examples and many other proteases have been isolated from the latex of the fruit. These examples all

Table 4.1 TPP isolation of enzymes showing increases in enzyme activity, t-butanol: homogenate ratio,percent ammonium sulfate, pH, fold purification and protein size.

Protein	Source	Butanol	Percent ammonium sulfate	pH	Fold purification	Percent activity	Reference
Alkaline proteases	*Pangasianodon gigas*	1 to 0.5	50 percent Na Citrate	8	5.04	219 percent	Ketnawa et al.,2014
α-amylase	*Aspergillus oryzae*	N/A	N/A	6	7 to 9	700 to 900	Lovrien et al.,1987
α-amylase	*Rhyzopus oryzae*	N/A	N/A	N/A	NA	168.80 percent	El-Okki et al., 2017
β-amylase	*Abrus precatorius*	1 to 0.87	49.46 percent	5.2	10.17	156 percent	Sagu et al.,2015
Catalase	*Solanum tuberosum*	1 to 1	40 percent	7	14.1	262 percent	Duman et al.,2013b
Cucumisin	*Cucumis melo* var. *reticulatus*	1 to 1.25	60 percent	8	4.61	156 percent	Gagaoua et al., 2014
Ficin	*Ficus carica*	1 to 0.75	40 percent	7	6.04	167 percent	Gagaoua et al., 2014
β-galactosidase	*Aspergillus oryzae*	N/A	N/A	N/A	5 to 10	500 to 1000 percent	Lovrien et al.,1987
β-galactosidase	*Cicer arietinum*	1 to 0.5	60 percent	6.8	10.1	133 percent	Duman et al.,2013a
Invertase	*Sacchromyces cereviscae*	N/A	N/A	4 to 6	50 to 100	5000 to 10,000 percent	Lovrien et al.,1987
Invertase	*Saccharomyces cerevisiae*	N/A	N/A	4 to 6	75	500 to 1000 percent	Dennison and Lovrien 1997
Invertase	*Sacchromyces cereviscae*	1 to 0.05	50 percent	4	3.6	363 percent	Akardere et al.,2010
Invertase	*Lycopersicon esculentum*	1 to 1	50 percent	4.5	8.6	190 percent	Özer et al., 2010
Lipase	*Candida cylindracea*	N/A	N/A	4 to 8		500 to 600 percent	Lovrien et al.,1987

(continued)

Protein	Source	Butanol	Percent ammonium sulfate	pH	Fold purification	Percent activity	Reference
Lipase	*Candida cylindracea*	N/A	N/A	>5	8	900 percent	Dennison and Lovrien 1997
Lipase	*Thermomyces lanuginosus*	1 to 1	various	7	2	200 percent	Kumar et al.,2015
Nattokinase	*Bacillus natto*	1 to 1.5	30 percent	8	5.6	130 percent	Garg and Thorat 2014
Papain	*Carica papaya*	1 to 0.75	40 percent	6	11.45	134 percent	Hafid et al.,2020
Peroxidase	*Ipomoea palmate*	1 to 1	30 percent	9	2	160 percent	Narayan et al.,2008
Peroxidase	*Amsonia orientalis*	1 to 1	20 percent	6	12.5	162 percent	Karakus et al.,2018
Protease	*Bacillus subtilis*	N/A	N/A	8	5	300 percent	Dennison and Lovrien 1997
Protease	*Calotropis procera*	1 to 0.5	30 percent	unbuffered	6.92	132 percent	Rawdkuen et al.,2010
Proteases	*Carica papaya*	1 to 0.5	20 percent	7	23.8	165 percent	Chaiwut et al.,2010
Proteinase K	Merck purified	1 to 2	30 percent	6	NA	210 percent	Singh et al.,2001
Trypsin inhibitor	*Phaseolus vulgaris*	1 to 1	30 percent	8.2	5	315 percent	Wati et al.,2009
Trypsin inhibitor	*Phaseolus vulgaris* L.	1 to 1	30 percent	8.2	14	441 percent	Wati et al.,2009
Trypsin inhibitor	*Vigna vulgaris*	1 to 1	30 percent	8.2	7	228 percent	Wati et al.,2009
Xylanase	*Bacillus oceanisediminis*	1 to 1.5	50 percent	8	3.48	107 percent	Boucherba et al., 2017
Zingibain	*Zingiber officinale*	1 to 1	50 percent	7	14.19	215 percent	Gagaoua et al.,2015

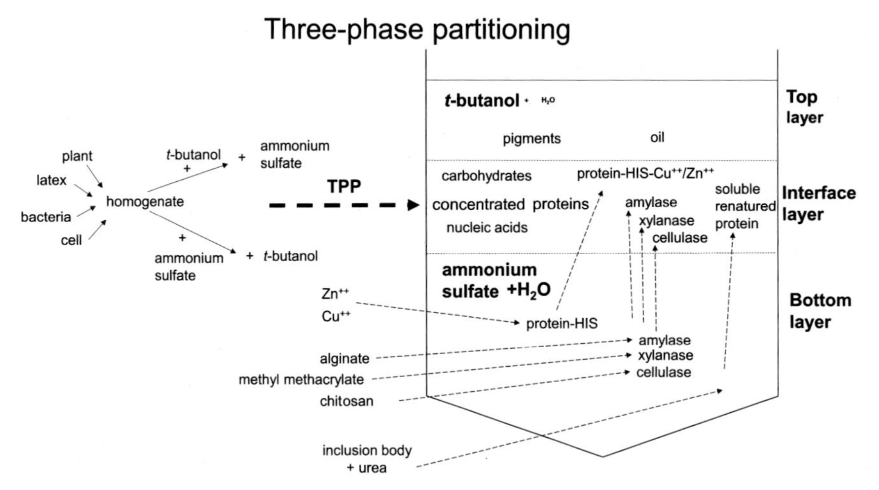

Fig. 4.1 ***Overview of three-phase partitioning.*** **Samples from different sources are homogenized and ammonium sulfate/*t*-butanol added. After equilibration samples are centrifuged to produce an upper *t*-butanol, interface and lower aqueous ammonium sulfate phase. TPP separates proteins, carbohydrates, pigments, nucleic acids and oils. TPP with affinity ligands can separate receptor proteins. TPP can renature proteins from inclusion bodies.**

show an increase in enzyme specific activity, due to the removal of unwanted protein, and there is considerable concentration of the isolated protein compared to the starting material. Isolating a protein in a single step is very important for routine laboratory work and can save considerable costs in an industrial application.

4.3 TPP concentrates proteins by decreasing the volume of water

During many protein isolation techniques it is necessary to increase the concentration of the desired protein by removing the aqueous solvent (Goldring, 2015; Goldring, 2019). Many protein isolation steps like the addition of buffers, dialysis and chromatography dilute protein solutions (Dennison, 2003; Pohl, 1990; Scopes, 1988). Enzymes at low concentrations tend to lose activity due to disaggregation of subunits, denaturation, loss of co-factors and loss of the crowding that stabilizes proteins (Goldring, 2019; Ma and Nussinov, 2013). When little enzyme activity is present, enzyme solutions are less convenient to work with and can be problematic to assay. Thus concentrating them is often an important component of any protein isolation protocol.

Lovrien et al. (1987) and Pike and Dennison (1989) observed that TPP concentrates proteins into small volumes in the interface layer separating proteins from the lower aqueous ammonium sulfate salt solution. Table 4.1 is not an exhaustive list but illustrates 29 instances where proteins from multiple sources were concentrated from dilute solutions. This is an advantage over all chromatographic methods which, typically, dilute proteins. TPP can be employed to concentrate proteins after any "diluting" step in a purification procedure (Dennison and Lovrien, 1997). TPP has been shown to concentrate proteins from the supernatants of cultures of *Corynebacterium pseudotuberculosis* during the purification of immunoreactive secreted proteins (Paule et al., 2004), laccase from *Ganoderma* culture supernatants (Rajeeva and Lele, 2011) and recombinant Trypanosome cysteine protease from *Pischia pastoris* culture supernatants (Pillay et al., 2010). An outline of TPP concentrating proteins is illustrated in Fig. 4.2.

Should the desired protein remain in the lower aqueous layer of TPP, it too has been concentrated by the addition of *t*-butanol and ammonium sulfate which have removed some water from around the protein and moved some proteins to the interface (Ward, 2009). Concentrating a protein with TPP takes place during a single step where enzyme activity and specific activity also increase as discussed below.

4.4 TPP concentrates individual proteins by removing unwanted proteins

The primary purpose of a protein purification technique is to obtain the desired protein whilst removing all or as many other "contaminating" proteins as possible. There is always a trade between losses of the desired protein, losses in the activity of the desired protein and the number of steps undertaken. To follow the progress of purification and to be able to compare each technique employed at any particular step in a purification protocol the amount of protein, the amount of enzyme activity and the volumes are carefully monitored and recorded in a "protein purification table" (Burgess, 2009;

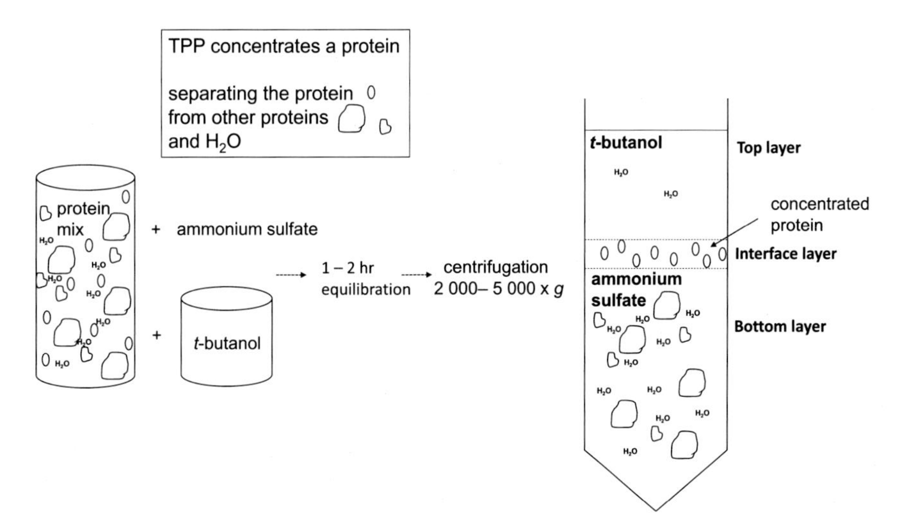

Fig. 4.2 ***Three-phase partitioning concentrates proteins.*** At an appropriate concentration of ammonium sulfate and the correct ratio of homogenate to t-butanol the desired protein is found in the interface layer. The protein in the interface is concentrated due to the removal of unwanted proteins which remain in the bottom layer and the presence of the protein in the interface layer with a decreased quantity of water.

Scopes, 1988). This record of protein and enzyme activity at each step enables informed decisions to be taken on the merit or otherwise of each step. The removal of undesired protein from the protein mixture is recorded as the changes in specific activity and yield of the enzyme. TPP has been shown to very successfully separate the desired protein into an insoluble intermediate phase, whilst leaving unwanted proteins in the aqueous ammonium sulfate lower layer, thus an increase in specific activity of the protein. As observed in the list in Table 4.1 different proteins from multiple sources are separated from unwanted proteins. Fig. 4.2 illustrates TPP concentrating proteins. Increases in specific enzyme activity are recorded for invertases, β-glycosidase, lipases and acid phosphatase (Lovrien et al., 1987), cucumisin (Gagaoua et al., 2017) trypsin inhibitor (Wati et al., 2009) protease/amylase inhibitor (Sharma and Gupta, 2001) and listed for many proteins by Dennison and Lovrien (Dennison and Lovrien, 1997).

Instead of designing the experiment to place the protein of interest in the interface layer a second approach is to remove unwanted proteins in the interface layer and leave the protein of interest in the lower aqueous phase. This approach was used in the isolation of recombinant Green Fluorescent Protein from *Escherichia coli* for a commercial operation (Ward, 2009) and zingibain from *Zingiber officinale* Roscoe rhizomes (Gagaoua et al., 2015). Rajeeva and Lele (Rajeeva and Lele, 2011) isolated laccase from *Ganoderma* where the protein was initially in the lower aqueous layer and unwanted proteins in the interface and lipids, carbohydrates and pigments in the upper t-butanol layer. A second TPP step precipitated laccase in the interface layer.

Whether the protein is in the interface phase or in the lower aqueous phase, some water and some proteins are removed from the desired protein as illustrated in Fig. 4.2.

4.5 TPP preserves and increases enzyme activity

Tan and Lovrien (1972) originally showed that *t*-butanol stabilized enzymes. Lovrien et al.,(1987) then showed with several enzymes that enzyme activity was increased in TPP. Fig. 4.3 represents this graphically.

Table 4.1 illustrates examples of several enzymes in this category listed in alphabetical order. The enzymes are from diverse organisms and include amylases, catalases, galactosidases, invertases, lipases, peroxidases, xylanases and a range of proteases. The activity of α-amylase from *Aspergillus oryzae* increased 7 to 9 fold or 700 to 900 percent (Lovrien et al., 1987) and to 168.8 percent from *Rhyzopus oryzae* (Ait Kaki El-Hadef El-Okki et al., 2017) for example. In Table 4.1 the median increase in enzyme activity is 213 percent and the range is from 107 to 10,000 percent. The highest increases include trypsin inhibitor 441 percent (Wati et al., 2009), α-amylase 900 percent (Lovrien et al., 1987), lipase 900 percent and invertase 1000 percent (Dennison and Lovrien, 1997) and invertase 10,000 percent (Lovrien et al., 1987). For those of us isolating and purifying proteins, these are very interesting observations.

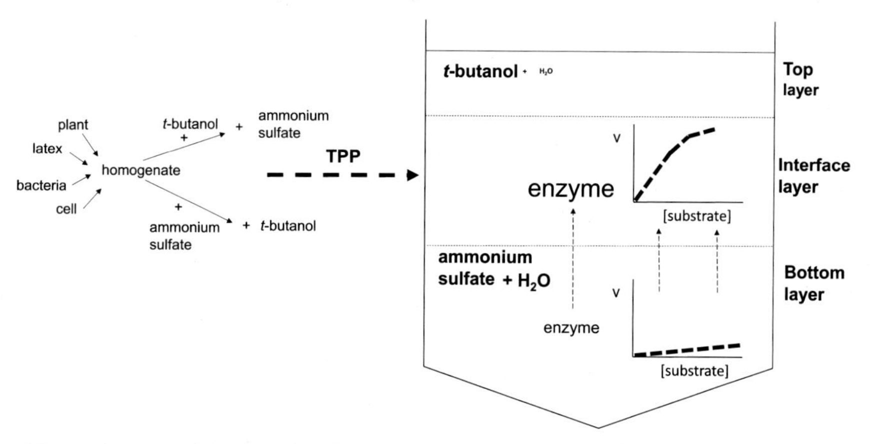

Fig. 4.3 ***Three-phase partitioning often leads to an increase in enzyme activity.*** Enzyme activity in the lower phase is illustrated as low and in the interface as high on a graph of velocity of enzyme activity against substrate concentration.

4.6 Explanations for TPP preserving and increasing enzyme activity

There are some twelve explanations for an increase in enzyme activity. Let's start with the initial homogenate solution. Whether this is from plant latex or mammalian tissue the original proteins mixture, described very nicely as a "gamisch" by Lovrien et al. (1987), contains multiple proteins, carbohydrates, nucleic acids, fats and metabolites. These components may participate in or inhibit enzyme activity and their removal then increases enzyme activity. Components of the homogenate may influence the enzyme reaction itself and their removal increases enzyme activity. Lastly *t*-butanol can influence the structure and architecture around the active site of an enzyme. Suggestions and examples for each of these possibilities are outlined below.

- Enzyme activity increases after TPP as the natural substrates of the enzyme being isolated are separated from the enzyme. For example if protease activity in an initial extract was measured by the azocasein substrate, the presence of endogenous substrate would compete with the azocasein and give a lower reading for enzyme activity. The removal of the endogenous substrate by TPP leads to an apparent increase in enzyme activity.
- Enzymes can be inhibited by products of their reactions that are present in the homogenate. Removal of these products increases enzyme activity. Fatty acids, for example, inhibit acetyl-coA carboxylase activity (Goldring and Read, 1993, 1994) and fatty acids could be separated from the enzyme by TPP.
- Enzyme activity increases because of the removal of components influencing enzyme activity. Niehous and Dilts (1982) (Jr. Niehaus and Jr Dilts, 1982) reported a

5 fold increase in mannitol dehydrogenase specific enzyme activity due to the removal of"unwanted components in the aqueous layer". Rajeeva and Lele (2011)report the removal of enzyme inhibitors by TPP. Dennison and Lovrien (1997). describe the removal of detergents by TPP.
- Enzyme activity increases because TPP prevents unwanted protein-protein interactions (Dennison et al., 2000).
- Enzyme activity increases because material that influence/interfere with the readings in an enzyme catalyzed reaction are removed. An example is the removal of haemoglobin when working with red blood cells (Dennison, 2012; Jacobs et al., 1989). Haemoglobin which absorbs at wavelengths (Takatani and Graham, 1979) used to measure the activity of several enzymes, including erythrocytic carbonic anhydrase activity at 405 nm, is removed in the TPP isolation of the enzyme from erythrocytes by Pol et al. (1990). Other pigments are removed by TPP (Lovrien et al., 1987; Rajeeva and Lele, 2011).
- Enzyme activity increases because the enzyme is converted from an active to a more active form of the enzyme. Roy and Gupta (2004) suggest this is not the case with the isolation of α-chymotrypsin based on the percent of active protein before and after TPP.
- Enzyme activity decreased by possible autolysis that is inhibited by the addition of *t*-butanol or ammonium sulfate. Roy and Gupta (2004) suggest this is not the case with α-chymotrypsin.
- TPP inhibits proteases that may digest the required enzyme. (Dennison et al., 2000)
- Enzyme activity increases due to a change in protein architecture around the catalytic site. Singh et al. (2001) support this with a X ray diffraction study on purified protease K.
- Enzyme activity increased due to increased flexibility of the enzyme as suggested by Gagaoua et al. (2017) when isolating cucumisin.
- Enzyme activity increases due to changes in conformational flexibility of an enzyme during TPP as indicated by Kumar et al. (2015).
- Enzyme activity may increase due to the crowding effect of *t*-butanol. Crowding stabilizes proteins (Ma and Nussinov, 2013) and crowding by the alcohol is described by Dennison and Lovrien (1997).

4.7 Purification and/or refolding of denatured enzymes with TPP

When recombinant proteins are expressed in *Escheriscia coli* host bacteria (or other bacteria, yeast or cell expression system) the expressed proteins are, relative to their natural environment, in an artificial host environment and, relative to host proteins, present as a novel foreign protein, and often present in higher concentrations than host proteins. These factors often result in the formation of insoluble inclusion bodies. One can spend considerable time attempting to express the proteins: with different fusion partners; under multiple conditions; with chaperones; and other manipulations to obtain soluble proteins. The end result is that one often works with the inclusion bodies (Rinas et al., 2017). Opinion and practice, according to García-Fruitós (2010) has changed and suggests that inclusion bodies are an advantage as they can readily be isolated from lysed bacteria exploiting their insoluble nature. Enzymes may be reversibly immobilized or loose activity due to immobilization (Roy et al., 2005). It is therefore desirable for laboratory studies and industrial applications to be able to renature and refold enzymes and recover enzyme activity.

Raghava et al. (2008) used TPP to refold 12 different proteins from inclusion bodies. The approach was to resolublise proteins trapped as inclusion bodies in 100 mM dithiothreitol and 8 M urea followed by a two-step TPP protocol starting with 5 percent ammonium sulfate followed by adding 35 percent ammonium sulfate to the aqueous phase. The initial ammonium sulfate removed unwanted proteins that are unfortunately often overlooked when isolating proteins from inclusion bodies. The second ammonium sulfate step placed the desired proteins in the interface layer as depicted in Fig. 4.4.

Each and every one of the proteins in the study recovered enzyme activity with yields as good as or better than conventional methods. Roy et al. (2005) denatured and renatured a commercial enzyme preparation with TPP and recovered 90 to 98 percent enzyme activity. An adaptation of this method was to include an affinity ligand with the TPP step to accompany refolding, known as macro-(affinity ligand) facilitated three-phase partitioning – MLFTPP (Gautam et al., 2012). The MLFTPP with recombinant

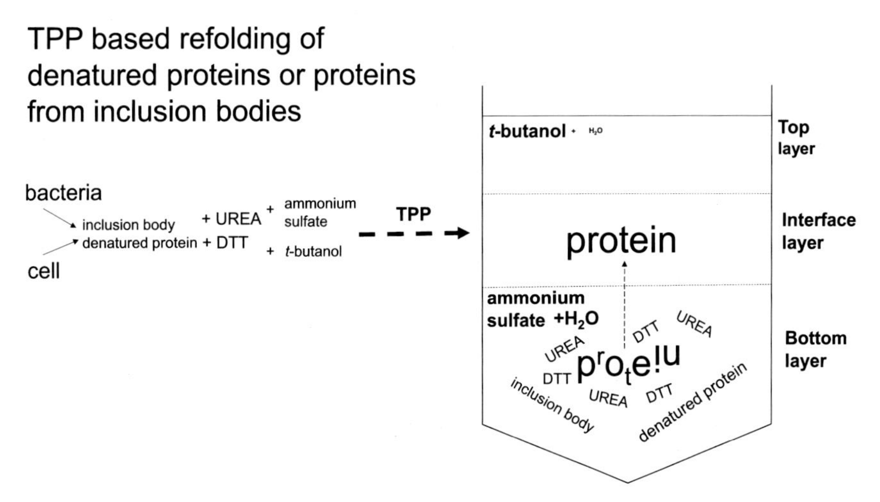

Fig. 4.4 ***Three-phase partitioning can be used to renature proteins from inclusion bodies.*** The denatured protein is depicted in the bottom layer in the presence of urea and DTT and the protein can be renatured and concentrated in the interface layer.

Green Fluorescent Protein, an MBP fusion protein and other protein examples produced concentrated proteins in a short time (Gautam et al., 2012). Like others, the author is disappointed not to have discovered the Raghava publication before spending several years attempting to solubilize an insoluble malarial protein with limited success (Raghava et al., 2008)!

4.8 TPP purification of recombinant HIS-Tag fusion proteins or metal binding proteins

The addition of a HIS-tag consisting of six or more histidine amino acid residues to a recombinant protein has enabled a single step purification of the HIS-tagged fusion protein. Histidine binds reversibly to nickel, cobalt or copper so these metals can be immobilized on an affinity matrix to isolate recombinant HIS-tagged proteins or histidine rich recombinant proteins from lysed bacterial host cells or histidine rich proteins from lysed tissues and cells (Gaberc-Porekar and Menart, 2001; Roy and Gupta, 2002; Terpe, 2003). Roy and Gupta (Roy and Gupta, 2002) explored the use of TPP where, instead of using ammonium sulfate as the salt, they added either $CuSO_4$ or $ZnCl_2$ to the aqueous phase then *t*-butanol. The trypsin inhibitor protein in the study has surface histidine residues that bind the metals and was precipitated by both of the two metallic salts at the TPP interface as shown in Fig. 4.5 (Roy and Gupta, 2002). The added advantage is that the TPP step may remove endotoxins, detergents and *E. coli* host proteins from the fusion protein.

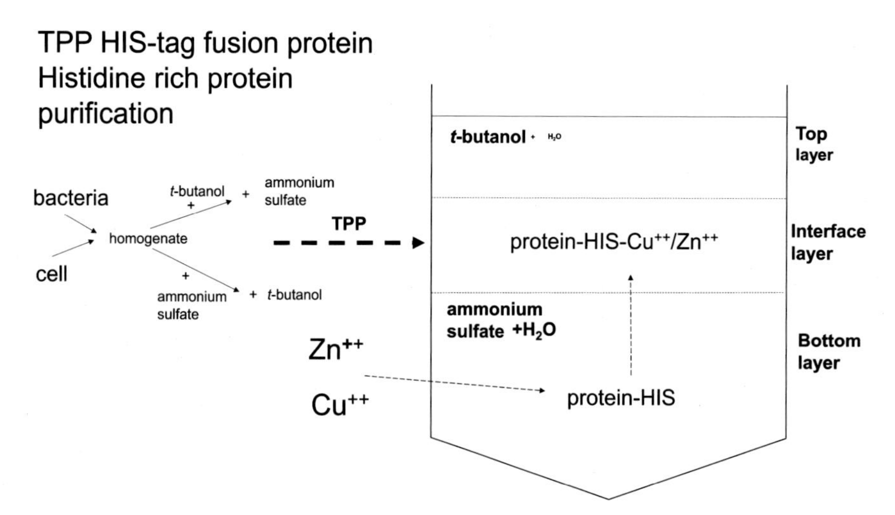

Fig. 4.5 ***Three-phase partitioning can isolate HIS-tagged fusion proteins or proteins with a histidine rich domain.*** The HIS-tagged recombinant protein or "histidine rich domain protein" can be isolated from other proteins by the addition of the appropriate metal ligand shown as copper or zinc. The ligand addition moves the His-proteins from the bottom aqueous layer to the interface layer.

4.9 TPP purification of proteins with affinity ligands

Three-phase partitioning has been used extensively to isolate proteins, but the approach by the nature of the technique is trial and error, exploring the ratio of *t*-butanol to homogenate volume, ammonium sulfate, pH and temperature. A more targeted approach was described with the addition of copper or zinc salts to the aqueous phase, which pushed trypsin inhibitor, which has a histidine rich domain that binds the metals, into the interface phase (Roy and Gupta, 2002). This approach was followed up when it was observed that the polymers methyl methacrylate, alginate and chitosan bind to and precipitate xylanase, α-amylase and cellulase respectively from solution (Homma et al., 1993; Mondal et al., 2003; Sharma and Gupta, 2002a). TPP with the polymers, known as micro-affinity ligand TPP selectively precipitated the partner enzymes into the insoluble interface TPP layer as illustrated in Fig. 4.6 (Sharma et al., 2003).

4.10 Ultrasound assisted TPP to isolate proteins, oils and polysaccharides

Ultrasound is an efficient method to break open a number of different cell types and has been combined with TPP to isolate enzymes, ursolic and oleanolic acid, oil and polysaccharides as described below. Ultrasound when combined with TPP improved the yield three fold and increased the activity of isolated polyphenol oxidase from potato peel, *Solanum tuberosum* by 20 percent (Niphadkar and Rathod, 2015). Mangiferin was isolated from *Mangifera indica* by TPP and when ultrasound was combined with TPP, the yield of mangiferin doubled and the combination reduced the time to execute the protocol

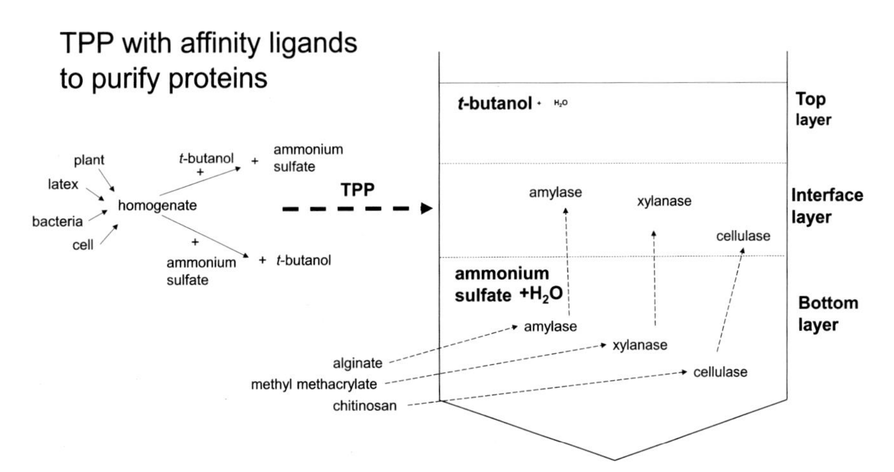

Fig. 4.6 ***Three-phase partitioning can isolate target proteins with appropriate affinity ligands.*** Three affinity ligands are added to the homogenate with ammonium sulfate and *t*-butanol. The ligands alginate, methyl methacrylate and chitosan bind to their receptors amylase, xylanase and cellulose and each receptor protein is moved to the interface layer under appropriate conditions.

from 2 h to 25 min (Kulkarni and Rathod, 2014). Utrasound was performed on *Bacillus sphaericus* bacterial cells suspended in an ammonium sulfate *t*-butanol mixture to isolate fibrinolytic enzymes (Avhad et al., 2014). Serratiopeptidase was isolated from *Serratia marcescens* with ultrasound assisted TPP (Pakhale and Bhagwat, 2016). Ultrasound was also used with TPP in the isolation of ursolic and oleanolic acid from *Ocimum sanctum* with (Vetal et al., 2014) and oil from custard apples (Panadare et al., 2020). The technique was used to isolate polysacharides from peas, *Pisum sativum* (Zhang et al., 2020).

4.11 Microwave assisted TPP

Microwave ovens are often used to rapidly heat samples in laboratories. Microwave ovens have been used with TPP to both improve yields and decrease the length of time to isolate material. Microwave assisted TPP was found reduce the isolation time for mangiferin from *Mangifera indica* from 5 h to 20 min (Kulkarni and Rathod, 2015). Laccase was isolated from white rot fungi, *Trametes hirsuta* where microwave assisted TPP increased yields nearly 6 fold over TPP alone and decreased the time from 60 min to 7 (Patil and Yadav, 2018).

4.12 Versatility of TPP. separating DNA, carbohydrates and oils and two-step TPP protocols

Part of the versatility of the TPP technique is that cellular components other than proteins can be isolated or removed as shown in Fig. 4.7. The range of proteins include

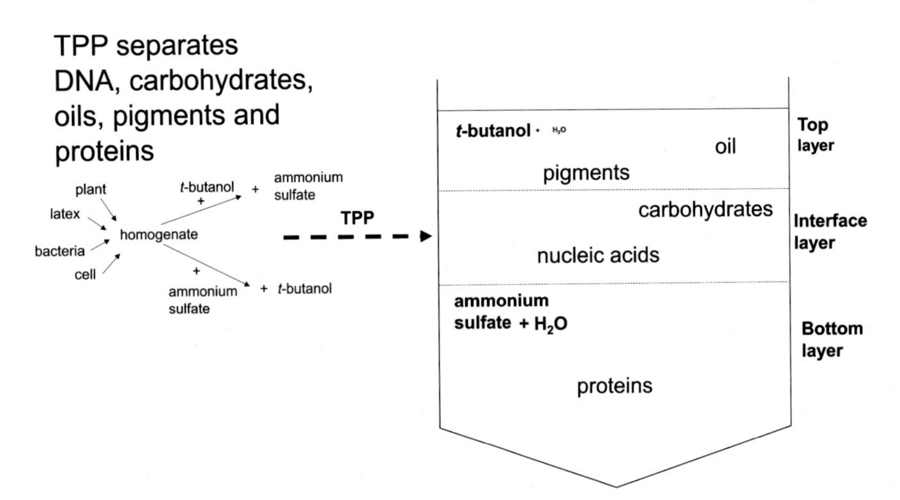

Fig. 4.7 ***Three-phase partitioning separates many cellular components.*** Examples in the figure include carbohydrates, nucleic acids, oils and pigments depicted in the different phases under appropriate conditions.

proteases from bacterial, plant and animal material (chapter 7), and other enzymes like amylases (chapter 9), peroxidases (chapter 8), β-galactosidases and laccases illustrated in Table 4.1. Chromosomal DNA and polysaccharides can be removed or isolated with TPP. Chromosomal DNA was found in the interface layer in the TPP based isolation of recombinant Green Fluorescent Protein from *E. coli* (Ward, 2009). Nucleic acids were found in the interface phase at 20, 30 and 40 percent ammonium sulfate and 1:0.2 *t*-butanol treatment of a lentil homogenate (Goldring personal observation). TPP has been described in the separation of carbohydrates (Sharma and Gupta, 2002b) and the separation of hydrolyzed levan (polysaccharide) from *Zymomonas mobilis* (Coimbra et al., 2010). Oils have been isolated with TPP from *Crotolaria juncea* seeds (Dutta et al., 2015). Edible oils were isolated from soya beans, rice bran and mango kernals by the inclusion of a protease treatment prior to TPP which lead to increased yields of the oils (Gaur et al., 2007). The oils partitioned in the *t*-butanol phase. TPP was found to separate fatty acids from albumin (Lovrien et al., 1987).

TPP is particularly useful as a single step in protein isolation to replace several alternative steps involving chromatography. TPP is not limited to a single step protocol. Two steps were used to isolate proteases from papaya skins (Chaiwut et al., 2010), β-galactosidase from *Lactobacillus acidophilus* (Choonia and Lele, 2013) and α-galactosidase and invertase from *Aspergillus oryzae* (Dhananjay and Mulimani, 2008), laccase from *Ganoderma* (Rajeeva and Lele, 2011).

4.13 Conclusion

Writing this review has reminded the author of the unexpected surprises in science that bring pleasure and wonder alongside practical applications. Three-phase partitioning is a technique that has led to interesting and remarkable advancements in a range of fields. TPP reduced the time from 6 months to an hour or two for isolating green fluorescent protein (Ward, 2009). TPP can replace multiple steps in the isolation of a large range of proteins as illustrated in Table 4.1. It is rare that one technique can both remove water from around proteins and remove unwanted proteins, lipids, nucleic acids and pigments in a single step. A single technique that can isolate protein, carbohydrate or oil. A technique that can be used to refold denatured enzymes and separate an individual protein from a complex mixture. A technique that can do all of the above and increase enzyme activity. One can therefore conclude that three-phase partitioning has multiple facets in the purification, concentration and preservation of proteins.

References

Akardere, E., Özer, B., Çelem, E.B., Önal, S., 2010. Three-phase partitioning of invertase from Baker's yeast. Sep. Purif. Technol. 72, 335–339.

Avhad, D.N., Niphadkar, S.S., Rathod, V.K., 2014. Ultrasound assisted three phase partitioning of a fibrinolytic enzyme. Ultrason. Sonochem. 21, 628–633.
Boucherba, N., Gagaoua, M., Bouanane-Darenfed, A., Bouiche, C., Bouacem, K., Kerbous, M.Y., Maafa, Y., Benallaoua, S., 2017. Biochemical properties of a new thermo- and solvent-stable xylanase recovered using three phase partitioning from the extract of Bacillus oceanisediminis strain SJ3. Bioresour. Bioprocess. 4, 29.
Burgess, R.R., 2009. Preparing a purification summary table. Methods Enzymol. 463, 29–34.
Çalci, E., Demir, T., Biçak Çelem, E., Önal, S., 2009. Purification of tomato (*Lycopersicon esculentum*) α-galactosidase by three-phase partitioning and its characterization. Sep. Purif. Technol. 70, 123–127.
Chaiwut, P., Pintathong, P., Rawdkuen, S., 2010. Extraction and three-phase partitioning behavior of proteases from papaya peels. Process. Biochem. 45, 1172–1175.
Choonia, H.S., Lele, S.S., 2013. Three phase partitioning of β-galactosidase produced by an indigenous *Lactobacillus acidophilus* isolate. Sep. Purif. Technol. 110, 44–50.
Coimbra, C.G., Lopes, C.E., Calazans, G.M., 2010. Three-phase partitioning of hydrolyzed levan. Bioresour. Technol. 101, 4725–4728.
Dennison, C., 2003. A Guide to Protein Isolation. Kluwer academic publishers, Dordrecht, The Netherlands.
Dennison, C., 2012. Three-phase partitioning. In: Tschesche, H. (Ed.), Methods in Protein Biochemistry. Walter de Gruyter GmbH & Co, Berlin/Boston.
Dennison, C., Lovrien, R., 1997. Three phase partitioning: concentration and purification of proteins. Protein Expr. Purif. 11, 149–161.
Dennison, C., Moolman, L., Pillay, C.S., Meinesz, R.E., 2000. Use of 2-methylpropan-2-ol to inhibit proteolysis and other protein interactions in a tissue homogenate: an illustrative application to the extraction of cathepsins B and L from liver tissue. Anal. Biochem. 284, 157–159.
Dhananjay, S.K., Mulimani, V.H., 2008. Purification of alpha-galactosidase and invertase by three-phase partitioning from crude extract of *Aspergillus oryzae*. Biotechnol. Lett. 30, 1565–1569.
Dogan, N., Tari, C., 2008. Characterization of three-phase partitioned exo-polygalacturonase from *Aspergillus sojae* with unique properties. Biochem. Eng. J. 39, 43–50.
Duman, Y., Kaya, E., 2013a. Purification, recovery, and characterization of chick pea (*Cicer arietinum*) β-galactosidase in single step by three phase partitioning as a rapid and easy technique. Protein Expr. Purif. 91, 155–160.
Duman, Y., Kaya, E., 2013b. Three-phase partitioning as a rapid and easy method for the purification and recovery of catalase from sweet potato tubers (*Solanum tuberosum*). Appl. Biochem. Biotechnol. 170, 1119–1126.
Dutta, R., Sarkar, U., Mukherjee, A., 2015. Process optimization for the extraction of oil from *Crotalaria juncea* using three phase partitioning. Ind. Crops. Prod. 71, 89–96.
El-Okki, A K E-H, Gagaoua, A., Bourekoua, M., Hafid, H., Bennamoun, K., Djekrif-Dakhmouche, L., El-Hadef El-Okki, S., Meraihi, Z., 2017. Improving bread quality with the application of a newly purified thermostable α-amylase from rhizopus oryzae FSIS4Foods6, 1.
Gaberc-Porekar, V., Menart, V., 2001. Perspectives of immobilized-metal affinity chromatography. J. Biochem. Biophys. Methods 49, 335–360.
Gagaoua, M., Boucherba, N., Bouanane-Darenfed, A., Ziane, F., Nait-Rabah, S., Hafid, K., Boudechicha, H.R., 2014. Three-phase partitioning as an efficient method for the purification and recovery of ficin from Mediterranean fig (*Ficus carica* L.) latex. Sep. Purif. Technol. 132, 461–467.
Gagaoua, M., Hoggas, N., Hafid, K., 2015. Three phase partitioning of zingibain, a milk-clotting enzyme from *Zingiber officinale* Roscoe rhizomes. Int. J. Biol. Macromol. 73, 245–252.
Gagaoua, M., Ziane, F., Nait Rabah, S., Boucherba, N., Ait Kaki El-Hadef El-Okki, A., Bouanane-Darenfed, A., Hafid, K., 2017. Three phase partitioning, a scalable method for the purification and recovery of cucumisin, a milk-clotting enzyme, from the juice of *Cucumis melo var.* reticulatus. Int. J. Biol. Macromol. 102, 515–525.
García-Fruitós, E., 2010. Inclusion bodies: a new concept. Microb. Cell Fact. 9, 80.
Garg, R., Thorat, B.N., 2014. Nattokinase purification by three phase partitioning and impact of t-butanol on freeze drying. Sep. Purif. Technol. 131, 19–26.
Gaur, R., Sharma, A., Khare, S.K., Gupta, M.N., 2007. A novel process for extraction of edible oils: enzyme assisted three phase partitioning (EATPP). Bioresour. Technol. 98, 696–699.

Gautam, S., Dubey, P., Singh, P., Varadarajan, R., Gupta, M.N., 2012. Simultaneous refolding and purification of recombinant proteins by macro-(affinity ligand) facilitated three-phase partitioning. Anal. Biochem. 430, 56–64.
Goldring, J.P., 2015. Methods to concentrate proteins for protein isolation, proteomic and peptidomic evaluation. Methods Mol. Biol. 1314, 15–18.
Goldring, J.P., Read, J.S., 1993. Insect acetyl-CoA carboxylase: activity during the larval, pupal and adult stages of insect development. Comp. Biochem. Physiol. B 106, 855–858.
Goldring, J.P., Read, J.S., 1994. Insect acetyl-CoA carboxylase: enzyme activity during adult development and after feeding in the tsetse fly, Glossina morsitans. Comp. Biochem. Physiol. Biochem. Mol. Biol. 108, 27–33.
Goldring, J.P.D., 2019. Concentrating proteins by salt, polyethylene glycol, solvent, SDS precipitation, three-phase partitioning, dialysis, centrifugation, ultrafiltration, lyophilization, affinity chromatography, immunoprecipitation or increased temperature for protein isolation, drug interaction, and proteomic and peptidomic evaluation. Methods Mol. Biol. 1855, 41–59.
Hafid, K., John, J., Sayah, T.M., Domínguez, R., Becila, S., Lamri, M., Dib, A.L., Lorenzo, J.M., Gagaoua, M., 2020. One-step recovery of latex papain from *Carica papaya* using three phase partitioning and its use as milk-clotting and meat-tenderizing agent. Int. J. Biol. Macromol. 146, 798–810.
Homma, T., Fujii, M., Mori, J., Kawakami, T., Kuroda, K., Taniguchi, M., 1993. Production of cellobiose by enzymatic hydrolysis: removal of beta-glucosidase from cellulase by affinity precipitation using chitosan. Biotechnol. Bioeng. 41, 405–410.
Jacobs, G.R., Pike, R.N., Dennison, C., 1989. Isolation of cathepsin D using three-phase partitioning in t-butanol/water/ammonium sulfate. Anal. Biochem. 180, 169–171.
Karakus, Y.Y., Acemi, A., isik, S., Duman, Y., 2018. Purification of peroxidase from *Amsonia orientalis* by three-phase partitioning and its biochemical characterisation. Sep. Sci. Technol. 53, 756–766.
Ketnawa, S., Benjakul, S., Martínez-Alvarez, O., Rawdkuen, S., 2014. Three-phase partitioning and proteins hydrolysis patterns of alkaline proteases derived from fish viscera. Sep. Purif. Technol. 132, 174–181.
Kulkarni, V.M., Rathod, V.K., 2014. Extraction of mangiferin from *Mangifera indica* leaves using three phase partitioning coupled with ultrasound. Ind. Crops Prod. 52, 292–297.
Kulkarni, V.M., Rathod, V.K., 2015. A novel method to augment extraction of mangiferin by application of microwave on three phase partitioning. Biotechnol. Rep. 6, 8–12.
Kumar, M., Mukherjee, J., Sinha, M., Kaur, P., Sharma, S., Gupta, M.N., Singh, T.P., 2015. Enhancement of stability of a lipase by subjecting to three phase partitioning (TPP): structures of native and TPP-treated lipase from *Thermomyces lanuginosa*. Sustain. Chem. Process. 3, 14.
Lovrien, R., Goldensoph, C., Anderson, P.C., Odegaard, B., 1987. Three phase partitioning (TPP) via t-butanol: enzymes separations from crudes. in: Burgess, R. (ed.) Protein purifi cation: micro to macro. New York: A.R. Liss, 131-148.
Ma, B., Nussinov, R., 2013. Structured crowding and its effects on enzyme catalysis. Top. Curr. Chem. 337, 123–137.
Mondal, K., Sharma, A., Lata, L., Gupta, M.N., 2003. Macroaffinity ligand-facilitated three-phase partitioning (MLFTPP) of alpha-amylases using a modified alginate. Biotechnol. Prog. 19, 493–494.
Narayan, A.V., Madhusudhan, M.C., Raghavarao, K.S., 2008. Extraction and purification of Ipomoea peroxidase employing three-phase partitioning. Appl. Biochem. Biotechnol. 151, 263–272.
Jr. Niehaus, W.G., Jr Dilts, R.P., 1982. Purification and characterization of mannitol dehydrogenase from Aspergillus parasiticus. J. Bacteriol. 151, 243–250.
Niphadkar, S.S., Rathod, V.K., 2015. Ultrasound-assisted three-phase partitioning of polyphenol oxidase from potato peel (*Solanum tuberosum*). Biotechnol. Prog. 31, 1340–1347.
Odegaard, B., Anderson, P.C., Lovrien, R., 1984. Resolution of the multienzyme cellulase complex of Trichoderma reesei QM9414. J. Appl. Biochem. 6, 158–183.
Özer, B., Akardere, E., Çelem, E.B., Önal, S., 2010. Three-phase partitioning as a rapid and efficient method for purification of invertase from tomato. Biochem. Eng. J. 50, 110–115.
Pakhale, S.V., Bhagwat, S.S., 2016. Purification of serratiopeptidase from *Serratia marcescens* NRRL B 23112 using ultrasound assisted three phase partitioning. Ultrason. Sonochem. 31, 532–538.

Panadare, D.C., Gondaliya, A., Rathod, V.K., 2020. Comparative study of ultrasonic pretreatment and ultrasound assisted three phase partitioning for extraction of custard apple seed oil. Ultrason. Sonochem. 61, 104821.

Patil, P.D., Yadav, G.D., 2018. Application of microwave assisted three phase partitioning method for purification of laccase from *Trametes hirsuta*. Process. Biochem. 65, 220–227.

Paule, B.J., Meyer, R., Moura-Costa, L.F., Bahia, R.C., Carminati, R., Regis, L.F., Vale, V.L., Freire, S.M., Nascimento, I., Schaer, R., Azevedo, V., 2004. Three-phase partitioning as an efficient method for extraction/concentration of immunoreactive excreted-secreted proteins of Corynebacterium pseudotuberculosis. Protein Expr. Purif. 34, 311–316.

Pike, R.N., Dennison, C., 1989. Protein fractionation by three phase partitioning (TPP) in aqueous/t-butanol mixtures. Biotechnol. Bioeng. 33, 221–228.

Pillay, D., Boulangé, A.F., Coetzer, T.H., 2010. Expression, purification and characterisation of two variant cysteine peptidases from Trypanosoma congolense with active site substitutions. Protein Expr. Purif. 74, 264–271.

Pohl, T., 1990. Concentrations of Proteins and Removal of Solutes. Acacemic Press, London.

Pol, M.C., Deutsch, H.F., Visser, L., 1990. Purification of soluble enzymes from erythrocyte hemolysates by three phase partitioning. Int. J. Biochem. 22, 179–185.

Raghava, S., Barua, B., Singh, P.K., Das, M., Madan, L., Bhattacharyya, S., Bajaj, K., Gopal, B., Varadarajan, R., Gupta, M.N., 2008. Refolding and simultaneous purification by three-phase partitioning of recombinant proteins from inclusion bodies. Protein Sci. 17, 1987–1997.

Rajeeva, S., Lele, S.S., 2011. Three-phase partitioning for concentration and purification of laccase produced by submerged cultures of *Ganoderma sp. WR-1*. Biochem. Eng. J. 54, 103–110.

Rawdkuen, S., Chaiwut, P., Pintathong, P., Benjakul, S., 2010. Three-phase partitioning of protease from *Calotropis procera* latex. Biochem. Eng. J. 50, 145–149.

Rinas, U., Garcia-Fruitós, E., Corchero, J.L., Vázquez, E., Seras-Franzoso, J., Villaverde, A., 2017. Bacterial inclusion bodies: discovering their better half. Trends Biochem. Sci. 42, 726–737.

Roy, I., Gupta, M.N., 2002. Three-phase affinity partitioning of proteins. Anal. Biochem. 300, 11–14.

Roy, I., Gupta, M.N., 2004. Preparation of highly active alpha-chymotrypsin for catalysis in organic media. Bioorg. Med. Chem. Lett. 14, 2191–2193.

Roy, I., Sharma, A., Gupta, M.N., 2005. Recovery of biological activity in reversibly inacivated proteins by three phase partitioning. Enzyme Microb. Technol. 37, 113–129.

Sagu, S.T., Nso, E.J., Homann, T., Kapseu, C., Rawel, H.M., 2015. Extraction and purification of beta-amylase from stems of *Abrus precatorius* by three phase partitioning. Food Chem. 183, 144–153.

Scopes, R.K., 1988. Protein Purification: Principles and Practice. Springer-Verlag, New York.

Sharma, A., Gupta, M.N., 2002a. Macroaffinity ligand-facilitated three-phase partitioning (MLFTPP) for purification of xylanase. Biotechnol. Bioeng. 80, 228–232.

Sharma, A., Gupta, M.N., 2002b. Three phase partitioning of carbohydrate polymers: separation and purification of alginates. Carbohydr. Polym. 48, 391–395.

Sharma, A., Mondal, K., Gupta, M.N., 2003. Separation of enzymes by sequential macroaffinity ligand-facilitated three-phase partitioning. J. Chromatogr. A 995, 127–134.

Sharma, A.I., Gupta, M.N., 2001. Three phase partitioning as a large-scale separation method for purification of a wheat germ bifunctional protease/amylase inhibitor. Process. Biochem. 37, 193–196.

Singh, R.K., Gourinath, S., Sharma, S., Roy, I., Gupta, M.N., Betzel, C., Srinivasan, A., Singh, T.P., 2001. Enhancement of enzyme activity through three-phase partitioning: crystal structure of a modified serine proteinase at 1.5 A resolution. Protein. Eng. 14, 307–313.

Takatani, S., Graham, M.D., 1979. Theoretical analysis of diffuse reflectance from a two-layer tissue model. IEEE Trans. Biomed. Eng. 26, 656–664.

Tan, K.H., Lovrien, R., 1972. Enzymology in aqueous-organic cosolvent binary mixtures. J. Biol. Chem. 247, 3278–3285.

Terpe, K., 2003. Overview of tag protein fusions: from molecular and biochemical fundamentals to commercial systems. Appl. Microbiol. Biotechnol. 60, 523–533.

Vetal, M.D., Shirpurkar, N.D., Rathod, V.K., 2014. Three phase partitioning coupled with ultrasound for the extraction of ursolic acid and oleanolic acid from *Ocimum sanctum*. Food Bioprod. Process. 92, 402–408.

Ward, W.W., 2009. Three phase partitioning for protein purification. Innov. Pharm. Technol.
Wati, R.K., Theppakorn, T., Benjakul, S., Rawdkuen, S., 2009. Three-phase partitioning of trypsin inhibitor from legume seeds. Process. Biochem. 44, 1307–1314.
Wilson, K., Walker, W., 2010. Principals and Techniques of Biochemsitry and Molecular Biology. Cambridge, University press, Cambridge.
Zhang, S.J., Hu, T.T., Chen, Y.Y., Wang, S., Kang, Y.F., 2020. Analysis of the polysaccharide fractions isolated from pea (Pisum sativum L.) at different levels of purification. J. Food Biochem. 44, e13248.

CHAPTER 5

Enzymes recovery by three phase partitioning

Mohammed Gagaoua
Food Quality and Sensory Science Department, Teagasc Ashtown Food Research Centre, Ashtown, Dublin, Ireland

Chapter outline

5.1 Introduction

To date, a number of techniques have been proposed for the extraction and recovery of macromolecules from different biological sources, among which aqueous systems are gaining specific interest and have the potential for high throughput purifications (Gagaoua, 2018; Grilo et al., 2016; Nadar et al., 2017). Generally, they represent a highly biocompatible and continuously operating purification process, which is easy to scale up for proteins from microbial, plant or animal sources using fluids, tissues or cells samples. Aqueous recovery systems are generally based on precipitation separations where two main systems can be distinguished, among which we can found the elegant Three Phase Partitioning (TPP) method (Dennison and Lovrien, 1997; Nadar et al., 2017). TPP, a technique first described serendipitously forty years ago by Lovrien's group (Dennison, 2011; Dennison and Lovrien, 1997; Lovrien et al., 1987; Tan and Lovrien, 1972), has been an emerging purification method intensively used to recover several target proteins, especially enzymes (Dennison, 2011; Gagaoua, 2018; Roy and Gupta, 2002; Yan et al., 2018) and nowadays has become a versatile, early-exploratory and a common bioseparation method with wide applications (Chew et al., 2018; Gagaoua, 2018; Yan et al., 2018). Briefly, TPP bioseparation approach is a three-stage recovery batch method, which is a hybrid of "salting out" and alcohol precipitations for extracting, dewatering, purifying and concentrating proteins, especially enzymes, for use in small or large

Three Phase Partitioning
DOI: https://doi.org/10.1016/B978-0-12-824418-0.00015-1

manufacturing operations as well as for research purposes. It is for its simplicity, fastness and efficiency that TPP system was intensively used in the last decade to recover proteins (Gagaoua, 2018), especially several enzymes (Fig. 5.1A) from myriad crude biological samples originating from microorganisms, plants and animals. The recovered enzymes, with varying success, belong to different groups such as glycosidases (Ait Kaki El-Hadef El-Okki et al., 2017b; Akardere et al., 2010; Bayraktar and Önal, 2013; Belligün and Demir, 2019; Bilen et al., 2019; Boucherba et al., 2017; Çalci et al., 2009; Dhananjay and Mulimani, 2009; Duman and Kaya, 2014; Mondal et al., 2003a; Şen et al., 2011; Sharma and Gupta, 2001a; 2002; Sharma et al., 2003); proteases (Chaiwut et al., 2010; de Melo Oliveira et al., 2020; Gagaoua et al., 2014, 2016, 2015; Gagaoua et al., 2017; Hafid et al., 2020; Kaur et al., 2019; Ketnawa et al., 2014; Rawdkuen et al., 2012; Senphan and Benjakul, 2014); oxidoreductases including laccase (Kumar et al., 2011b; Liu et al., 2015; Nadaroglu et al., 2019; Rajeeva and Lele, 2011; Wasak et al., 2018), peroxidase (Narayan et al., 2008; Panadare and Rathod, 2017; Vetal and Rathod, 2015; Karakus et al., 2018), polyphenol oxidases (Alici and Arabaci, 2016; Noori et al., 2020; Karakus and Kocak, 2020), catalase (Duman and Kaya, 2013b; Kaur et al., 2020); lipases (Dobreva et al., 2019; Kuepethkaew et al., 2017; Roy and Gupta, 2005; Saifuddin and Raziah, 2008; Sharma and Gupta, 2001b) and other hydrolases acting on carbon-nitrogen bonds other than peptide bonds such as asparaginase (Biswas et al., 2014; Shanmugaprakash et al., 2015a) or acting on ester bonds like esterase (Dong et al., 2020), etc. The TPP system consists in the sequential addition of a sufficient amount of solid salt (ammonium sulfate, potassium phosphate or sodium citrate) which is mostly ammonium sulfate $[(NH_4)_2SO_4]$ and an organic solvent (*tert*-butnaol, 2-butanol, 1-propanol, dimethyl carbonate or 2-propanol), which is usually *tert*-butanol (*t*-BuOH) to a crude protein slurry in order to obtain three phases. Thus, the process is based on the ability of the salt to separate a *tert*-butanol – water mixture, an otherwise miscible solution, into two phases, an upper organic phase and a lower aqueous phase. In the presence of proteins such as enzymes, an intermediate protein layer is formed, in which the target enzyme precipitates and can be easily extracted with high purity and enhanced activity. However, it is important to highlight that in certain cases and as a function of the p*I* of the protein, the enzyme may precipitate in the bottom phase referred also as the aqueous phase (Çamurlu et al., 2020; Dong et al., 2020; Duman and Kaya, 2014; Gagaoua et al., 2016, 2015; Hafid et al., 2020; Vetal and Rathod, 2015; 2016).

The process of TPP involves collective operation of principles including salting out, cosolvent precipitation, isoionic precipitation, kosmotropic precipitation, osmolytic, protein hydration shifts and conformation tightening, which all interact in a complex manner to often result in the recovery and precipitation of the target enzyme with high yield and purity. Therefore, this chapter focuses and gathers all the enzymes of various biological sources recovered using conventional TPP or its variants in combination with other methods such as ultrasounds, microwaves or macroaffinity ligands. This chapter

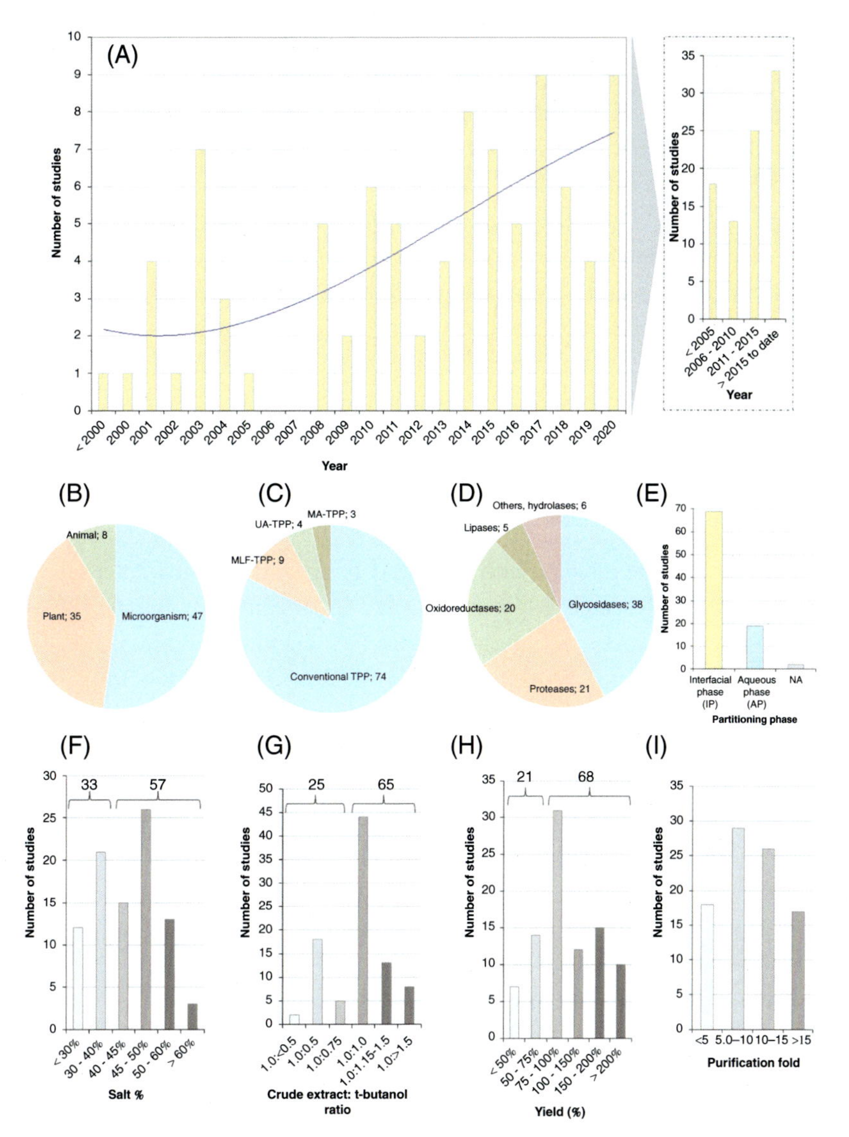

Fig. 5.1 ***Data about the recovery of enzymes from different sources using Three Phase Partitioning (TPP) or its variants.*** (A) Number of the related works (total $n = 90$) reported over the past 20 years in the literature for the recovery of enzymes using TPP or its variants. The frequency of use each five years was further shown on the right side of the graph. (B) Main biological sources used to purify the 90 enzymes using TPP system or its variants. (C) Number of the studies that used to recover the 90 enzymes from different sources by TPP system or its variants: MLF-TPP: Macroaffinity ligand-facilitated TPP; UA-TPP: Ultrasound-assisted TPP and MA-TPP: Microwave-assisted TPP. (D) Main enzymes families to which the recovered 90 enzymes belong. (E) Frequency of the partitioning of the 90 enzymes in the aqueous (bottom) phase (AP) or interfacial (middle) phase (IP) of TPP system. Two studies (NA) did not specify the recovery phase. (F) Percentage of salt used for the recovery of the 90 enzymes and number of studies corresponding to each range. (G) Crude extract: *tert*-butanol ratio and number of corresponding studies used to recover the 90 enzymes. (H) Distribution of the 90 enzymes recovered by TPP or its variants by yield percentage. (I) Distribution of the 90 enzymes recovered by TPP or its variants by purification fold.

further details the main enzymes and objectives of their purification or use, conditions of recovery in the aqueous or interfacial phases, overall purification results such as the yields, the purification folds and the activities achieved for each enzyme. The studies used in this chapter to identify the enzymes recovered by TPP and its variants (Table 5.1 and Table 5.2) were selected based a comprehensive systematic computerized search using Google Scholar, Scopus, Pubmed and Web of Science databases. The main key words used for searches were three phase partitioning, TPP, enzyme, protease, proteins, recovery, or purification and a combination of these words. All the eligible studies that investigated the purification and concentration of enzymes by the means of TPP system and its variants from original papers published up to October 2020 were identified, and organized into a spreadsheet reference manager, analyzed and discussed in this chapter.

5.2 Overview of the studies and conditions of use of three phase partitioning and its variants for the recovery of enzymes

The comprehensive systematic search followed in this study allowed to create a database from 81 eligible research papers, published in 46 different journals, dealing with TPP and its variants to purify or extract around 90 enzymes (Table 5.1 and Table 5.2). The enzymes belonging to different groups were mainly extracted from microorganisms (n = 47 enzymes) followed by plant protein extracts (n = 35) and a few number of enzymes (n = 8) were from animal sources (Fig. 5.1B). Most of the studies were published during the last decade for a total of 65 enzymes reported in 63 papers among which 40 papers were published in the last five years, hence confirming the emerging and great interest of using TPP system and its variants by numerous scientists and laboratories (Fig. 5.1A). In fact, more than 75 percent of the studies on TPP have focused on the extraction and recovery of proteins including enzymes, protein inhibitors and other proteins while the 25 percent remaining part focused on carbohydrates, lipids, oils, small-molecule organic compounds, DNA and other molecules. In the current database, 69 studies applied the conventional TPP system to recover 74 enzymes (Fig. 5.1C and Table 5.1) while 12 studies proposed TPP-based methods such as macro-affinity ligand-facilitated TPP (MLF-TPP), ultrasound-assisted TPP (UA-TPP) and microwave-assisted TPP (MA-TPP) to recover 9, 4 and 3 enzymes, respectively (Fig. 5.1C and Table 5.2). The TPP based methods were applied to improve the efficiency of the recovery as well as for reducing the extraction time, and consequently maximizing the activity yield (mostly by preventing enzyme inactivation) and purification folds of the recovered enzymes. Out of the 90 enzymes, 38 were glycosidases (42.2 percent) followed by 21 proteases (23.3 percent), 20 oxidoreductases (22.2 percent), 5 lipases and 6 other enzymes being hydrolases (Fig. 5.1D, Table 5.1 and Table 5.2) acting on carbon-nitrogen bonds other than peptide bonds (EC 3.5.), those acting on ester bonds (EC 3.1.) or those acting on ether bonds (EC 3.3.).

Table 5.1 List of the enzymes ($n = 74$, organized by family) recovered using conventional Three Phase Partitioning (TPP) system from different sources.

Source	Enzymes	CE: t-butanol ratio [a]	Salt, %percent [b]	pH	Temp. (C)	Phase location [c]	Purification fold	Yield (percent)	References
Glycosidases ($n = 29$)									
Aspergillus oryzae	α-galactosidase	1.0:1.0	60	–	37	IP	12	92	(Dhananjay and Mulimani, 2009)
Aspergillus niger	α-galactosidase	1.0:1.5	60	5.5	25	IP	6.27	97.2	(Gu et al., 2012)
Aspergillus Lentulus	α-galactosidase	1.0:1.0	55	5.5	25	AP	5.3	178	(Çamurlu et al., 2020)
Aspergillus oryzae	α-galactosidase	1.0:1.0	30	4.3	40	IP	15	50	(Dhananjay and Mulimani, 2008)
Tomato (*Lycopersicon esculentum*)	α-galactosidase	1.0:1.0	50	4.5	25	IP	4.3	81	(Çalci et al., 2009)
Pepino (*Solanum muricatum*)	α-galactosidase	1.0:1.5	50	5.3	25	IP	6.2	127	(Şen et al., 2011)
Watermelon (*Citrullus vulgaris*)	α-galactosidase	1.0:1.0	50	5.5	25	IP	2.7	76.7	(Bayraktar and Önal, 2013)
Chick pea (*Cicer arietinum*)	β-galactosidase	1.0:0.5	60	6.8	37	IP	10.1	133	(Duman and Kaya, 2013a)
Lactobacillus acidophilus	β-galactosidase	1.0:0.5	30	–	25	IP	7.5	78	(Choonia and Lele, 2013)
Tomato (*Lycopersicon esculentum*)	Invertase	1.0:1.0	50	4.5	25	IP	8.6	190	(Özer et al., 2010)
Momordica charantia (bitter melon)	Invertase	1.0:0.5	70	5.0	–	IP	10.48	20.28	(Belligün and Demir, 2019)
Potato tubers (*Solanum tuberosum*)	Invertase	1.0:1.0	25	–	37	AP	12.8	156	(Duman and Kaya, 2014)

(continued)

Table 5.1 (Cont'd)

Source	Enzymes	CE: t-butanol ratio [a]	Salt, %percent [b]	pH	Temp. (C)	Phase location [c]	Purification fold	Yield (percent)	References
Saccharomyces cerevisiae	Invertase	1.0:0.5	50	4.0	25	AP	15	363	(Akardere et al., 2010)
Aspergillus oryzae	Invertase	1.0:1.0	30	5.0	40	IP	12	54	(Dhananjay and Mulimani, 2008)
Aspergillus niger	Xylanase	1.0:1.0	30	5.6	40	AP	21	93	(Roy et al., 2004)
Bacillus oceanisediminis SJ3	Xylanase	1.0:1.5	50	8.0	10	IP	3.48	107	(Boucherba et al., 2017)
Trichoderma citrino-viride AUKAR04	Xylanase	1.0:0.5	55	5.0	25	IP	5.7	99.8	(Periyasamy et al., 2017)
Rhizopus oryzae FSIS4	α-amylase	1.0:1.5	50	5.5	25	IP	14.97	168.8	(Ait Kaki El-Hadef El-Okki et al., 2017b)
Stems of *Abrus precatorius*	β-amylase	1.0:1.15	49.5	5.2	4	IP	10.17	156.2	(Sagu et al., 2015)
Aspergillus niger	Pectinase	1.0:1.0	30	5.0	25	IP	10	76	(Sharma and Gupta, 2001a)
Tomato (*Lycopersicon esculentum*)	Pectinase	1.0:2.0	20	–	25	AP	9	183	(Sharma and Gupta, 2001a)
Aspergillus sojae ATCC 20,235	Exo-polygalacturonase	1.0:1.0	30	6.6	25	–	6.7	25	(Dogan and Tari, 2008)
Bitter almond	β-glucosidase	1.0:1.5	50	5.0	25	–	5.97	85.7	(Wei et al., 2016)
Corn (*Zea mays indurata*)	β-glucosidase	1.0:0.75	50	4.5	25	IP	3.8	60	(Bilen et al., 2019)
Trichoderma citrino-viride AUKAR04	Carboxymethyl cellulase	1.0:0.5	55	5.0	25	IP	5.5	96.5	(Periyasamy et al., 2017)

Bacillus methylotrophicus	Cellulase	1.0:1.0	40	7.0	25	IP	5.8	155	(Duman et al., 2020)
Aspergillus niger	Exo-inulinase	1.0:0.5	30	4	25	IP	10.2	88	(Kumar et al., 2011a)
Aspergillus brasiliensis MTCC 1344	Naringinase	1.0:1.0	60	5.5	28	IP	4.2	21.44	(Shanmugaprakash et al., 2015b)
Trichoderma citrino-viride AUKAR04	β−1,3-Glucanase	1.0:0.5	55	5.0	25	IP	5.6	98.4	(Periyasamy et al., 2017)
Proteases (n = 19)									
Carica papaya latex	Papain	1.0:0.75	40	6.0	25	AP	11.45	134	(Hafid et al., 2020)
Dried papaya leaves	Papain	1.0:0.5	20	7.0	25	AP	15.8	253.5	(Chaiwut et al., 2010)
Zingiber officinale roscoe rihizome	Zingibain	1.0:1.0	50	7.0	25	AP	14.91	215	(Gagaoua et al., 2016, 2015)
Bacillus pumilus MP27	Alkaline protease	1.0:1.0	20	9.0	4.0	IP	10.65	186	(Baweja et al., 2017)
Bacillus sp. ICTF2	Alkaline Protease	1.0:1.0	35	10	20	IP	12	148	(Kaur et al., 2019)
Bacillus natto NRRL-3666	Nattokinase	1.0:1.5	30	8.0	37	IP	5.6	129.5	(Garg and Thorat, 2014)
Viscera of farmed giant catfish	Alkaline proteases	1.0:0.5	50	8.0	25	IP	6	220	(Ketnawa et al., 2014)
Calotropis procera	Calotropain	1.0:1.5	50	6.5	25	IP	8.4	124	(Gagaoua and Hafid, 2016)
Peacock bass *(Cichla ocellaris)*	Collagenolytic protease	1.0:0.5	30	–	–	IP	5.07	239.2	(de Melo Oliveira et al., 2020)

(continued)

Table 5.1 (Cont'd)

Source	Enzymes	CE: t-butanol ratio [a]	Salt, %percent [b]	pH	Temp. (C)	Phase location [c]	Purification fold	Yield (percent)	References
Juice of *Cucumis melo* var. *reticulatus*	Cucumisin	1.0:1.25	60	8.0	20	IP	4.62	156	(Gagaoua et al., 2017)
Pichia pastoris	Cysteine peptidase CP_{SHN}	1.0:0.3	30	6.5	25	IP	2.8	28.4	(Pillay et al., 2010)
Pichia pastoris	Cysteine peptidase CP_{SYN}	1.0:0.3	30	6.5	25	IP	11	85.6	(Pillay et al., 2010)
Fig. (*Ficus carica* L.) latex	Ficin	1.0:0.75	40	6.5	25	IP	6.04	167	(Gagaoua et al., 2014)
Latex of *Calotropis procera*	Protease	1.0:0.5	50–65	7.0	25	IP	6.92	132	(Rawdkuen et al., 2010)
Viscera of *Pangasianodon gigas*	Proteases	1.0:0.5	50	8.0	25	IP	5	163.1	(Rawdkuen et al., 2012)
Hepatopancreas of *Litopenaeus vannamei*	Proteases	1.0:1.0	30	8.0	4.0	IP	2.6	76	(Senphan and Benjakul, 2014)
Wrightia tinctoria stem	Proteases [d]	1.0:1.0	60	5.5	4.0	IP	22.6	89.49	(Rajagopalan and Sukumaran, 2018)
NA [e]	Proteinase K	1.0:2.0	30	6.0	25	IP	2.1	210	(Singh et al., 2001)
NA	α-chymotrypsin	1.0:2.0	50	7.8	25	IP	3.58	119	(Roy and Gupta, 2004)
Oxidoreductases (*n* = 17)									
Pleurotus ostreatus	Laccase	1.0:1.8	50–60	4.0	42 - 45	IP	7.22	184	(Kumar et al., 2011b)
Ganoderma sp. WR-1	Laccase	1.0:0.5	90	7.0	35	IP	13.2	60	(Rajeeva and Lele, 2011)
Coriolopsis trogii	Laccase	1.0:1.0	40	5.0	20	IP	20	75	(Liu et al., 2015)
Trametes versicolor	Laccase	1.0:1.0	50	7.0	25	IP	24	73	(Wasak et al., 2018)

Weissella viridescens LB37	Laccase	1.0:0.25	40	6.0	25	AP	12.69	83.07	(Nadaroglu et al., 2019)
Ipomoea palmata (leaves)	Peroxidase	1.0:1.0	30	9.0	37	AP	18.36	80.95	(Narayan et al., 2008)
Orange peels (*Citrus sinenses*)	Peroxidase	1.0:1.5	50	6.0	30	AP	18.2	93.96	(Vetal and Rathod, 2015)
Amsonia orientalis	Peroxidase	1.0:1.0	20	6.0	25	AP	12.5	162	(Yuzugullu Karakus and Kocak, 2020)
Momordica charantia (bitter gourd)	Peroxidase [f]	1.0:0.75	20	7.0	30	AP	4.84	177	(Panadare and Rathod, 2017)
Potato (*Solanum tuberosum*) peel	Polyphenol oxidase (PPO)	1.0:1.0	45	7.0	30	IP	9.19	68.89	(Noori et al., 2020)
Trachystemon orientalis L.	Polyphenol oxidase (PPO)	1.0:1.0	15	6.5	25	IP	3.59	68.75	(Alici and Arabaci, 2016)
Rosmarinus officinalis L.	Polyphenol oxidase (PPO)	1.0:1.0	50	6.5	25	IP	14	230	(Yuzugullu Karakus and Kocak, 2020)
Sweet potato (*Solanum tuberosum*)	Catalase	1.0:1.0	40	7.0	35	IP	14.1	262	(Duman and Kaya, 2013b)
Potato leaf (*Solanum tuberosum*L.)	Catalase	1.0:1.0	40	7.0	25	IP	3.81	179	(Kaur et al., 2020)
Pleurotus ostreatus	Aryl alcohol oxidase	1.0:1.0	30	6.0	30	IP	10.19	10.95	(Kumar and Rapheal, 2011)
Aspergillus parasiticus	Mannitol dehydrogenase	1.0:0.5	50	6.8	4.0	IP	4.8	77	(Niehaus and Dilts, 1982)
Kluyveromyces marxianus	Superoxide dismutase	1.0:1.0	20	–	25	IP	9.8	80	(Simental-Martinez et al., 2014)

(continued)

Table 5.1 (Cont'd)

Source	Enzymes	CE: t-butanol ratio [a]	Salt, %percent [b]	pH	Temp. (C)	Phase location [c]	Purification fold	Yield (percent)	References
Lipases (*n* = 4)									
Chromobacterium viscosum	Lipase	1.0:2.0	50	7.8	25	IP	3.24	320	(Roy and Gupta, 2005)
Hepatopancreas of *Litopenaeus vannamei*	Lipase	1.0:1.0	50	–	–	IP	3.49	87.41	(Kuepethkaew et al., 2017)
Rhizopus arrhizus	Lipase	1.0:0.5	30	7.0	25	IP	19.1	71	(Dobreva et al., 2019)
Dacus carota	Phospholipase D	1.0:1.0	30	7.0	25	IP	13	72	(Sharma and Gupta, 2001b)
Others, hydrolases (*n* = 5)									
Capsicum annuum L.	Asparaginase [g]	1.0:0.75	50	9.0	40	IP	6.83	567.4	(Shanmugaprakash et al., 2015a)
E. coli k12	Asparaginase [g]	1.0:1.0	50	6.0	37	IP	11	36.21	(Biswas et al., 2014)
Bacillus cereus EMB20	β-Lactamase [g]	1.0:1.0	40	8.0	25	AP	7.5	25	(Sadaf et al., 2018)
Wheat flour	Esterase [h]	1.0:1.0	50	4.0	35	AP	11.35	–	(Dong et al., 2020)
Chicken intestine	Alkaline phosphatase [h]	1.0:1.0	60	7.0	25	AP	23	88	(Sharma et al., 2000)

[a]Crude extract: *tert*-butanol ratio.
[b]Salt (Ammonium sulfate)percent unless stated.
[c]Aqueous phase (AP) or interfacial phase (IP).
[d]TPP recovery profile of proteases from *W. tinctoria* stem was based on caseinolytic activity.
[e]The exact source was not specified.
[f]Na-citrate and dimethyl carbonate were used in this case.
[g]Other hydrolases: Acting on carbon-nitrogen bonds, other than peptide bonds.
[h]Other hydrolases: Acting on ester bonds.

Table 5.2 List of the enzymes ($n = 16$) recovered using TPP-based variants from different sources.

Source	Enzymes	Enzyme family	CE: t-butanol ratio [a]	Salt, %percent [b]	pH	Temp. (°C)	Phase location [c]	Purification fold	Yield (percent)	Conditions	References
MLF-TPP: Macroaffinity ligand-facilitated TPP ($n = 9$)											
Bacillus amyloliquefaciens	α-Amylase	Glycosidases	1.0:1.0	20	5.6	37	IP	5.5	74	Alginate ester 0.5 percent	(Mondal et al., 2003a)
Wheat germ	α-Amylase		1.0:1.0	20	5.6	37	IP	55	77	Alginate ester 0.5 percent	(Mondal et al., 2003a)
Porcine pancreas	α-Amylase		1.0:1.0	20	6.8	37	IP	10	92	Alginate ester 0.5 percent	(Mondal et al., 2003a)
Pectinex Ultra SP-L [d]	Pectinases		1.0:2.0	30	3.8	37	IP	4	100	Esterified alginate 0.5 percent	(Sharma et al., 2004)
Pectinex Ultra SP-L	Pectinases		1.0:1.0	45	5.0	38	IP	13	96	Alginate 1 percent	(Sharma et al., 2003)
Pectinex Ultra SP-L	Cellulase		1.0:1.0	45	4.8	37	IP	16	92	Chitosan 0.2 percent	(Sharma et al., 2003)
Aspergillus niger	Xylanase		1.0:1.0	40	4.5	40	IP	95	60	Eudragit S-100 1 percent	(Sharma and Gupta, 2002)
Aspergillus niger	Glucoamylase		1.0:2.0	30	4.5	37	IP	20	83	Esterified alginate 0.5 percent	(Mondal et al., 2003b)
Bacillus acidopullulyticus	Pullulanase		1.0:1.0	20	5.0	37	IP	38	89	Esterified alginate 0.5 percent	(Mondal et al., 2003b)

(continued)

Table 5.2 (Cont'd)

Source	Enzymes	Enzyme family	CE: t-butanol ratio [a]	Salt, %percent [b]	pH	Temp. (°C)	Phase location [c]	Purification fold	Yield (percent)	Conditions	References
UA-TPP: Ultrasound-assisted TPP ($n = 4$)											
Bacillus sphaericus	Fibrinolytic enzyme	Proteases	1.0:0.5	80	9.0	30	AP	16.15	65	25 kHz frequency, 150 W ultrasonication power, 40 percent duty cycle, 5 min	(Avhad et al., 2014)
Serratia marcescens	Serratiopeptidase	Proteases	1.0:1.5	30	7.0	30	IP	9.4	96	25 kHz frequency, intensity 0.05 W/cm^2, duty cycle 20 percent, 5 min	(Pakhale and Bhagwat, 2016)
Citrus sinensis	Peroxidase	Oxidoreductases	1.0:1.5	50	6.0	30	AP	24.28	91.84	25 kHz frequency, 150 W ultrasonication power, 40 percent duty cycle, 6 min	(Vetal and Rathod, 2016)
Solanum tuberosum	Polyphenol oxidase (PPO)	Oxidoreductases	1.0:1.0	40	7.0	30	AP	19.7	98.3	25 kHz frequency, 150 W ultrasonication power, 40 percent duty cycle, 5 min	(Niphadkar and Rathod, 2015)

MA-TPP: Microwave-assisted TPP (n = 3)											
Trametes hirsuta	Laccase	Oxidoreductases	1.0:1.0	40	4.8	30	IP	13.9	59.8	Irradiation time (7 min), duty cycle (20 percent), and power (20 W)	(Patil and Yadav, 2018)
Candida rugosa	Lipase	Lipases	1.0:2.0	50	7.8	25	IP	5.4	123	Irradiation time 10 s, frequency 2.45 GHz, power 100 W, 100 °C	(Saifuddin and Raziah, 2008)
Glycine max	Epoxide hydrolase [e]	Hydrolases [f]	1.0:1.5	40	6.0	45	IP	2.38	142.6	Irradiation time (8 min), duty cycle (50 percent), and power (20 W)	(Salvi and Yadav, 2020)

[a]Crude extract: t-butanol ratio.
[b]Salt (Ammonium sulfate)percent unless specified;.
[c]Aqueous phase (AP) or interfacial phase (IP);.
[d]A commercial mixture of enzymes (pectinase and cellulase) from *Aspergillus niger*;.
[e]Dimethyl carbonate was used in this case.
[f]Other hydrolases: Acting on ether bonds.

Regardless the TPP methods used for the extraction/purification of the 90 enzymes, 69 of them were recovered (precipitated) in the interfacial phase, while 19 being in the aqueous phase (Fig. 5.1E, Table 5.1 and Table 5.2). The mechanisms behind the recovery phase are not fully understood, but they are mainly admitted to be related to the pH of the protein and the system pH of the buffer in addition to the interplay with salt amount, temperature, *tert*-butanol, etc. For example, if the system pH is above the isoelectric point (p*I*) of the enzyme, hydrophilic amino acid residues are charged negatively and the enzyme will acquire net negative charge moving it to the bottom of the aqueous phase (Gagaoua, 2018). Under these conditions, the enzyme remains in the lower aqueous phase, which was for example the case of α-galactosidase of *Aspergillus Lentulus* (Çamurlu et al., 2020), invertase of *Saccharomyces cerevisiae* (Akardere et al., 2010), xylanase of *Aspergillus niger* (Roy et al., 2004), pectinase from tomato (Sharma and Gupta, 2001a), papain from both the latex and dried papaya leaves (Chaiwut et al., 2010; Hafid et al., 2020), zingibain from ginger rhizomes (Gagaoua et al., 2016, 2015) and interestingly all the plant peroxidases irrespective of the biological source and TPP methods used (Narayan et al., 2008; Panadare and Rathod, 2017; Vetal and Rathod, 2015; 2016; Yuzugullu Karakus et al., 2018). However, if the system pH is below the p*I* of the target enzyme, there is maximum precipitation in the interfacial precipitate phase (refer to Table 5.1 and Table 5.2 for the large list of the enzymes exemplifying this case). This is may be due to the development of a positive charge on the surface of the enzyme at a low pH that perhaps facilitates the binding of the salt to the cationic sites, thus significantly increasing the impact of electrostatic screening and minimizing the electrostatic repulsion between like-charged enzyme molecules.

The bioseparation of enzymes by the TPP methods is affected by numerous important parameters such as salt concentration (ammonium sulfate) and co-solvent types (*tert*-butanol) as well as other physical conditions likely the temperature and pH of the medium (Table 5.1 and Table 5.2). Therefore, the best results in terms of the recovery yield and purification folds can be achieved by varying these parameters. Among the first parameters, salt at various concentrations, 20 percent, 30 percent, 40 percent, 50 percent, 60 percent, 70 percent and 80 percent (w/v), during several partitioning experiments are performed. Salt saturation, mostly ammonium sulfate as stated above, plays a crucial part in TPP as it is accountable for protein-protein interaction and precipitation. For the 90 enzymes, the optimal amount of salt concentration used for the extraction varied from less than 30 percent to more than 60 percent (Fig. 5.1F, Table 5.1 and Table 5.2). Overall, 57 enzymes, which account for more than 63 percent of the database, were found to precipitate using more than 40 percent salt concentration. Focusing on conventional TPP only and by comparing glycosidases (n = 29 enzymes), proteases (n = 19) and oxidoreductases (n = 17), the first group needed on average 5 percent more salt than the two other enzyme groups. *tert*-Butanol was used as the best organic solvent for TPP and its variants to recover enzymes. This was mainly due to the ability of *tert*-butanol to bind easily to the enzymes, hence, increasing their buoyancy and efficiency of separation without causing denaturation (Dennison and Lovrien, 1997; Dennison et al., 2000; Sardar et al., 2007). Further, *tert*-butanol was described to allow significant crowding effects at room temperature due to its kosmotropic properties which not only intensifies the partitioning process but also makes it convenient (Dennison, 2011; Dennison et al., 2000; Gagaoua, 2018). The effect of slurry (crude protein extract) to *tert*-butanol ratio needed to recover the enzymes with high yields was generally optimized at a constant concentration of salt and by using various ratios: 1.0:0.25, 1.0:0.5, 1.0:0.75, 1.0:1.0, 1.0:1.25, 1.0:1.5, 1.0:1.75 and 1.0:2.0 (v/v). The analysis of the findings show that the best partitioning results were achieved mostly at 1.0:≥1.0 ratio for 65 enzymes among which 44 were recovered with a ratio of 1.0:1.0 (Fig. 5.1G) and only 25 enzymes needed low *tert*-butanol volumes (1.0:≤0.75) and 8 needed very high ratios (1.0:>1.5–2.0). The 8 enzymes that were recovered using very high ratios were pectinase from tomato (Sharma and Gupta, 2001a), proteinase K (Singh et al., 2001), α-chymotrypsin (Roy and Gupta, 2004), laccase from *Pleurotus ostreatus* (Kumar et al., 2011b), lipase from *Chromobacterium viscosum* (Roy and Gupta, 2005), all under conventional TPP and pectinase and glucoamylase from *A. niger* (Mondal et al., 2003b; Sharma et al., 2004) using MLF-TPP and the lipase of *Candida rugosa* using MA-TPP (Saifuddin and Raziah, 2008). For pH and temperature, the optimal conditions used depend mainly on the properties of the enzymes and the families to which they belong (Table 5.1 and Table 5.2). For example, glycosidases were mainly recovered by conventional TPP at low and acidic pH values (on average 5.3 ± 0.94) compared to proteases (7.3 ± 1.15) and oxidoreductases (6.5 ± 1.08). On another hand, proteases were recovered at lower temperatures, on average 21 ± 9 °C, using conventional TPP compared to glycosidases (27 ± 8 °C) and oxidoreductases (28 ± 9 °C).

Under optimal conditions, the enzymes were extracted, concentrated and recovered by using conventional TPP and its variants with various activity recovery and purification fold degrees (Table 5.1 and Table 5.2). Overall, the findings evidenced that TPP systems are effective and useful methods to obtain concentrated enzymes in desirable amounts and activities compared to chromatography methods. In terms of the overall recovery yields, the enzymes were mostly recovered with more than 75 percent for 68 enzymes and only 7 were recovered with less than 50 percent and 14 were between 50 and 75 percent (Fig. 5.1H, Table 5.1 and Table 5.2). The highest number of enzymes (n = 31) were recovered with yields varying between 75 and 100 percent (Fig. 5.1H). Around 40 percent of the enzymes included in the database (n = 36) were recovered with yields > 100 percent, these being 15 proteases, 11 glycosidases, 6 oxidoreductases that were all purified by conventional TPP using a one-step or two-step process and the 4 remaining enzymes were 2 lipases and 2 other hydrolases recovered by conventional TPP or MA-TPP process (Table 5.1 and Table 5.2). Overall, proteases were recovered with high yields using conventional TPP, on average 151 percent, compared to glycosidases (111 percent) and oxidoreductases (116 percent). It is well known from earlier studies that large quantities of enzymes sustained their activities in *tert*-butanol/water mixtures (Tan and Lovrien, 1972), a peculiarity that is confirmed in this unique large database in agreement with the previous reports (Chew et al., 2018; Gagaoua, 2018; Gagaoua and Hafid, 2016; Nadar et al., 2017). The purification folds achieved by TPP and its variants were also high and varied from 2.1 to 24 for conventional TPP (Table 5.1) and from 2.4 to 95 for TPP-based methods (Table 5.2). The enzymes that were recovered with very high purification folds were all glycosidases and by means of MLF-TPP: 38-fold for pullulanase of *Bacillus acidopullulyticus* (Mondal et al., 2003b), 55-fold for α-amylase of wheat germ (Mondal et al., 2003a) and 95-fold for *A. niger* xylanase (Sharma and Gupta, 2002). Overall, 55 enzymes (61 percent) were recovered with purification folds from 5 to 15 (Fig. 5.1I). However, it is worth noting from the database that the average purification folds of oxidoreductases (11.8 ± 6.0) using conventional TPP is slightly higher to those of proteases (8.3 ± 5.4) and glycosidases (8.6 ± 4.3).

From the above, TPP based separations can be regarded as a potential non-chromatography method in industrial enzymology due the numerous superiorities as compared to the conventional recovery methods (de Melo Oliveira et al., 2020; Gagaoua et al., 2014, 2015, 2017; Hafid et al., 2020). These advantages of TPP include selective partition and concentration of desired enzymes to one phase in shorter time periods, the low costs of the process, the ability to reuse *tert*-butanol, involving less steps and its operations at room temperatures (52 percent of them at 20–25 °C and 35 percent at 30–40 °C) for most of the enzymes without impact on their activity. Beside this, it is possible to recover the enzymes from crude protein extracts of several biological sources with good activity yields and purification folds than traditional methods.

5.3 Three phase partitioning for the recovery of glycosidases

Glycosidases or glycoside hydrolases are a broad family of ubiquitous intracellular and extracellular enzymes (EC 3.2.1.-) responsible for the hydrolysis of glycosidic linkages. The CAZy database (Carbohydrate-Active Enzymes at http://www.cazy.org) has divided its members into 153 glycosidase families based on amino acid sequence and structural similarities (Lombard et al., 2013). As presented above, glycosidase family members were the main enzymes (42%percent, n = 38 enzymes) recovered by TPP methods from various biological sources, especially from microorganisms (25 from 38) followed by plant (n = 12) and animal (n = 1) sources (Table 5.1 and Table 5.2). Glycosidases can be exo- (catalyzing cleavage of terminal bonds) or endo- (catalyzing cleavage of internal bonds) O-, N- and S-glycosidic bonds in carbohydrate chains and various glycoconjugates (Herscovics, 1999). These enzymes are specific for the anomeric configuration, and hydrolyze glycosidic bonds with either retention or inversion (Slámová et al., 2018). They are known to play various functions, with the vast majority of them being required as degradative enzymes for the digestion of extracellular carbohydrates to monosaccharides. It is for these properties that around 16 glycosidases were mainly extracted and purified by TPP systems to be used in several food and pharmaceutical industry purposes likely to produce sugar syrups or other products.

Among the glycosidases, galactosidases (7 α-galactosidases and 2 β-galactosidases) which catalyze the hydrolysis of galactosides into monosaccharides are the main group of enzymes recovered from microorganisms and plants by conventional TPP (Table 5.1). α-Galactosidases (α-D-galactoside galactohydrolase, EC 3.2.1.22) are enzymes of interest catalyzing the hydrolysis of α-linked terminal non-reducing galactose residues from galactose disaccharides/oligosaccharides, galactomannans and galactolipids. They are used in several industrial and biotechnological applications in sugar industry to increase the yield of crystallized sugar; in pulp and paper industry to improve pulp bleaching; in animal nutrition as feed additives to improve the nutritional value; for enzymatic synthesis and also in clinical applications (Bayraktar and Önal, 2013; Çamurlu et al., 2020; Dhananjay and Mulimani, 2009; Wang et al., 2016). α-Galactosidases were recovered from different *Aspergillus* mould species, which were *A. oryzae* (Dhananjay and Mulimani, 2008; 2009), *A.* niger (Gu et al., 2012) and *A. Lentulus* (Çamurlu et al., 2020). The α-galactosidase from *A. Lentulus* isolated from sand of a hot spring was the only one recovered in the aqueous phase using 55 percent salt, 1.0:1.0 ratio with optimised pH and temperature of 5.5 and 25 °C, respectively. The resulted yield and purification factor were high, being 178 percent and 5.3, respectively. α-Galactosidases were further efficiently recovered by TPP from plants with yields ranging from 77 to 127 percent (Table 5.1) mainly from tomato (*Lycopersicon esculentum*) (Çalci et al., 2009), pepino (*Solanum muricatum*) (Şen et al., 2011) and watermelon (*Citrullus vulgaris*) (Bayraktar and Önal, 2013). On another hand, β-galactosidases (EC 3.2.1.23) which catalyze transgalactosylation of several substrates including mono-, di-, and oligosaccharides of both anomeric configurations, methyl/aryl glycosides of both primary and secondary alcohols and phenols were for instance recovered in the interfacial phase of TPP from chick pea (*Cicer arietinum*) (Duman and Kaya, 2013a) and *Lactobacillus acidophilus* (Choonia and Lele, 2013). In both studies, an optimized 1.0:0.5 crude extract: *tert*-butanol ratio was used to achieve yields of 133 and 78 percent, respectively.

Invertases are the second largest group of enzymes that were efficiently purified five times by using conventional TPP in both of the aqueous and interfacial phases (Table 5.1). Invertase also known as β-fructofuranosidase (EC 3.2.1.26) is the enzyme involved in the hydrolysis of sucrose into an equimolar mixture of glucose and fructose, also known as invert sugar (Akardere et al., 2010). It is an enzyme widely used in myriad food applications mainly to produce sugar syrup and in drugs, beverage, cosmetic and paper industries (Kotwal and Shankar, 2009). From plants, it was recovered by TPP from the crude extract of tomato (*Lycopersicon esculentum*) (Özer et al., 2010) with a yield of 195 percent using 1.0:1.0 ratio with *tert*-butanol, 50 percent salt and under pH and temperature of 4.5 and 25 °C; from bitter melon (*Momordica charantia*) (Belligün and Demir, 2019) but with a low yield of 20 percent; and in the aqueous phase of potato tubers (*Solanum tuberosum*) with yield and purification fold of 156 percent and 12.8, respectively (Table 5.1). From microorganisms, invertase was successfully purified by Akardere et al. (Akardere et al., 2010) in the aqueous phase from Baker's yeast (*Saccharomyces cerevisiae*) and the findings indicated a very high recovery yield of 363 percent and a purification yield of 15 by using the optimized parameters of 1.0:0.5 ratio to *tert*-butanol, 50 percent ammonium sulfate at pH and temperature of 4.0 and 25 °C, respectively. An earlier study by Dhananjay and Mulimani (Dhananjay and Mulimani, 2008) used *A. oryzae* to achieve 54%percent activity recovery in the interfacial phase by 30%percent salt and temperature and pH of 40 °C and 5.0, respectively.

Four xylanases were also successfully recovered using TPP and MLF-TPP based methods (Table 5.1 and Table 5.2). These enzymes are among the microbial enzymes of great interest in the last two decades due to their biotechnological potential in many industrial processes including food industry, bioethanol production and animal nutrition as used to improve the digestibility of some feeds (Boucherba et al., 2014; Chakdar et al., 2016). They further represent one of the largest percentages of the world enzymes market. Xylanases were purified from *A. niger* protein extracts in the aqueous phase by conventional TPP (Roy et al., 2004) and in the interfacial phase by MLF-TPP (Sharma and Gupta, 2002). Both protocols used 1.0:1.0 crude extract: *tert*-butanol ratio and a temperature of 40 °C. MLF-TPP at 1 percent eudragit S-100 and pH of 4.5 allowed achieving a purification fold of 95 and yielding recovery of 60%percent, while conventional TPP resulted in 93 percent recovery yield and a purification fold of 21. In another study, an extracellular thermostable xylanase from the *Bacillus oceanisediminis* strain SJ3 newly isolated from Algerian soil (Boucherba et al., 2017) was recovered in the interfacial precipitate phase using the optimum parameters of 50 percent salt saturation with 1.0:1.5 ratio of crude extract: *tert*-butanol at pH of 8.0 and temperature of 10 °C. The xylanase was recovered with 3.48 purification fold and 107 percent activity recovery with stability over a broad pH range of 5.0–10. Another extremophile, a strain of *Trichoderma citrinoviride* AUKAR04, was found to produce a cocktail of enzymes from which a xylanase was purified with a yield of 99.8 percent using 55 percent salt, a ratio of 1.0:0.5 and at temperature and pH of 25 and 5.0, respectively (Periyasamy et al., 2017).

Five amylases, both α- and β, were efficiently recovered in the interfacial phases by TPP methods from both microorganisms and plants (Table 5.1 and Table 5.2). α-Amylase (1,4-α-D-glucanglucanohydrolase, EC 3.2.1.1) is the most important carbohydrate-degrading enzyme for starch-based industries. A thermostable α-amylase was recovered with a purification fold of 14.94 and a recovery yield of 168.8 percent using one-step TPP from *R. oryzae* FSIS4 strain isolated from the wheat seed cultivated in an arid area from Algerian Sahara (Ait Kaki El-Hadef El-Okki et al., 2017b). The authors indicated clearly that the recovered enzyme has potential application in the bread-making industry and in other food biotechnology applications due to its interesting physicochemical characteristics (Ait Kaki El-Hadef El-Okki et al., 2017a). By using MLF-TPP, Mondal and co-workers recovered under optimized parameters of 1.0:1.0 crude extract: *tert*-butanol ratio, 20 percent salt at 37 °C and under alginate ester 0.5 percent as ligand; three α-amylases from *B.* amyloliquefaciens (5.5-fold, pH 5.6), wheat germ (55-fold, pH 5.6), and porcine pancreas (10-fold, pH 6.8) with recovery yields of 74 percent, 77 percent and 92 percent, respectively (Mondal et al., 2003a). The only β-amylase recovered by conventional TPP was from the stems of *Abrus precatorius* (Sagu et al., 2015). The optimal conditions of 49.5 percent ammonium sulfate, 1.0:1.15 crude extract: *tert*-butanol ratio at pH 5.2 and temperature of 4 °C fitted in a second-order polynomial model allowed to recover the enzyme in the interfacial phase with a recovery yield of 156.3 percent and a purification fold of 10.17 (Table 5.1). The enzyme was found to be stable up to 65 °C (Sagu et al., 2015).

Pectinases are a complex group of enzymes involved in the biological degradation of pectin, an acidic polysaccharide whose basic structural repeats are α−1,4-linked-D-galacturonic acid, present in plant tissues (Anand et al., 2020). Pectinases are used for several industrial objectives including the clarification of fruit juices, treatment of pectin wastewater, coffee and tea leaf fermentation, retting of natural fibers, oil extraction, virus purifications and also for developing functional foods (Anand et al., 2020; Garg et al., 2016). Both conventional TPP and MLF-TPP were used to extract with varying success five pectinases (Table 5.1 and Table 5.2). From *A. niger*, pectinases were recovered in the interfacial phases with 76 percent yield and 10-fold by conventional TPP (Sharma and Gupta, 2001a) and with 100%percent and 4-fold (Sharma et al., 2004) or 96 percent and 13-fold (Sharma et al., 2003) by MLF-TPP using esterified alginate [0.5 percent w/v] or alginate [1 percent w/v] as ligands, respectively (Table 5.2). Another earlier study by Sharma and Gupta (Sharma and Gupta, 2001a), obtained in the aqueous phase of conventional TPP a high yield of 183 percent and a 9-fold purification of pectinases extracted from tomato (*Lycopersicon esculentum*) under optimal conditions of 20 percent salt, a crude extract: *tert*-butanol ratio of 1.0:2.0 and a temperature of 25 °C. Dogan and Tari (Dogan and Tari, 2008) worked with a strain of *A. sojae* to extract an exo-polygalacturonase, which catalyzes the hydrolytic release of one saturated galacturonic acid residue from the non-reducing end of homogalacturonan (Anand et al., 2020). The authors recovered the enzyme with a low yield of 25%percent compared to the other pectinases using optimal conditions of 30 percent salt, pH 6.6 and a crude extract: *tert*-butanol ratio of 1.0:1.0 (Dogan and Tari, 2008).

The other glycosidases were recovered by conventional TPP or MLF-TPP (Table 5.1 and Table 5.2) in the interfacial phases, likely β-glucosidases from bitter almond (Wei et al., 2016) and corn (*Zea mays indurata*) (Bilen et al., 2019). β-Glucosidases are key enzymes for cellulose degradation with potential roles in the production of glucose syrup and maltooligosaccharides as foodstuff, in brewing and starch processing. Cellulases as a multienzyme system that hydrolyses cellulose has well-known roles in different areas of industry including food, beverages, biofuel, feed, paper, textile, pharmaceutical and agriculture (Behera et al., 2017). Cellulases were also efficiently recovered from microbial sources such as *Trichoderma citrinoviride* AUKAR04 (Periyasamy et al., 2017), *Bacillus methylotrophicus* (Duman et al., 2020) and *A. niger* (Sharma et al., 2003). From *Aspergillus* species, other enzymes were also purified, *e.g.* exo-inulinase from *A. niger* (Kumar et al., 2011a) which hydrolyses inulin into pure fructose, which is an excellent alternative for the production of fructose syrup. Naringinase specifically hydrolyses naringin (4,5,7-trihydroxy flavanone 7-rhamnoglucoside) to release rhamnose and prunin and the latter product is further hydrolyzed to yield naringenin and glucose. Naringinase was effectively recovered from a strain of *A. brasiliensis* (Shanmugaprakash et al., 2015b); glucoamylase was extracted with 20-fold and 80 percent yield recovery by MLF-TPP (using esterified alginate 0.5 percent as a ligand) from *A. niger* (Mondal et al., 2003b). In this latter study, the authors also efficiently recovered with a 38-fold a pullulanase (EC 3.2.1.41), also known as α-dextrin 6-glucanohydrolase, from *B. acidopullulyticus* with a recovery yield of 89 percent (Mondal et al., 2003b). Another enzyme recovered with a 5.6-fold and a yield recovery of 98.4 percent using conventional TPP from a strain of *Trichoderma citrinoviride* was a β−1,3-glucanase (Periyasamy et al., 2017). This preparation consisted of two enzymes, namely exo-glucanase (EC 3.2.1.58) and endo-glucanase (EC 3.2.1.39), that act on (1, 3) and (1, 6) positions of β-D-glucan to release glucose (Levy et al., 2007).

5.4 Three phase partitioning for the recovery of proteases

Proteases (EC 3.4.x.x) also known as proteolytic enzymes are the second major group of enzymes (n = 21) from microbial, plant and animal sources efficiently recovered by TPP methods with high purity and recovery yields (Table 5.1 and Table 5.2). They constitute one of the most important classes of enzymes used in biotechnology and industrial enzymology, especially in the food sector as meat tenderizing (Gagaoua et al., 2021) or milk-clotting agents as well as for producing bioactive peptides and protein hydrolysates, therefore constituting around 60 percent of the total industrial market (Dhillon et al., 2017). In the current database (Table 5.1 and Table 5.2), 8 of them were recovered by TPP methods from plants, 8 from microorganisms and 5 were from animal sources and especially from the marine digestive organs. Although the bulk of proteases come from microbial and animal sources, vegetable proteases extracted from higher plant organs, have been extensively investigated in the last two decades as potential proteolytic enzymes (Gagaoua and Hafid, 2016) in food industry for cheese-making (Gagaoua et al., 2014, 2015, 2017; Shah et al., 2014; Zikiou et al., 2020) as well as for meat tenderization (Collados et al., 2020; Hafid et al., 2020; Shah and Mir, 2019; Gagaoua et al., 2021). The use of plant proteases can be considered as an emerging sustainable option and as an alternative to reduce food industry waste in the frame of circular economy (Banerjee et al., 2018; Campos et al., 2020). The examples to cite for this latter objective are the recovery of bromelain from peels and core of pineapple processing waste (Banerjee et al., 2020; Seguí Gil and Fito Maupoey, 2018) or cucumisin from *Cucumis melo* (Gagaoua et al., 2017). Further, the interest for the use of natural plant proteases is augmented by technical and societal reasons. Technically, plant proteases have unique properties and activity over a wide range of pH and temperature (Gagaoua et al., 2021), as well as high stability and substrate specificity that make them suitable for several food processes or other purposes. Religiously, they conform for example to the halal consumption requirements by avoiding products from animal origin that are prohibited or the use of growth media for microbial protease fermentation for similar reasons (Ermis, 2017). The use of genetically modified organisms to produce proteolytic enzymes has also been controversial among some consumers (Zhang et al., 2019).

Among the plant proteases, papain (EC 3.4.22.2) was recovered in the aqueous phase of conventional TPP from the latex (Hafid et al., 2020) and dried papaya leaves (Chaiwut et al., 2010) with yield recoveries of 134 and 253.5 percent, respectively (Table 5.1). The latex papain showed activity at temperatures ranging from 40 to 80 °C (maximum activity at 60 °C) using casein as substrate, and it retained 70 percent of its activity when incubated at 40 °C for 1 h (Hafid et al., 2020). For these interesting properties, the recovered latex papain was successfully applied as a meat-tenderizing agent (Gagaoua et al., 2021) including on tough camel meat (Hafid et al., 2020). As milk-clotting proteolytic agents, zingibain (EC 3.4.22.67) from *Zingiber officinale* roscoe rhizomes (Gagaoua et al., 2016, 2015), ficin known also as ficain (EC 3.4.22.3) from Mediterranean *Ficus carica* latex (Gagaoua et al., 2014), calotropain from *Calotropis procera* leaves (Gagaoua and Hafid, 2016), a protease from the latex of *Calotropis procera* (Rawdkuen et al., 2010), proteases of *Wrightia tinctoria* stem (Rajagopalan and Sukumaran, 2018) and Cucumisin (EC 3.4.21.25) from the juice of *Cucumis melo* var. *reticulatus* (Gagaoua et al., 2017) were all recovered with excellent yields by one-step TPP process and demonstrated promising cheese-making applications (Table 5.1). Under optimized conditions, these plant proteases were extracted and recovered with various activity recoveries and purification fold degrees such as cucumisin in the interfacial phase with 156 percent activity recovery and 4.62-fold; zingibain at the aqueous phase with 215 percent recovery and 14.91-fold, ficin in the interfacial phase with 167 percent activity recovery and 6.04-fold and calotropain with 124 percent recovery and 8.4-fold in the interfacial phase.

Among the microbial proteases, *Bacillus* species were used to extract different proteases (Table 5.1). Among them, *B. pumilus* MP27 isolated from Southern ocean water samples and the *Bacillus.* sp. ICTF2 allowed the recovery of alkaline proteases in the interfacial phases (Baweja et al., 2017; Kaur et al., 2019) with activity yields of 186 percent and 148 percent, respectively. The 53 kDa protease from *B. pumilus* MP27 showed an exceptional activity along with tested detergents, with 98 percent stability against SDS and 99 percent stability against Tide detergent (Baweja et al., 2017), making it compatible with surfactants and commercial detergents. Nattokinase, a fibrinolytic enzyme isolated from a traditional fermented food "Natto" in Japan (Sumi et al., 1987) was further effectively purified from *B. natto* NRRL-3666 in the middle phase by Garg et al. (Garg and Thorat, 2014) with 129.5 percent recovery and 5.6-fold using 1.0:1.5 crude extract: *tert*-butanol ratio, 30 percent salt, pH and temperature of 8.0 and 37 °C, respectively. Ultrasound-assisted TPP was also tested for its suitability to recover in the aqueous phase another fibrinolytic enzyme from *B. sphaericus* using 25 kHz frequency, 150 W ultrasonication power during 5 min irradiation time and 40 percent duty cycle (Avhad et al., 2014). The authors concluded that UA-TPP increased fold purity and recovery yield of the fibrinolytic enzyme from 7.98 to 16.15 and from 45 to 65 percent, respectively, with reducing time of operation from 1 h to 5 min as compared to conventional TPP. Using the same method, serratiopeptidase from *Serratia marcescens* was recently purified with the similar advantages by Pakhale and Bhagwat (Pakhale and Bhagwat, 2016). The authors reported an efficient selective partitioning of the extracellular metalloprotease in the interfacial phase leaving contaminant proteins in the bottom phase. The optimum conditions of 30 percent ammonium sulfate saturation, pH 7.0, broth to *tert*-butanol ratio 1.5 (v/v) and under a temperature of 30 °C, the conventional TPP achieved a 4.2-fold purification with 83 percent recovery in 1 h However, UA-TPP allowed increasing the fold purity to 9.4 and the yield to 96 percent at optimized conditions of irradiation of 25 kHz frequency, intensity of 0.05 W/cm^2 during 5 min, therefore reducing the time of purification to 5 min (Pakhale and Bhagwat, 2016). It seemed from these two studies that UA-TPP not only decreases the time of operation of TPP but also allow an increase in the recovery yield and purity of the enzymes. The TPP system was further successfully used as an initial concentration method to remove contaminating proteins and improve the purification yield of cysteine peptidases from *Pichia pastoris* (Table 5.1) in combination with chromatography techniques (Pillay et al., 2010).

From animal sources, most of the proteases were recovered using conventional TPP only from the digestive and viscera content of marine species (Table 5.1). The proteases of fish viscera are of interest and were described to help in the processing of seafood likely to increase the rate of fish sauce fermentation or the production of protein hydrolysates (Klomklao et al., 2012). Among the successful studies, Ketnawa et al. (Ketnawa et al., 2014) used the viscera of farmed giant catfish to recover alkaline proteases in the interfacial phase with 220%percent recovery yield and a 6-fold purification using optimal conditions of 50 percent salt and broth to *tert*-butanol ratio of 0.5 (v/v) at a temperature of 25 °C and pH of 8.0. The same research group achieved 163 percent and 76 percent recovery yields by extracting proteases from the viscera of *Pangasianodon gigas* (Rawdkuen et al., 2012) and hepatopancreas of *Litopenaeus vannamei* (Senphan and Benjakul, 2014). Another very recent study targeted the collagenolytic proteases of peacock bass (*Cichla ocellaris*) that were extracted with a very high yield of 239 percent and a 5-fold purification with low salt and *tert*-butanol amounts (de Melo Oliveira et al., 2020).

5.5 Three phase partitioning for the recovery of oxidoreductases

Oxidoreductases comprise a large class of enzymes that catalyze the biological oxidation/reduction reactions that occur within the cell (May 1999) and have myriad industrial applications (for a review: (Xu, 2005)). In the current database, they are the third largest group of enzymes that were recovered 17 times by conventional TPP (Table 5.1) and three times using MA-TPP and UA-TPP based methods (Table 5.2). Oxidoreductases were extracted from plant (n = 11) and microbial sources (n = 9). Among the members of oxidoreductases, laccase also known as 1,4-benzenediol: oxygen oxidoreductases (EC 1.10.3.2) is the enzyme frequently recovered by TPP methods. Laccase catalyzes the oxidation of some monomers, polymers, diamines, phenolic compounds and aromatic amines (Aktaş and Tanyolaç, 2003) and used in industrial enzymology for delignification, degradation, paper pulp bleaching and design of biosensors (Fernández-Fernández et al., 2013) and by food industries for baking, fruit juice and wine processing (Rodríguez Couto and Toca Herrera, 2006). Using conventional TPP, laccase was recovered mostly in the interfacial phase using submerged cultures of *Ganoderma* sp. WR-1 (Rajeeva and Lele, 2011), *Pleurotus ostreatus* (Kumar et al., 2011b, 2012), *Trametes versicolor* (Wasak et al., 2018), *Coriolopsis trogii* (Liu et al., 2015) and *Weissella viridescens* LB37 (Nadaroglu et al., 2019). MA-TPP has been also applied for laccase purification from *Trametes hirsute* (Patil and Yadav, 2018). The authors reported that there is a significant lowering of processing time (7 min) compared to conventional TPP (60 min). The authors achieved a recovery yield of 59.8 percent and a 13.9-fold purification (Table 5.2).

Plant peroxidases (E.C 1.11.1.7) were also efficiently recovered by TPP methods and they seemed to precipitate, for instance, exclusively in the aqueous phase only (Table 5.1 and Table 5.2). From the leaves of *Ipomoea palmate*, Narayan et al. (Narayan et al., 2008) used 30 percent ammonium sulfate and broth to *tert*-butanol ratio of 1.0 (v/v) at a temperature of 37 °C and pH of 9.0 to achieve a activity recovery and purification fold of 80.95 percent and 18.36, respectively. Better activity recovery and by using 20 percent salt only were obtained from the medicinal plant *Amsonia orientalis* (Yuzugullu Karakus et al., 2018) and bitter gourd *Momordica charantia* (Panadare and Rathod, 2017). For the peroxidase of bitter gourd Na-citrate and dimethyl carbonate were used instead of ammonium sulfate and *tert*-butanol (Table 5.1). From the orange peel (*Citrus sinenses*) waste produced by orange processing industry, Vetal and Rathod (Vetal and Rathod, 2015) were able a recover a peroxidase by using 50 percent ammonium sulfate, a broth to *tert*-butanol ratio of 1.5 (v/v) at a temperature of 30 °C and pH of 6.0. These optimal conditions allowed obtaining a recovery yield of 93.96 percent and 18.2-fold purification. From another study by the same authors on *C. sinensis* but by using UA-TPP (Vetal and Rathod, 2016), 91.84 percent recovery and 24.28-fold purification were obtained in a significant reduced time of 6 min compared to 80 min in conventional TPP (Table 5.2).

Polyphenol oxidases (PPOs) are metalloenzymes containing copper which catalyze the oxidation of mono-, di-, and polyhydric phenols to quinones; they are bifunctional in nature and recovered with high efficiency using TPP methods (Table 5.1 and Table 5.2). From potato (*Solanum tuberosum*), PPOs were recovered by means of conventional TPP (Noori et al., 2020) and UA-TPP (Niphadkar and Rathod, 2015). In TPP method, the enzyme was concentrated in the interfacial phase and in UA-TPP it was in the aqueous phase. The best results were achieved using UA-TPP, with recovery yields of 98.3 percent *versus* 68.89%percent for TPP and a fold purification of 19.7 *versus* 9.19 using the same optimal conditions of salt, temperature, pH and *tert*-butanol volume (Table 5.1 and Table 5.2). Furthermore, Niphadkar and Rathod (Niphadkar and Rathod, 2015) concluded that UA-TPP reduced the operation time from 40 min to 5 min when compared with conventional TPP. Alici and Arabaci in 2016 (Alici and Arabaci, 2016) tested the possibility to recover a PPO from borage *Trachystemon orientalis* L. and they achieved preliminary results in the interfacial phase of yield and purification fold of 68.75 percent and 3.59, respectively. The authors used 15 percent salt, a broth to *tert*-butanol ratio of 1.0 (v/v) at room temperature (25 °C) and pH 6.5. Very recently, Karakus et al. (Yuzugullu Karakus and Kocak, 2020) used the same conditions and increased only the ammonium sulfate concentration to 50 percent and achieved a very high recovery of 230 percent and 14-fold purification by using rosemary (*Rosmarinus officinalis*) as raw material.

Other oxidoreductases were recovered in the interfacial phase of conventional TPP (Table 5.1). Among them, plant catalases were purified sweet potato (*Solanum tuberosum*) (Duman and Kaya, 2013b) and leaves (Kaur et al., 2020). Very high recovery yields were obtained in both studies and the higher was in the first study compared to the leaves of potato. The conditions used were *tert*-butanol in equal volume, salt (40 percent) and pH (7.0) were all same but there was a 10 °C temperature difference between the first (35 °C) and second study (25 °C). From microorganisms, aryl alcohol oxidase (AAO) was purified from *P. ostreatus* (also used for laccase (Kumar et al., 2011b)) by Kumar and Rapheal, (Kumar and Rapheal, 2011) but with low recovery yield (10.95 percent) and fold purification (10.19). From other fungi, mannitol dehydrogenase and superoxide dismutase were recovered with interesting results of recovery yield > 75 percent (Table 5.1) from *A. parasiticus* (Niehaus and Dilts, 1982) and *Kluyveromyces marxianus* (Simental-Martinez et al., 2014).

5.6 Three phase partitioning for the recovery of lipases

Besides the success of TPP for the other enzyme groups, the number of lipases (EC 3.1.1.3) also known as triacylglycerol acylhydrolases, recovered by TPP methods is low although high recovery yields and purification folds have been reported for those recovered to date from plant, microbial and animal sources (Table 5.1 and Table 5.2). The purified lipases using TPP or MA-TPP were all concentrated in the interfacial phases with recovery yields > 70%percent. The lipase recovered with great success was that from *Chromobacterium viscosum* (Roy and Gupta, 2005) with a recovery yield of 320 percent and 3.24-fold purification using the optimal parameters of 50 percent salt, pH 7.8, temperature of 25 °C and crude protein extract: *tert*-butanol ratio of 1.0:2.0. From other microbial sources, *Rhizopus arrhizus* using conventional TPP (Dobreva et al., 2019) and *Candida rugosa* using MA-TPP (Saifuddin and Raziah, 2008), were used to extract lipases with 72 percent (19.1-fold purification) and 123 percent (5.4 -fold purification) recovery activities, respectively. The optimal parameters used by the two studies were somewhat similar in terms of pH and temperature (Table 5.1 and Table 5.2). Another lipase was recovered from the pacific white shrimp (*Litopenaeus vannamei*) hepatopancreas (Kuepethkaew et al., 2017) using a broth to *tert*-butanol ratio of 1.0 (v/v) and 50 percent ammonium sulfate that allowed recovery activity of 87.71 percent and 3.49-fold purification. The biochemical characterization of the recovered TPP lipase has shown its potential for use in various applications. From plants, use of a single step TPP resulted in the purification of phospholipase D from *Dacus carota* (carrot) (Sharma and Gupta, 2001b) with an activity recovery of 72 percent and 13-fold purification.

5.7 Three phase partitioning for the recovery of other hydrolases

TPP methods were also used to recover other hydrolases (Table 5.1 and Table 5.2), like those acting on carbon-nitrogen bonds other than peptide bonds (E.C 3.5.x.x) such as asparaginases of *Capsicum annuum* L. (Shanmugaprakash et al., 2015a) and *Escherichia coli* k12 (Biswas et al., 2014) or β-Lactamase of *B. cereus* EMB20 (Sadaf et al., 2018). Among the three enzymes, asparaginase of *C. annuum* was recovered in the interfacial phase with a very high activity recovery of 567.4 percent. The authors further tested the recovered TPP enzyme for anti-proliferative activity in cancer cell lines (Shanmugaprakash et al., 2015a). Among the enzymes acting on ester bonds (E.C 3.1.x.x), esterase from wheat flour (Dong et al., 2020) and alkaline phosphatase from chicken intestine (Sharma et al., 2000) were recovered in the aqueous phase of conventional TPP. Epoxide hydrolase (EC 3.3.2.3) was recovered very recently by MA-TPP from *Glycine* max (soybean) (Salvi and Yadav, 2020) and it belongs to the category of the enzymes acting on ether bonds. For its recovery, dimethyl carbonate was used as a solvent and allowed a recovery yield of 142.6 percent and 2.38-fold purification under optimal conditions of 40 percent salt, crude extract: dimethyl carbonate ratio 1.0:1.5, pH 6.0, temperature 45 °C, irradiation time of 8 min, duty cycle of 50 percent, and a power of 20 W (Salvi and Yadav, 2020).

5.8 Conclusion

To sum up, this chapter describes the applications of three phase partitioning and its variants to successfully recover myriad enzymes, including glycosidases, proteases, oxidoreductases, lipases and other hydrolases, from microbial, plant and animals sources. The use of TPP in routine or as an early-stage exploratory research method would allow in the near future identifying, purifying and studying more enzymes from various biological sources. The data and its analysis described in this chapter should help such explorations. The combination of TPP with other methods not yet explored for enzymes recovery might be another alternative in furthering more research on three phase partitioning.

References

Ait Kaki El-Hadef El-Okki, A., Gagaoua, M., Bennamoun, L., Djekrif, S., Hafid, K., El-Hadef El-Okki, M., Meraihi, Z., 2017a. Statistical optimization of thermostable α-amylase production by a newly isolated rhizopus oryzae strain FSIS4 using decommissioned dates. Waste Biomass Valori. 8 (6), 2017–2027.

Ait Kaki El-Hadef El-Okki, A., Gagaoua, M., Bourekoua, H., Hafid, K., Bennamoun, L., Djekrif-Dakhmouche, S., El-Hadef El-Okki, M., Meraihi, Z., 2017b. Improving bread quality with the application of a newly purified thermostable α-amylase from rhizopus oryzae FSIS4. Foods 6 (1), 1.

Akardere, E., Özer, B., Çelem, E.B., Önal, S., 2010. Three-phase partitioning of invertase from Baker's yeast. Sep. Purif. Technol. 72 (3), 335–339.

Aktaş, N., Tanyolaç, A., 2003. Kinetics of laccase-catalyzed oxidative polymerization of catechol. J. Mol. Catal. B Enzym. 22 (1–2), 61–69.

Alici, E.H., Arabaci, G., 2016. Purification of polyphenol oxidase from borage (Trachystemon orientalis L.) by using three-phase partitioning and investigation of kinetic properties. Int. J. Biol. Macromol. 93 (Pt A), 1051–1056.

Anand, G., Yadav, S., Gupta, R., Yadav, D., 2020. 14 - Pectinases: from microbes to industries. In: Chowdhary, P., Raj, A., Verma, D., Akhter, Y. (Eds.), Microorganisms for Sustainable Environment and Health. Elsevier, pp. 287–313.

Avhad, D.N., Niphadkar, S.S., Rathod, V.K., 2014. Ultrasound assisted three phase partitioning of a fibrinolytic enzyme. Ultrason. Sonochem. 21 (2), 628–633.

Banerjee, S., Arora, A., Vijayaraghavan, R., Patti, A.F., 2020. Extraction and crosslinking of bromelain aggregates for improved stability and reusability from pineapple processing waste. Int. J. Biol. Macromol. 158, 318–326.

Banerjee, S., Ranganathan, V., Patti, A., Arora, A., 2018. Valorisation of pineapple wastes for food and therapeutic applications. Trends Food Sci. Technol. 82, 60–70.

Baweja, M., Singh, P.K., Sadaf, A., Tiwari, R., Nain, L., Khare, S.K., Shukla, P., 2017. Cost effective characterization process and molecular dynamic simulation of detergent compatible alkaline protease from Bacillus pumilus strain MP27. Process Biochem. 58, 199–203.

Bayraktar, H., Önal, S., 2013. Concentration and purification of α-galactosidase from watermelon (Citrullus vulgaris) by three phase partitioning. Sep. Purif. Technol. 118 (0), 835–841.

Behera, B.C., Sethi, B.K., Mishra, R.R., Dutta, S.K., Thatoi, H.N., 2017. Microbial cellulases – Diversity & biotechnology with reference to mangrove environment: a review. J. Genet. Eng. Biotechnol. 15 (1), 197–210.

Belligün, N.K., Demir, B.S., 2019. Partial purification of invertase from Momordica charantia (bitter melon) by three phase partitioning (TPP) method. Afr. J. Biochem. Res. 13 (5), 56–62.

Bilen, B., Bayraktar, H., Onal, S., 2019. Partial purification and biochemical characterization of α-glucosidase from corn by three-phase partitioning. Hacettepe J. Biol. Chem. 46 (4), 481–494.

Biswas, T., Kumari, K., Singh, H., Jha, S.K., 2014. Optimization of three-phase partitioning system for enhanced recovery of L-asparaginase from Escherichia coli k12 using design of experiment (DOE). Int. J. Adv. Res. 2 (5), 1142–1147.

Boucherba, N., Gagaoua, M., Bouanane-Darenfed, A., Bouiche, C., Bouacem, K., Kerbous, M.Y., Maafa, Y., Benallaoua, S., 2017. Biochemical properties of a new thermo- and solvent-stable xylanase recovered

using three phase partitioning from the extract of Bacillus oceanisediminis strain SJ3. Bioresour Bioprocess 4 (1), 29.
Boucherba, N., Gagaoua, M., Copinet, E., Bettache, A., Duchiron, F., Benallaoua, S., 2014. Purification and characterization of the xylanase produced by Jonesia denitrificans BN-13. Appl. Biochem. Biotechnol. 172 (5), 2694–2705.
Çalci, E., Demir, T., Biçak Çelem, E., Önal, S., 2009. Purification of tomato (Lycopersicon esculentum) α-galactosidase by three-phase partitioning and its characterization. Sep. Purif. Technol. 70 (1), 123–127.
Campos, D.A., Gómez-García, R., Vilas-Boas, A.A., Madureira, A.R., Pintado, M.M., 2020. Management of fruit industrial by-products—A case study on circular economy approach. Molecules 25 (2), 320.
Çamurlu, D., Bayraktar, H., Özdemir, S.Ç., Uzel, A., Önal, S., 2020. Three-Phase partitioning of α-galactosidase from aspergillus lentulus: optimization of system and characterization of enzyme. Hacettepe J. Biol. Chem. 48 (1), 83–98.
Chaiwut, P., Pintathong, P., Rawdkuen, S., 2010. Extraction and three-phase partitioning behavior of proteases from papaya peels. Process Biochem. 45 (7), 1172–1175.
Chakdar, H., Kumar, M., Pandiyan, K., Singh, A., Nanjappan, K., Kashyap, P.L., Srivastava, A.K., 2016. Bacterial xylanases: biology to biotechnology. 3 Biotech 6 (2), 150.
Chew, K.W., Ling, T.C., Show, P.L., 2018. Recent developments and applications of three-phase partitioning for the recovery of proteins. Sep. Purif. Rev., 1–13.
Choonia, H.S., Lele, S.S., 2013. Three phase partitioning of β-galactosidase produced by an indigenous Lactobacillus acidophilus isolate. Sep. Purif. Technol. 110 (0), 44–50.
Collados, A., Conversa, V., Fombellida, M., Rozas, S., Kim, J.H., Arboleya, J.-.C., Román, M., Perezábad, L., 2020. Applying food enzymes in the kitchen. Int. J. Gastron. Food Sci. 21, 100212.
de Melo Oliveira, V., Carneiro da Cunha, M.N., Dias de Assis, C.R., Matias da Silva Batista, J., Nascimento, T.P., dos Santos, J.F., de Albuquerque Lima, C., de Araújo Viana Marques, D., de Souza Bezerra, R., Figueiredo Porto, A.L., 2020. Separation and partial purification of collagenolytic protease from peacock bass (Cichla ocellaris) using different protocol: precipitation and partitioning approaches. Biocatal. Agric. Biotechnol. 24, 101509.
Dennison, C., 2011. Three-phase partitioning. In: Tschesche, H. (Ed.), Methods in Protein Biochemistry. Walter de Gruyter, Berlin, Germany, pp. 1–5.
Dennison, C., Lovrien, R., 1997. Three phase partitioning: concentration and purification of proteins. Protein Expr. Purif. 11 (2), 149–161.
Dennison, C., Moolman, L., Pillay, C.S., Meinesz, R.E., 2000. t-Butanol: nature's gift for protein isolation. S. Afr. J. Sci. 96 (4), 159–160.
Dhananjay, S.K., Mulimani, V.H., 2008. Purification of alpha-galactosidase and invertase by three-phase partitioning from crude extract of Aspergillus oryzae. Biotechnol. Lett. 30 (9), 1565–1569.
Dhananjay, S.K., Mulimani, V.H., 2009. Three-phase partitioning of α-galactosidase from fermented media of Aspergillus oryzae and comparison with conventional purification techniques. J. Ind. Microbiol. Biotechnol. 36 (1), 123–128.
Dhillon, A., Sharma, K., Rajulapati, V., Goyal, A., 2017. 7 - Proteolytic Enzymes A2 - Pandey, Ashok. In: Negi, S., Soccol, C.R. (Eds.), Current Developments in Biotechnology and Bioengineering. Elsevier, pp. 149–173.
Dobreva, V., Zhekova, B., Dobrev, G., 2019. Use of aqueous two-phase and three-phase partitioning systems for purification of lipase obtained in solid-state fermentation by rhizopus arrhizus. Open Biotechnol. J. 13 (1).
Dogan, N., Tari, C., 2008. Characterization of three-phase partitioned exo-polygalacturonase from Aspergillus sojae with unique properties. Biochem. Eng. J. 39 (1), 43–50.
Dong, L., He, L., Huo, D., 2020. Three phase partitioning as a rapid and efficient method for purification of plant-esterase from wheat flour. Pol. J. Chem. Technol. 22 (2), 42–49.
Duman, Y., Kaya, A.U., Yağci, Ç., 2020. Three phase partitioning and immobilization of bacillus methylotrophicus Y37cellulase into organo-bentonite and its kinetic and thermodynamic properties. Clay Miner., 1–37.
Duman, Y., Kaya, E., 2014. Purification and recovery of invertase from potato tubers (Solanum tuberosum) by three phase partitioning and determination of kinetic properties of purified enzyme. Turkish J. Biochem./Turk Biyokimya Dergisi 39 (4).

Duman, Y.A., Kaya, E., 2013a. Purification, recovery, and characterization of chick pea (Cicer arietinum) beta-galactosidase in single step by three phase partitioning as a rapid and easy technique. Protein Expr. Purif. 91 (2), 155–160.

Duman, Y.A., Kaya, E., 2013b. Three-phase partitioning as a rapid and easy method for the purification and recovery of catalase from sweet potato tubers (Solanum tuberosum). Appl. Biochem. Biotechnol. 170 (5), 1119–1126.

Ermis, E., 2017. Halal status of enzymes used in food industry. Trends Food Sci. Technol. 64, 69–73.

Fernández-Fernández, M., Sanromán, M.Á., Moldes, D., 2013. Recent developments and applications of immobilized laccase. Biotechnol. Adv. 31 (8), 1808–1825.

Gagaoua, M., 2018. Aqueous methods for extraction/recovery of macromolecules from microorganisms of atypical environments: a focus on three phase partitioning Methods in Microbiology. Academic Press, pp. 203–242.

Gagaoua, M., Boucherba, N., Bouanane-Darenfed, A., Ziane, F., Nait-Rabah, S., Hafid, K., Boudechicha, H.R., 2014. Three-phase partitioning as an efficient method for the purification and recovery of ficin from Mediterranean fig (Ficus carica L.) latex. Sep. Purif. Technol. 132 (0), 461–467.

Gagaoua, M., Dib, A.L., Lakhdara, N., et al., 2021. Artificial meat tenderization using plant cysteine proteases. Current Opinion in Food Science 38, 177–188. doi:10.1016/j.cofs.2020.12.002.

Gagaoua, M., Hafid, K., 2016. Three phase partitioning system, an emerging non-chromatographic tool for proteolytic enzymes recovery and purification. Biosens. J. 5 (1), 1–4.

Gagaoua, M., Hafid, K., Hoggas, N., 2016. Data in support of three phase partitioning of zingibain, a milk-clotting enzyme from Zingiber officinale Roscoe rhizomes. Data Brief 6, 634–639.

Gagaoua, M., Hoggas, N., Hafid, K., 2015. Three phase partitioning of zingibain, a milk-clotting enzyme from Zingiber officinale Roscoe rhizomes. Int. J. Biol. Macromol. 73, 245–252.

Gagaoua, M., Ziane, F., Nait Rabah, S., Boucherba, N., Ait Kaki El-Hadef El-Okki, A., Bouanane-Darenfed, A., Hafid, K., 2017. Three phase partitioning, a scalable method for the purification and recovery of cucumisin, a milk-clotting enzyme, from the juice of Cucumis melo var. reticulatus. Int. J. Biol. Macromol. 102, 515–525.

Garg, G., Singh, A., Kaur, A., Singh, R., Kaur, J., Mahajan, R., 2016. Microbial pectinases: an ecofriendly tool of nature for industries. 3 Biotech 6 (1), 47.

Garg, R., Thorat, B.N., 2014. Nattokinase purification by three phase partitioning and impact of t-butanol on freeze drying. Sep. Purif. Technol. 131 (0), 19–26.

Grilo, A.L., Raquel Aires-Barros, M., Azevedo, A.M, 2016. Partitioning in aqueous two-phase systems: fundamentals, applications and trends. Sep. Purif. Rev. 45 (1), 68–80.

Gu, F., Gao, J., Xiao, J., Chen, Q., Ruan, H., He, G., 2012. Efficient methods of purification of α-galactosidase from Aspergillus niger: aqueous two-phase system *versus* three-phase partitioning. Rom. Biotechnol. Lett. 17, 7853–7862.

Hafid, K., John, J., Sayah, T.M., Dominguez, R., Becila, S., Lamri, M., Dib, A.L., Lorenzo, J.M., Gagaoua, M., 2020. One-step recovery of latex papain from Carica papaya using three phase partitioning and its use as milk-clotting and meat-tenderizing agent. Int. J. Biol. Macromol. 146, 798–810.

Herscovics, A., 1999. 3.02 - Glycosidases of the asparagine-linked oligosaccharide processing pathway. In: Barton, S.D., Nakanishi, K., Meth-Cohn, O. (Eds.), Comprehensive Natural Products Chemistry. Pergamon, Oxford, pp. 13–35.

Kaur, G., Sharma, S., Das, N., 2020. Comparison of catalase activity in different organs of the potato (Solanum tuberosum L.) cultivars grown under field condition and purification by three-phase partitioning. Acta Physiologiae Plantarum 42 (1), 10.

Kaur, N., Gat, Y., Panghal, A., 2019. Cost-Effective purification and characterization of an industrially important alkaline protease from a newly isolated strain of bacillus sp. ICTF2. Ind. Biotechnol. 15 (1), 20–24.

Ketnawa, S., Benjakul, S., Martínez-Alvarez, O., Rawdkuen, S., 2014. Three-phase partitioning and proteins hydrolysis patterns of alkaline proteases derived from fish viscera. Sep. Purif. Technol. 132 (0), 174–181.

Klomklao, S., Benjakul, S., Simpson, B.K., 2012. Seafood enzymes: biochemical properties and their impact on quality. Food Biochem. Food Proc., 263–284.

Kotwal, S.M., Shankar, V., 2009. Immobilized invertase. Biotechnol. Adv. 27 (4), 311–322.

Kuepethkaew, S., Sangkharak, K., Benjakul, S., Klomklao, S., 2017. Use of TPP and ATPS for partitioning and recovery of lipase from Pacific white shrimp (Litopenaeus vannamei) hepatopancreas. J. Food Sci. Technol. 54 (12), 3880–3891.
Kumar, V.V., Premkumar, M.P., Sathyaselvabala, V.K., Dineshkirupha, S., Nandagopal, J., Sivanesan, S., 2011a. Aspergillus niger exo-inulinase purification by three phase partitioning. Eng. Life Sci. 11 (6), 607–614.
Kumar, V.V., Rapheal, V.S., 2011. Induction and purification by three-phase partitioning of aryl alcohol oxidase (AAO) from Pleurotus ostreatus. Appl. Biochem. Biotechnol. 163 (3), 423–432.
Kumar, V.V., Sathyaselvabala, V., Kirupha, S.D., Murugesan, A., Vidyadevi, T., Sivanesan, S., 2011b. Application of response surface methodology to optimize three phase partitioning for purification of laccase from pleurotus ostreatus. Sep. Sci. Technol. 46 (12), 1922–1930.
Kumar, V.V., Sathyaselvabala, V., Premkumar, M.P., Vidyadevi, T., Sivanesan, S., 2012. Biochemical characterization of three phase partitioned laccase and its application in decolorization and degradation of synthetic dyes. J. Mol. Catal. B-Enzym. 74 (1–2), 63–72.
Levy, A., Guenoune-Gelbart, D., Epel, B.L., 2007. β-1,3-Glucanases. Plant Signal Behav 2 (5), 404–407.
Liu, Y., Yan, M., Huang, J., 2015. Three-phase partitioning for purification of laccase produced by coriolopsis trogii under solid fermentation. Am. J. Food Technol. 10 (3), 127–134.
Lombard, V., Golaconda Ramulu, H., Drula, E., Coutinho, P.M., Henrissat, B., 2013. The carbohydrate-active enzymes database (CAZy) in 2013. Nucleic. Acids. Res. 42 (D1), D490–D495.
Lovrien, R.E., Goldensoph, C., Anderson, P.C., Odegaard, B., 1987. Three phase partitioning (TPP) via t-butanol: enzyme separation from crudes. Protein Purification: Micro to Macro, 131–148.
May, S.W., 1999. Applications of oxidoreductases. Curr. Opin. Biotechnol. 10 (4), 370–375.
Mondal, K., Sharma, A., Lata, L., Gupta, M.N., 2003a. Macroaffinity ligand-facilitated three-phase partitioning (MLFTPP) of alpha-amylases using a modified alginate. Biotechnol. Prog. 19 (2), 493–494.
Mondal, K., Sharma, A., Nath Gupta, M., 2003b. Macroaffinity ligand-facilitated three-phase partitioning for purification of glucoamylase and pullulanase using alginate. Protein Expr. Purif. 28 (1), 190–195.
Nadar, S.S., Pawar, R.G., Rathod, V.K., 2017. Recent advances in enzyme extraction strategies: a comprehensive review. Int. J. Biol. Macromol. 101, 931–957.
Nadaroglu, H., Mosber, G., Gungor, A.A., Adıguzel, G., Adiguzel, A., 2019. Biodegradation of some azo dyes from wastewater with laccase from Weissella viridescens LB37 immobilized on magnetic chitosan nanoparticles. J. Water Process Eng. 31, 100866.
Narayan, A.V., Madhusudhan, M.C., Raghavarao, K.S., 2008. Extraction and purification of Ipomoea peroxidase employing three-phase partitioning. Appl. Biochem. Biotechnol. 151 (2–3), 263–272.
Niehaus, W.G., Dilts, R.P., 1982. Purification and characterization of mannitol dehydrogenase from Aspergillus parasiticus. J. Bacteriol. 151 (1), 243–250.
Niphadkar, S.S., Rathod, V.K., 2015. Ultrasound-assisted three-phase partitioning of polyphenol oxidase from potato peel (Solanum tuberosum). Biotechnol. Prog. 31 (5), 1340–1347.
Noori, R., Perwez, M., Mazumder, J.A., Sardar, M., 2020. Development of low-cost paper-based biosensor of polyphenol oxidase for detection of phenolic contaminants in water and clinical samples. Environ. Sci. Pollut. Res. 27 (24), 30081–30092.
Özer, B., Akardere, E., Çelem, E.B., Önal, S., 2010. Three-phase partitioning as a rapid and efficient method for purification of invertase from tomato. Biochem. Eng. J. 50 (3), 110–115.
Pakhale, S.V., Bhagwat, S.S., 2016. Purification of serratiopeptidase from Serratia marcescens NRRL B 23112 using ultrasound assisted three phase partitioning. Ultrason. Sonochem. 31, 532–538.
Panadare, D.C., Rathod, V.K., 2017. Extraction of peroxidase from bitter gourd (Momordica charantia) by three phase partitioning with dimethyl carbonate (DMC) as organic phase. Process Biochem. 61, 195–201.
Patil, P.D., Yadav, G.D., 2018. Application of microwave assisted three phase partitioning method for purification of laccase from Trametes hirsuta. Process Biochem. 65, 220–227.
Periyasamy, K., Santhalembi, L., Mortha, G., Aurousseau, M., Guillet, A., Dallerac, D., Sivanesan, S., 2017. Production, partial purification and characterization of enzyme cocktail from trichoderma citrinoviride AUKAR04 through solid-state fermentation. Arab. J. Sci. Eng. 42 (1), 53–63.
Pillay, D., Boulange, A.F., Coetzer, T.H., 2010. Expression, purification and characterisation of two variant cysteine peptidases from Trypanosoma congolense with active site substitutions. Protein Expr. Purif. 74 (2), 264–271.

Rajagopalan, A., Sukumaran, B.O., 2018. Three phase partitioning to concentrate milk clotting proteases from Wrightia tinctoria R. Br and its characterization. Int. J. Biol. Macromol. 118 (Pt A), 279–288.

Rajeeva, S., Lele, S.S., 2011. Three-phase partitioning for concentration and purification of laccase produced by submerged cultures of Ganoderma sp. WR-1. Biochem. Eng. J. 54 (2), 103–110.

Rawdkuen, S., Chaiwut, P., Pintathong, P., Benjakul, S., 2010. Three-phase partitioning of protease from Calotropis procera latex. Biochem. Eng. J. 50 (3), 145–149.

Rawdkuen, S., Vanabun, A., Benjakul, S., 2012. Recovery of proteases from the viscera of farmed giant catfish (Pangasianodon gigas) by three-phase partitioning. Process Biochem. 47 (12), 2566–2569.

Rodríguez Couto, S., Toca Herrera, J.L., 2006. Industrial and biotechnological applications of laccases: a review. Biotechnol. Adv. 24 (5), 500–513.

Roy, I., Gupta, M.N., 2002. Three-phase affinity partitioning of proteins. Anal. Biochem. 300 (1), 11–14.

Roy, I., Gupta, M.N., 2004. α-Chymotrypsin shows higher activity in water as well as organic solvents after three phase partitioning. Biocatal. Biotransformation 22 (4), 261–268.

Roy, I., Gupta, M.N., 2005. Enhancing reaction rate for transesterification reaction catalyzed by Chromobacterium lipase. Enzyme Microb. Technol. 36 (7), 896–899.

Roy, I., Sharma, A., Gupta, M.N., 2004. Three-phase partitioning for simultaneous renaturation and partial purification of Aspergillus niger xylanase. Biochimica et Biophysica Acta (BBA) - Proteins and Proteomics 1698 (1), 107–110.

Sadaf, A., Sinha, R., Khare, S.K., 2018. Structure and functional characterisation of a distinctive beta-lactamase from an environmental strain EMB20 of bacillus cereus. Appl. Biochem. Biotechnol. 184 (1), 197–211.

Sagu, S.T., Nso, E.J., Homann, T., Kapseu, C., Rawel, H.M., 2015. Extraction and purification of beta-amylase from stems of Abrus precatorius by three phase partitioning. Food Chem. 183 (0), 144–153.

Saifuddin, N., Raziah, A.Z., 2008. Enhancement of lipase enzyme activity in non-aqueous media through a rapid three phase partitioning and microwave irradiation. E-J. Chem. 5, 864–871.

Salvi, H.M., Yadav, G.D., 2020. Extraction of epoxide hydrolase from Glycine max using microwave-assisted three phase partitioning with dimethyl carbonate as green solvent. Food Bioprod. Process. 124, 159–167.

Sardar, M., Sharma, A., Gupta, M.N., 2007. Refolding of a denatured α-chymotrypsin and its smart bioconjugate by three-phase partitioning. Biocatal. Biotransformation 25 (1), 92–97.

Seguí Gil, L., Maupoey, F., 2018. An integrated approach for pineapple waste valorisation. Bioethanol production and bromelain extraction from pineapple residues. J. Clean Prod. 172, 1224–1231.

Şen, A., Eryılmaz, M., Bayraktar, H., Önal, S., 2011. Purification of α-galactosidase from pepino (Solanum muricatum) by three-phase partitioning. Sep. Purif. Technol. 83 (0), 130–136.

Senphan, T., Benjakul, S., 2014. Use of the combined phase partitioning systems for recovery of proteases from hepatopancreas of Pacific white shrimp. Sep. Purif. Technol. 129 (0), 57–63.

Merillon, J.M. , M.A.Shah, S.A.Mir, 2019. Plant proteases in food processing. In: Ramawat, K.G. (Ed.), Bioactive Molecules in Food. Springer International Publishing, Cham, pp. 443–464.

Shah, M.A., Mir, S.A., Paray, M.A., 2014. Plant proteases as milk-clotting enzymes in cheesemaking: a review. Dairy Sci. Technol. 94 (1), 5–16.

Shanmugaprakash, M., Jayashree, C., Vinothkumar, V., Senthilkumar, S.N.S., Siddiqui, S., Rawat, V., Arshad, M., 2015a. Biochemical characterization and antitumor activity of three phase partitioned L-asparaginase from Capsicum annuum L. Sep. Purif. Technol. 142 (0), 258–267.

Shanmugaprakash, M., Vinothkumar, V., Ragupathy, J., Reddy, D.A., 2015b. Biochemical characterization of three phase partitioned naringinase from Aspergillus brasiliensis MTCC 1344. Int. J. Biol. Macromol. 80 (0), 418–423.

Sharma, A., Gupta, M.N., 2001a. Purification of pectinases by three-phase partitioning. Biotechnol. Lett. 23 (19), 1625–1627.

Sharma, A., Gupta, M.N., 2002. Macroaffinity ligand-facilitated three-phase partitioning (MLFTPP) for purification of xylanase. Biotechnol. Bioeng. 80 (2), 228–232.

Sharma, A., Mondal, K., Gupta, M.N., 2003. Separation of enzymes by sequential macroaffinity ligand-facilitated three-phase partitioning. J. Chromatogr. A 995 (1–2), 127–134.

Sharma, A., Roy, I., Gupta, M.N., 2004. Affinity precipitation and macroaffinity ligand facilitated three-phase partitioning for refolding and simultaneous purification of urea-denatured pectinase. Biotechnol. Prog. 20 (4), 1255–1258.

Sharma, A., Sharma, S., Gupta, M.N., 2000. Purification of alkaline phosphatase from chicken intestine by three-phase partitioning and use of phenyl-Sepharose 6B in the batch mode. Bioseparation 9 (3), 155–161.

Sharma, S., Gupta, M.N., 2001b. Purification of phospholipase D from Dacus carota by three-phase partitioning and its characterization. Protein Expr. Purif. 21 (2), 310–316.

Simental-Martinez, J., Rito-Palomares, M., Benavides, J., 2014. Potential application of aqueous two-phase systems and three-phase partitioning for the recovery of superoxide dismutase from a clarified homogenate of Kluyveromyces marxianus. Biotechnol. Prog. 30 (6), 1326–1334.

Singh, R.K., Gourinath, S., Sharma, S., Roy, I., Gupta, M.N., Betzel, C., Srinivasan, A., Singh, T.P., 2001. Enhancement of enzyme activity through three-phase partitioning: crystal structure of a modified serine proteinase at 1.5 Å resolution. Protein Eng. Des. Sel. 14 (5), 307–313.

Slámová, K., Kapešová, J., Valentová, K., 2018. Sweet flavonoids": glycosidase-catalyzed modifications. Int. J. Mol. Sci. 19 (7), 2126.

Sumi, H., Hamada, H., Tsushima, H., Mihara, H., Muraki, H., 1987. A novel fibrinolytic enzyme (nattokinase) in the vegetable cheese Natto; a typical and popular soybean food in the Japanese diet. Experientia 43 (10), 1110–1111.

Tan, K.H., Lovrien, R., 1972. Enzymology in aqueous-organic cosolvent binary mixtures. J. Biol. Chem. 247 (10), 3278–3285.

Vetal, M.D., Rathod, V.K., 2015. Three phase partitioning a novel technique for purification of peroxidase from orange peels (Citrus sinenses). Food Bioprod. Process. 94 (0), 284–289.

Vetal, M.D., Rathod, V.K., 2016. Ultrasound assisted three phase partitioning of peroxidase from waste orange peels. Green Process. Synth.

Wang, C., Wang, H., Ma, R., Shi, P., Niu, C., Luo, H., Yang, P., Yao, B., 2016. Biochemical characterization of a novel thermophilic α-galactosidase from Talaromyces leycettanus JCM12802 with significant transglycosylation activity. J. Biosci. Bioeng. 121 (1), 7–12.

Wasak, A., Drozd, R., Grygorcewicz, B., Jankowiak, D., Rakoczy, R., 2018. Purification and recovery of laccase produced by submerged cultures of Trametes versicolor by three-phase partitioning as a simple and highly efficient technique. Pol. J. Chem. Technol. 20 (4), 88–95.

Wei, S., Qian, W., Meng, N., Zhu, B., Zhou, Q., Li, Q., 2016. Preparation of b-glucosidase from bitter almond by three-phase partitioning in Chinese. Fine Chem. 33, 530–535.

Xu, F., 2005. Applications of oxidoreductases: recent progress. Ind. Biotechnol. 1 (1), 38–50.

Yan, J.K., Wang, Y.Y., Qiu, W.Y., Ma, H., Wang, Z.B., Wu, J.Y., 2018. Three-phase partitioning as an elegant and versatile platform applied to nonchromatographic bioseparation processes. Crit. Rev. Food Sci. Nutr. 58 (14), 2416–2431.

Karakus, Y.Y., Işık, S., A., Duman, Y, 2018. Purification of peroxidase from Amsonia orientalis by three-phase partitioning and its biochemical characterization. Sep. Sci. Technol. 53 (5), 756–766.

Karakus, Y.Y., Kocak, G., B., Acemi, A., 2020. Application of three-phase partitioning to the purification and characterization of polyphenol oxidase from antioxidant rosemary (Rosmarinus officinalis L.). Int. J. Food Eng. 0 (0), 20200118.

Zhang, Y., Geary, T., Simpson, B.K., 2019. Genetically modified food enzymes: a review. Curr. Opin. Food Sci. 25, 14–18.

Zikiou, A., Esteves, A.C., Esteves, E., Rosa, N., Gomes, S., Louro Martins, A.P., Zidoune, M.N., Barros, M., 2020. Algerian cardoon flowers express a large spectrum of coagulant enzymes with potential applications in cheesemaking. Int. Dairy J. 105, 104689.

CHAPTER 6

Emulsion gel formation in three phase partitioning

R. Borbás*, É. Kiss[1]
Laboratory of Interfaces and Nanostructures, Institute of Chemsitry, Eötvös Loránd University, Budapest, Hungary
current: Szent István Secondary Grammar School, Budapest

[1]Support of grant VEKOP-2.3.2–16–2017–00014 from the European Union and the State of Hungary, co-financed by the European Regional Development Fund is acknowledged.

Chapter outline

6.1 Introduction

Purification of proteins and other macromolecules is fundamentally an essential part of chemistry and chemical industry. The different purification procedures add a significant sum to economy, and their cost comprises a considerable part of the commercial price of proteins. There is a high demand for pure proteins, especially enzymes, for research, chemical and food industrial uses, sensors, pharmacy, etc. Three phase partitioning (TPP) was introduced as a low-cost, economical and simple alternative among the purification processes by Anderson, Odegaard, Wilson and Lovrien (1981) applied by Niehaus and Dilts (1982). Following the early descriptions the method was further developed by Lovrien, Pike, Dennison and their co-workers (Lovrien, 1987; Pike and Dennison, 1989; Dennison and Lovrien, 1997).

TPP combines the traditional salting out using ammonium sulfate and the beneficial properties of *tert*-butanol in order to separate and purify proteins even from crude extracts without the necessity of pre-treatment. Water and *tert*-butanol are fully miscible, but in the presence of an appropriate amount of ammonium sulfate, the system separates into two liquid phases. The ternary diagram (phase diagram) of the ammonium sulfate – *tert*-butanol – water system is shown in Fig. 6.1. The corners of the triangle indicate pure substances, the area with symbol L shows the compositions belonging to a homogeneous liquid, the L1 / L2 region indicates such three component systems which

DOI: https://doi.org/10.1016/B978-0-12-824418-0.00003-5

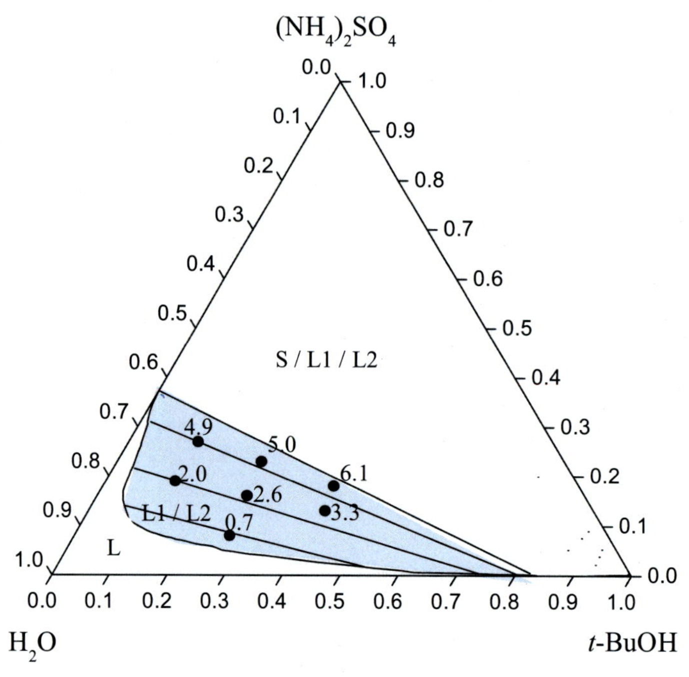

Fig. 6.1 ***Ternary phase diagram of ammonium sulfate,* tert*-butanol and water presenting the liquid (L), two-liquid (L1/L2) and solid-liquid-liquid (S/L1/L2) phase ranges.*** The numbers next to marked points show the interfacial tension between the two liquid phases measured by pendant drop technique (Kiss et al., 1998; Borbás, 2003b).

separate into two liquid phases with a composition shown by the end points of the tie line, while S/L1/L2 means that besides the two liquid phases, solid salt is present, as well (Kiss et al., 1998). Adding of the aforementioned components at appropriate ratios to crude protein extracts, following a rigorous mixing, their emulsification results in a three-phase system in which the components of the initial mixture are separated by polarity. The lower aqueous phase of the system is polar with a high ionic strength due to the presence of ammonium sulfate, and contains polar substances. The upper phase is rich in *tert*-butanol, thus it dissolves the non-polar components. For the composition of appropriate systems, see Fig. 6.1 and for more data on the ternary diagram, refer to Kiss et al. (1998). The proteins are enriched in the middle phase, from where they can be regained by simple steps like dissolution. The potential of TPP to separate the components of mixtures based on their polarity implemented its adaptation for the purification of oils, carbohydrates and small organic molecules (Yan et al., 2017). In these cases, the target molecules were found either in the upper or the lower phase, while the contaminants in the other phases. The main use of TPP however, is protein, particularly enzyme purification (Dennison, 2011; Szamos et al., 1998; Mondal et al., 2006; Yan et al., 2017; Ketnawa et al., 2017; Chew et al., 2018) (Fig. 6.2) even with the variants of TPP (Mondal et al.,

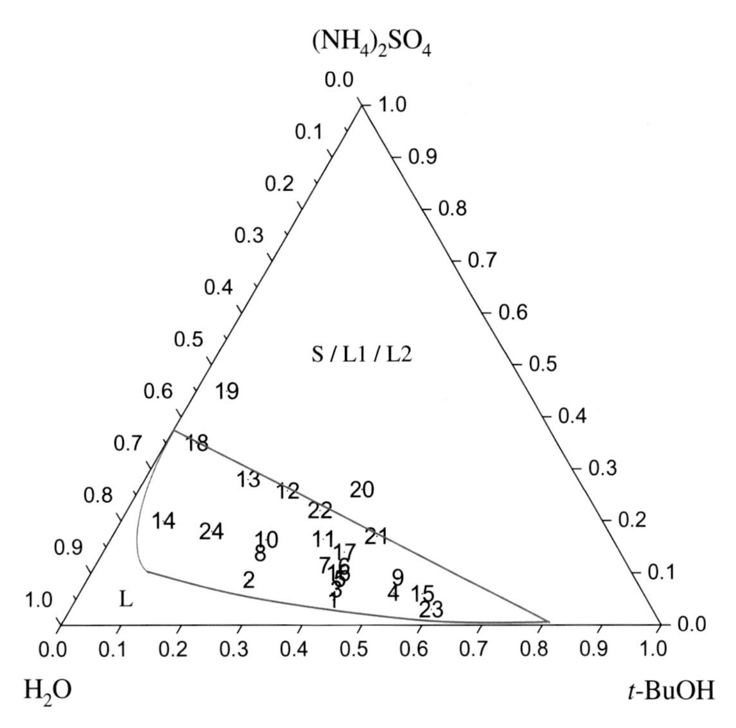

Fig. 6.2 ***Composition of systems used for purification purposes:*** 1. and 16. α-amylases, 2. and 10. β-galactosidase, 3. aryl alcohol oxidase, 4. cholesterol oxidase, 5. catalase, 6. exo-polygalacturonase, 7. β-amylase, 8. alkaline proteases, 9. α-galactosidase, 11. cholesterol oxidase, 12. fibrinolytic enzyme, 13. mannitol dehydrogenase, 14. and 17. cellulases, 15. carboxylester lipase, 18. horseradish peroxidase, 19. exo-pectinase, 20. protease/amylase inhibitor, 21. alginates, 22. soy trypsin inhibitor, 23. proteinase K, 24. pork DNA (Borbás, 2003b; Yan et al., 2017).

2006; Gagaoua, 2014). According to researchers using TPP for purification, it is mentioned that the method retains the activity of enzymes or increases it, causes only a mild change in the protein conformation, which is usually reversible . Moreover, it is run at room temperature, with recyclable, not highly volatile components. Another important characteristic of the process is that it is easily up- and down-scaled.

Although several reviews have appeared on TPP (see for example Mondal et al., 2006; Dennison, 2011; Yan et al., 2017; Ketanawa et al., 2017; Chew et al., 2018; Gagaoua, 2018), its complete mechanism is not well understood, hence it may still be considered an empirical method. A systematic research (Pike and Dennison, 1989) revealed that pH, molecular mass, and temperature are relevant parameters influencing the efficiency of TPP. The formation of the middle phase was seen as a *tert*-butanol – protein co-precipitate with such a density that the complex floats between the two liquid phases.

It has already been proven that the combined effect of ammonium sulfate and *tert*-butanol is needed for TPP (Dennison and Lovrien, 1997). Ammonium sulfate is a common agent in salting out. It increases the ionic strength in the solution, thus screens the charge of the dissolved protein. Consequently, the repulsion distance between the protein molecules decreases, speeding up the aggregation. Sulfate ions are kosmotropic ions (Moelbert et al., 2004) affecting the entropy of the system and changing the structure of water, hence the structure of protein as well, blocking certain conformations, and dehydrating the proteins through osmosis (Dennison and Lovrien, 1997). In addition, the bivalent sulfate ions bind to certain cationic groups of the proteins changing their net charge, and probably also their conformation. *Tert*-butanol has similar effect as ammonium sulfate in changing the water structure, but it does not denature proteins. Due to its size and branched structure, *tert*-butanol does not easily permeate inside the folded protein molecules and hence does not cause denaturation (Dennison and Lovrien, 1997). On the other hand, applying them together there is an enhanced effect. Less ammonium sulfate is needed for the formation of the protein rich middle phase in the presence of *tert*-butanol than the amount needed for salting out. Therefore, several authors (Dennison and Lovrien, 1997; Yan et al., 2017) suggest that TPP involves collective operation of principles such as salting out, isoionic and cosolvent precipitations, osmolytic and kosmotropic precipitation.

Pike and Dennison (1989) observed that the TPP systems do not provide conventional precipitation curves, rather a family of curves depending on the initial protein concentration. Our research group further investigated the solubility curves and found that the concentration of dissolved protein increases with increasing initial protein concentration. This finding led to the definition of the partitioning ratio in the TPP systems, since the relative amount of protein in the middle phase is a constant function of the initial protein concentration. In addition, it was also concluded that the amount of protein precipitated in the interphase was delineated as a function of the composition of the partitioning system as well as the initial protein concentration.

The increasing number of studies on the topic recognizes the fact that TPP works only under such conditions when the ammonium sulfate and *tert*-butanol form a heterogeneous mixture with water (see Fig. 6.2), however, this fact was only moderately investigated, and hardly correlated to the process itself. Our aim is to review fundamental research and findings on TPP systems, and propose a mechanism for TPP based on our investigations and outcome. Providing a thorough explanation to the driving forces of the process would lead to further understanding of TPP, and help in predicting the application of the process for a specific protein, even for industrial use. It is going to be demonstrated that the following phenomena play a crucial role in the process of TPP: protein adsorption at the interface of the two liquids, the formation of a viscoelastic protein layer at the interface accompanied by conformational changes of the protein, aggregation, coagulation, network forming, the emulsion gel formation as a middle phase and its stability. Therefore, each step of the TPP method is analyzed, with focusing on how the different components may affect the crucial processes.

6.2 The mechanism of TPP

6.2.1 The effect of ammonium sulfate and *tert*-butanol

When performing a TPP process, initially ammonium sulfate is mixed with the aqueous, macromolecule containing system. The target molecule is in a crude extract, containing different salts, small molecules, proteins, and other cell debris as well. The addition of ammonium sulfate may already precipitate some components at this stage, but the formation of middle phase does not occur until the addition of *tert*-butanol. The process can also be run by adding tert-butanol first, but the mixture still will not separate into the typical different fluid phases until the addition of the salt. It is worth noting that if the system is not shaken well (i.e. emulsified), the simultaneous addition of salt and *tert*-butanol does not produce three phases, just two; the protein rich middle phase is missing. The kosmotropic effect of sulfate ions and tert-butanol has already been discussed, as well as the screening of protein charges due to the ionic strength. That allows solvation by *tert*-butanol thus enhancing the hydrophobic interactions between the protein molecules.

The appropriate ratio of ammonium sulfate, *tert*-butanol and water for efficient TPP falls in the heterogeneous, two-liquid-phase region of the ternary system (as shown in Fig. 6.1). The forming upper and lower phases differ in the ratio of salt, alcohol and water, their composition is described by a tie line in the ternary diagram. The interfacial tension of these two-liquid-phase systems were determined and found to be in the interval of 0 to 7 mN/m. The relationship between the tie line length (L) and the interfacial tension, γ (mN/m) is given by the empirical formula of $\ln\gamma = 11.21\ L - 6.15$. This reflects that the longer the tie line, i.e. the higher the difference between the compositions, consequently the polarity of the two phases; the higher the interfacial tension. It was also observed that the protein yield increased with increasing interfacial tension (Kiss et al., 1998). That finding underlines the fact that the heterogeneity of the partitioning system is crucial for the success of TPP process.

6.2.2 Protein adsorption at the interface

The previously described heterogeneous two-liquid-system containing the macromolecules is emulsified. The emulsification is quick and dynamic, resulting in a turbulent flow of the phases, and producing a large new interfacial area available for the adsorption of amphiphilic proteins. The size of the droplets created depends on the energy of emulsification, the viscosity of the phases and the interfacial tension. The formation of a polydisperse system is expected, the typical drop size is between 20 and 50 μm by shaking, and 5 to 10 μm by laboratory mixers (Langevin, 2000). The mixing of liquids can be intensified by high speed homogenizer (e.g. Ultraturrax) or application of ultra sound (Kulkarni and Rathod, 2014; Yan et al., 2018) which leads to smaller average droplet size. Under these conditions, thermodynamic equilibrium cannot be reached, especially with proteins.

Proteins are generally amphipathic; therefore they have an affinity to the interface between two fluid phases with different polarity even though this difference is small. The process of protein adsorption comprises the diffusion of protein from the bulk to the layer right under the interface, the adsorption of protein at the interface, the rearrangement of molecules adsorbed (intramolecular as well as intermolecular), the formation of intermolecular bonds, the construction of multiple adsorption layers, and desorption. Naturally, not all steps occur in each case but the adsorption of proteins can be sensitively followed by interfacial tension measurement, for example by pendant drop method. Fig. 6.3 shows how different proteins change the interfacial tension of a definite TPP system.

Even though the interfacial tension between the two liquid phases of the TPP systems is very low, Fig. 6.3 shows clearly that proteins have a definite affinity towards this liquid-liquid interface. Three out of the above six model proteins considerably and one moderately decreased the interfacial tension in consequence to their adsorption. The changes in absolute values were small (up to 0.6–0.7 mN/m), but significant (from 3 to 35 percent). A steady-state condition was reached for two of the proteins in less than 10 min, a much shorter time than the adsorption time of proteins at the air/water or oil/water interfaces (Benjamins, 2000; Bos and van Vliet, 2001). Also the rate of change in the cases of the other proteins decreases noticeably during this time interval. The interfacial tension decreases without lag time indicating the lack of induction period of the interfacial adsorption at this protein concentration. However, if the concentration of protein is different, the curves are slightly altered, but their general profile is similar (Kiss and Borbás, 2003). At higher concentrations, the interfacial tension decreases more, reaching the steady-state value in a shorter period of time. At lower concentrations, a

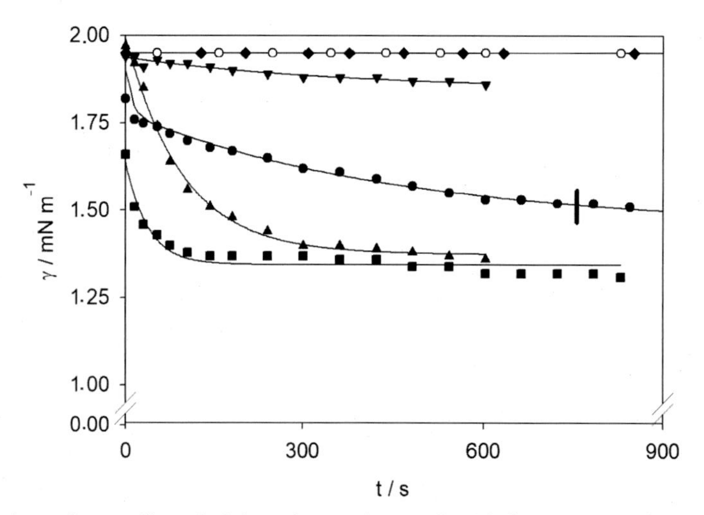

Fig. 6.3 ***Time dependence of interfacial tension, γ of a two-liquid phase system of TPP in the presence of various proteins at a concentration of 0.1 g × dm^{-3}.*** The studied proteins: bovine serum albumin (■), β–lactoglobulin (▲), ovalbumin (●), horseradish peroxidase (▼), lysozyme (◆), trypsin (O). The bar represents the average standard deviation of the data.

lag time may appear which shortens with increasing protein content of the solution. The initially unchanging interfacial tension is not necessarily indicative of zero surface concentration, as it was proven from independent measurements of adsorbed amount of protein and surface pressure (Tripp et al., 1995; Lucassen-Reynders and Benjamins, 1999; Benjamins, 2000). Even in the case of the two proteins not showing change in the interfacial tension during the first 15 min (see Fig. 6.3), the surface coverage may not be zero, which was supported by compression experiments (Borbás, 2003b).

Besides the sensitive detection of protein adsorption by measuring the change of interfacial tension, the adsorbed amount can not be calculated on the basis of Ward-Tordai equation (Ward and Tordai, 1946) since the conditions are not fulfilled, thermodynamic equilibrium could not be reached. Furthermore, the structure of the adsorbed layer might also be different. Separate protein molecules and smaller or larger aggregates might be present in the interfacial layer; moreover, it can also happen that only a segment of the protein is hooked to the interface, and the interfacial tension decreases slowly but gradually due to the conformational changes of the protein.

During the process of TPP, nevertheless an emulsion with large specific interfacial area is formed, and proteins are adsorbed at this interface. The amount of protein accumulated in the middle phase could be an indirect measure of the adsorbed amount. To check this possible relation, the change of interfacial tension ($\Delta\gamma$) by different proteins after 600 s of adsorption time is compared to the protein yield in Fig. 6.4. Yield is the relative amount of protein obtained in the middle phase after emulsification and separation.

The data show that the adsorption of bovine serum albumin, ovalbumin and β-lactoglobulin at the given concentration leads to significant reduction of the interfacial

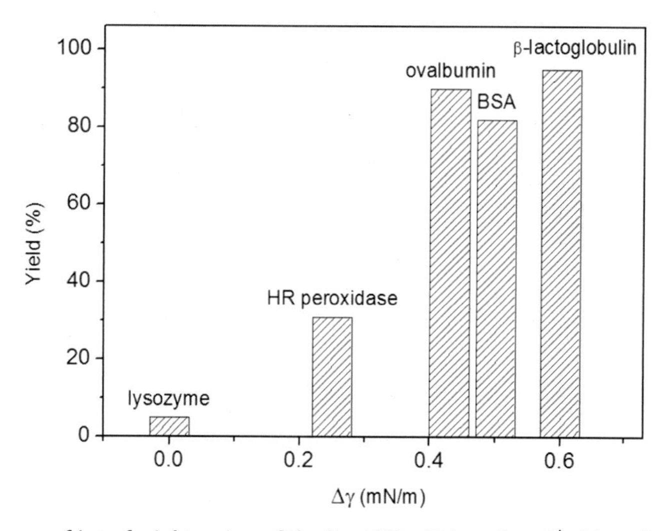

Fig. 6.4 ***The change of interfacial tension of the liquid/liquid interface ($\Delta\gamma$) by adsorption of different proteins ($c = 0.1\ g \times dm^{-3}$) and the yield of protein (percent) achieved in the TPP system*** *(Data from Borbás, 2003b).*

tension and at the same time produces rather high partitioning of the protein to the middle phase (yield > 80 percent). In the case of horseradish peroxidase the extent of interfacial tension reduction is much smaller along with a moderate yield. Adsorption of lysozyme does not cause measurable change in the interfacial tension connected to a very low yield obtained at high concentration relative to the others. From this comparison we can conclude that the proteins that produced a higher change in interfacial tension were apparently extracted in higher proportions into the middle phase.

This relation supports that interfacial adsorption is crucially important during the extraction process of TPP, suggesting that the partition may occur between the bulk of the lower phase and the interface between the lower and the upper phase.

6.2.3 The shear rheological properties of interfacial protein layer

After the initial adsorption of proteins at the interface, several other processes may start. One of those is the development of multiple adsorption layers, another is the conformational change of the protein, and also interfacial aggregation and intermolecular lateral bond formation can occur. The examination of the interfacial rheological behavior can help in obtaining structural information on the layer.

The liquid/liquid interface of TPP systems is between the high-ionic-strength lower phase and the organic-solvent-rich upper phase, but still each phase contains all three components, namely ammonium sulfate, *tert*-butanol, and water. For this reason, there is no competition for the interfacial spaces between the protein and the cosolvent. But there could be competition between water and tert-butanol for the solvation of protein molecule. In addition, the adsorbed protein (or parts of it) can penetrate the upper phase as well as the lower phase (the protein is almost always soluble in the lower phase). Therefore, the upper phase can promote the conformational changes of the protein in the interfacial region. According to Dickinson and Matsamura (1994), the adsorbed proteins reach a so called "molten globule" state, which is a state between the native state and the completely unfolded form. Thus the protein gets into an activated state, in which the secondary structure is conserved, while the tertiary is affected to a varying degree. This is supported for example by Singh et al. (2001) who studied serine proteinase after separation by TPP, and found that the structure of the protein after separation is similar to that of the native state in aqueous solution. The hydrogen-bonding system of the active center of the enzyme was untouched, but the structure of water had changed in this region: some water molecules were substituted or missing. They also found that the mobility and the number of possible configurations of certain side chains increased, and the protein reached a certain activated state.

The interfacial shear rheological parameters, the interfacial shear elasticity, G' and interfacial shear viscosity, η' reflect the interactions between the adsorbed protein molecules. They are sensitive to the conformation of the protein, and the structure of the layer formed from it, the packaging of the interfacial layer, the interparticle interactions within the layer, as well as its history (Murray, 1998). The adsorbed protein layer often behaves as a very thin protein gel, presenting a considerable elasticity and viscosity. The

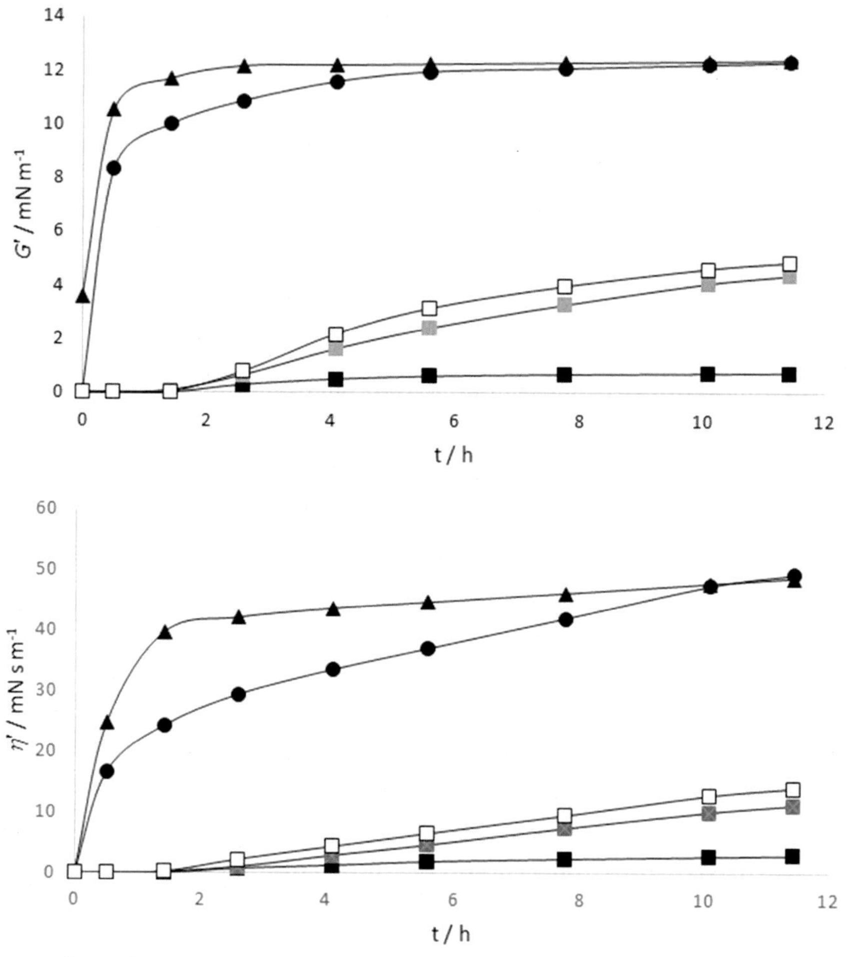

Fig. 6.5 ***Interfacial shear elastic modulus (G′) and viscosity (η′) of bovine serum albumin (■), β–lactoglobulin (▲), ovalbumin (●) as a function of time (t).*** The protein concentration: 0.05 (filled), 0.1 (grey), 0.5 (open symbol) $g \times dm^{-3}$.

examined proteins, ovalbumin, β-lactoglobulin and bovine serum albumin adsorbing at the liquid/liquid interface of TTP system developed into a viscoelastic gel with a moderately high apparent shear elastic modulus (*G*') and viscosity (*η*'), as shown in Fig. 6.5.

Examining first the behavior of bovine serum albumin, which proved to be the most surface active (Fig. 6.3) among the model proteins already mentioned (Borbás et al., 2003a; 2003c), both the interfacial shear elastic modulus (*G*') and interfacial shear viscosity (*η*') of bovine serum albumin started to increase only after a pronounced lag time. The evolution of a viscoelastic protein film takes 0.6 to 1.5 h depending on the concentration. In the case of the lowest protein concentration after ca. 5 h both *G*'and *η*' reached a steady-state value, but gradually increased for higher concentration. Although the change of interfacial tension was quick, approaching a steady-state value in less than

0.5 h as a result of bovine serum albumin adsorption, the evolution of a viscoelastic protein film takes longer time. This implies that not only a certain interfacial coverage but also an advanced molecular rearrangement is needed for the formation of a viscoelastic film, and this latter process takes longer time in the case of bovine serum albumin.

On the contrary, the changes in the shear rheological parameters are abrupt for adsorbed ovalbumin and β-lactoglobulin even though their effect in reducing the interfacial tension is less pronounced compared to bovine serum albumin. The above suggests that the molecular rearrangements of ovalbumin and β-lactoglobulin are notably faster (Bos and van Vliet, 2001) due to their greater flexibility, and strong intermolecular interactions are formed. The continued rise in G' and η' indicates the formation and strengthening of interparticle interactions, the continuous formation of lateral bonds between the adsorbed protein molecules and the possible multilayer formation (Borbás et al., 2003c). The thickening of protein adsorption layer is efficiently detected by interfacial shear rheology sensitive to the regions of the bulk in the vicinity of the interface (Dickinson and Stainsby, 1982).

The influence of the composition of the TPP system on the formation of viscoelastic protein layer was analyzed for bovine serum albumin (Fig. 6.6). Three TPP systems with 2, 3 and 4 interfacial tensions were compared. There was a rather long (at least 1.5 h) induction period for the development of measurable shear rheological parameters in all cases. The alteration of the initial interfacial tension from 2 to 3 and finally 4 mN/m means the increase in the polarity difference between the lower and upper phases, as well as the increase of the ionic strength of the lower phase. These result in the interfacial protein layer becoming more rigid: the increases in elasticity and the viscosity with increasing initial interfacial tension. This phenomenon can be explained by the higher salt content of the lower phase with an increased screening effect, further assisting the interfacial association of proteins. It is also possible, that due to the high ionic strength, the protein is already aggregated, and these aggregates adsorb at the interface, already bound to each other via crosslinks. The formation of further lateral bonds between the hydrophobic patches needs an expanded time, thus increasing the lag time. During 12 h, only the elasticity of the protein layer formed at interface of the lowest interfacial tension system reached a steady value. The constant increase in elasticity and viscosity is explained by the slow but dynamic processes at the interface (adsorption, unfolding, molecular rearrangements, formation of interparticle bonds).

The above detailed interfacial shear rheological properties of the protein layers explain the formation of multilayers and strong viscoelastic protein films, which may contribute to the long-term stability of the emulsion formed during the TPP process. The thick layer made of proteins around the emulsion droplets hinders the coalescence of emulsion droplets, as the droplets cannot approach each other. The gel-like protein layer would resist deformation of the droplets, also acting against coalescence.

The dilatational rheological properties of interfacial protein layer contribute to emulsion stability – The rheological measurements indicate that the adsorbed proteins form a viscoelastic layer at the liquid/liquid interface, which is highly enlarged during emulsification. The emulsion is stabilized by the proteins covering the droplet surfaces.

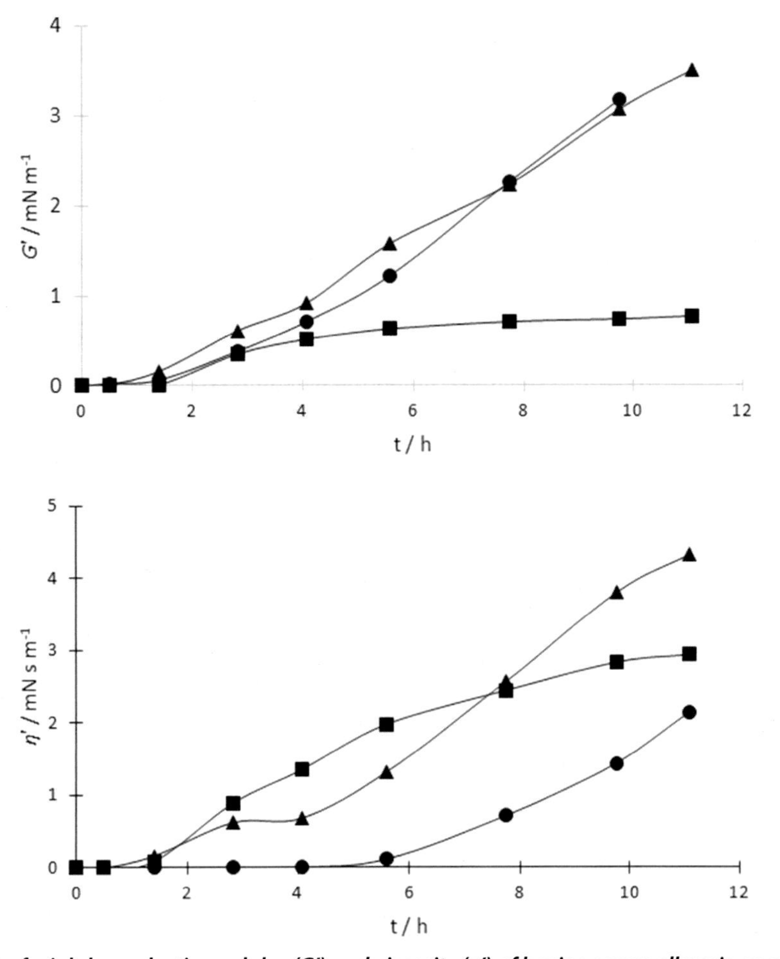

Fig. 6.6 ***Interfacial shear elastic modulus (G′) and viscosity (η′) of bovine serum albumin as a function of time (t):*** in different TPP systems with interfacial tension of 2 (■), 3 (▲), and 4 mN × m^{-1} (●). The protein concentration was 0.05 *g* × dm^{-3}.

Even aggregated protein precipitates could contribute to preserving the metastable state of the emulsion, similarly to Pickering emulsions stabilized by solid particles (Murray, 2019; Hoeben et al., 2009; Calabrese et al., 2018). Not only the interfacial shear rheology of the protein films affects emulsion stability, but also the dilatational rheology which contributes to the short-term stability (Murray and Dickinson, 1996; Dickinson, 1998; Petkov et al., 2000; Pezennec et al., 2000). The dilatational rheological properties can be determined by measuring interfacial tension while changing the interfacial area. Dilatational modulus (ε) can be obtained by the differentiation of the surface pressure (Π) as a function of the logarithm of the area (A_{rel}):

$$\varepsilon = -\frac{d\Pi}{d\ln A_{rel}}$$

where A_{rel} is the ratio of the actual (A) and the initial area (A_0) of the drop during compression while the surface pressure ($\Pi = \gamma_0 - \gamma$) is the change of interfacial tension due to protein adsorption.

The dilatational rheological properties of the adsorbed protein layer, besides other parameters that influence adsorption as well, are sensitive to the protein conformation at the interface and hence depend on the adsorption time and the age of the protein layer (Cao et al., 2014). An interfacial protein network with many lateral bonds can act against dilatation, lessen the interfacial fluctuations, slowing down the thinning of the liquid film in contact with the interfacial layer (Petkov et al., 2000). An interfacial protein layer can act against the Gibbs-Marangoni effect as well, thus stabilizing the emulsion by preventing the enlargement of the surface through convection.

The dilatational viscoelastic behavior of several model proteins was studied in detail (Kiss and Borbás, 2003). The change of interfacial pressure as a function of relative interfacial area for various proteins (bovine serum albumin, ovalbumin, β-lactoglobulin and trypsin) is shown in Fig. 6.7. The dilatational moduli were derived from compression measurements at small (maximum 20 percent) deformations. By compressing the interfacial protein layers, basically two different behaviors were observed. One was typical for the protein layer formed by bovine serum albumin, or ovalbumin or β-lactoglobulin with low-concentration and short adsorption time. In these cases, when the interfacial protein layer was compressed, the interfacial pressure was increasing with an almost constant or slightly increasing slope,

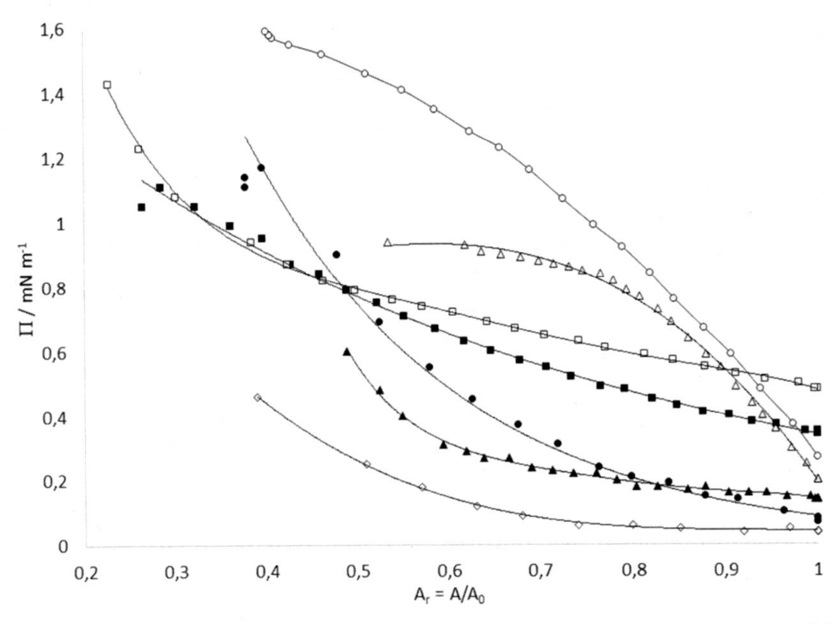

Fig. 6.7 *The interfacial pressure, Π as a function of relative area, A_{rel} upon compression of the interface of TPP system with 2 mN/m initial interfacial tension for bovine serum albumin (■), ovalbumin (●), β-lactoglobulin (▲), and trypsin (◆) after adsorption time of 120 s (filled) or 600 s (open symbols). The protein concentration was 0.05 g × dm^{-3}, except trypsin with 1.0 g × dm^{-3}.*

demonstrating that the interfacial layer is compressible containing only a few lateral bonds. At the beginning of the adsorption time, bovine serum albumin showed somewhat stronger lateral interactions than the other two proteins. The character of the bovine serum albumin film does not change in the time interval studied here. Its compression curve corresponding to 600 s adsorption time is quite similar to that belonging to the film at 120 s.

For ovalbumin and β-lactoglobulin however, as the adsorption proceeds, the dilatational behavior of the film has significantly been altered. The increase in the interfacial pressure upon compression occurs with a higher slope than in the previous cases. The higher dilatational moduli show (at 600 s adsorption time) the intermolecular interactions between adsorbed ovalbumin or β-lactoglobulin molecules became stronger with time, and those films behave as a mechanically strong protein layer at the interface made of coagulated molecules (Dickinson and Stainsby, 1982). Moreover, these rigid protein layers collapsed upon further compression indicated by the decreasing slope of the $\Pi - A_r$ curves. This kind of change indicates strong lateral interactions inside the adsorption layer, clearly showing intermolecular crosslinks, which leads to highly viscoelastic protein films, strongly resistant to dilatation.

It is worth noticing the behavior of trypsin, which exhibited negligible change in the interfacial tension (see Fig. 6.3). However, the compression of the interfacial layer showed that there are protein molecules present at the interface, but the interfacial coverage is so low that the molecules do not interact without compression. The slope of the Π- A_{rel} curve, thus the dilatational modulus increases at higher degree of compression.

Considering the dilatational moduli of different proteins, their behavior is consistent with their known properties, i.e. both ovalbumin and β-lactoglobulin possessing high conformational mobility easily unfold upon adsorption, forming a strongly cohesive, rigid, solid-like film (Petkov et al., 2000; Peron et al., 2000) interacting via secondary bonds, and probably also through disulfide bridges. On the other hand, bovine serum albumin and horseradish peroxidase form layers with weakly interacting molecules represented in the low values of dilatational modulus, which do not change significantly with increasing interfacial pressure. The results of interfacial rheological measurements performed with model proteins including the relevant data, i.e. the interfacial pressure, the parameters describing the viscoelasticity of the interfacial layers and the yield using TPP determined are given in Table 6.1. It seems to be a general trend that the dilatational and shear rheological properties of proteins adsorbed at the interface of TPP systems develop with a shift in time, the evolution of a viscoelastic layer resistant to shear strains takes longer time. Consequently, the dilatational viscoelasticity probably has a significant role in the short, while the shear viscoelasticity in the long-term stabilization of the emulsions of the TPP systems (Murray, 1998; Lucassen-Reynders and Benjamins, 1999; Makievsky et al., 1999).

In conclusion, based on the values in Table 6.1, it was found that in order to purify and extract a protein efficiently by TPP, either it has to have a high affinity towards the interface or it has to rearrange into a viscoelastic interfacial layer, or both. If a protein shows none of these properties, the finding of a suitable pH or other adjustment of the

Table 6.1 Interfacial rheological parameters and the yield of four model proteins relevant to their behavior in the TPP system with 2.0 mN × m^{-1} interfacial tension.

	Bovine serum albumin	Ovalbumin	β-lactoglobulin	Horseradish peroxidase
Π/mN × m^{-1}				
t = 120 s	0.37 ± 0.04	0.12 ± 0.05	0.31 ± 0.10	0.04 ± 0.01
t = 600 s	0.58 ± 0.03	0.26 ± 0.06	0.49 ± 0.08	0.09 ± 0.02
ε/ mN × m^{-1}				
t = 120 s	0.58	0.30	0.30	0.06
t = 600 s	0.45	2.5*	2.6*	0.19
G'/mN × m^{-1}; t = 8 h	0.65 ± 0.04	12.4 ± 0.7	11.9 ± 0.6	(no data)
η'/mN × s × m^{-1}; t = 8h	2.20 ± 0.29	42.3 ± 3.2	46.5 ± 2.5	(no data)
estimated characteristics of the protein at the interface	fast adsorption, slow unfolding, weak interparticle interactions, build-up of multiple layers	slow adsorption, fast unfolding, strong interparticle interactions, build-up of multiple layers	fast adsorption, fast unfolding, strong interparticle interactions, build-up of multiple layers	slow adsorption, slow or no unfolding, weak or no interparticle interactions
protein yield in the middle phase (percent)	82	90	95	31

(Data taken from Borbás, 2003b) The initial protein concentration was 0.05 g × dm^{-3}. Π : interfacial pressure; ε : interfacial dilatational modulus; G': interfacial shear modulus; η': interfacial shear viscosity; *collapse at high deformations.

conditions may help. In those cases when two-step TPP was needed for the expected purification, the competitive adsorption and molecular rearrangement arise. However, these latter suggestions have not been proven yet.

6.2.4 The role of emulsion stability in the formation of gel as the middle phase

In the course of TPP after rigorous mixing of the system phase separation occurs and the target component, proteins can be obtained as the middle phase. That is formed as the result of destabilization of the two-liquid dispersed system, the emulsion. Emulsion is not stable thermodynamically, but its kinetic stability is crucial regarding their applications. The kinetic stability is related to sedimentation or creaming of droplets, the isotherm dissolution due to various droplet sizes (Ostwald ripening) and the adhesion and coalescence of drops. Definitions of systems and processes related to TPP are collected in Table 6.2.

Type of an emulsion is defined whether the oil or water is the continuous phase. When the dispersed phase, the droplet is the oil or other nonpolar liquid and the dispersion medium is water, the type of the emulsion is oil-in-water, O/W. Conversely, when the dispersed phase is water and the medium is the oil, the type of the emulsion

Table 6.2 Definition of some systems and processes involved in TPP.

Emulsion	Disperse system formed from two immiscible (or partly miscible) liquids with a typical droplet size of 0.1–10 μm.
Gel	A cohesive three-dimensional cross-linked system of a polymeric (or particulate) network with a high liquid content
Emulsion gel	A cohesive system containing the interacting emulsion droplets
Ostwald ripening	Increase of size of the colloidal particles (solid or liquid) due to size dependent solubility or vapor pressure
Creaming	Concentration of emulsion droplets due to smaller density than that of the medium

is water-in-oil, W/O. Since we experienced both creaming and sedimentation of the emulsion droplets in various TPP systems, we can conclude that the type of the emulsion is not determined by the solubility of the stabilizing agent, the proteins. According to our observations the emulsion formed in TPP can happen to be O/W or W/O mainly depending on the volume ratio of the two liquid phases.

Once the emulsion is formed, it is left to stand and after a predetermined period of time the system is centrifuged during the TPP process. Meanwhile the drops can fuse together, which is called coalescence, driven by the effort to reach the energy minimum of the system. The occurrence of coalescence is fairly hindered by steric repulsion of adsorbed layer of proteins on the surface of the drops, even more efficiently in the case of multiple adsorption layers. Also a layer with high interfacial dilatational modulus can avoid droplet coalescence, as the adsorbed layer is less easily distorted when pressed together.

Coalescence and creaming can be followed by the volume change of the emulsion. Comparing the stability of the emulsions using the model proteins, the relationship between the properties of the interfacial layers and the emulsion stability was investigated. The volume of the three phases formed in TPP is displayed for systems containing various proteins in Fig. 6.8. Lysozyme and trypsin, which have low interfacial pressure, not forming strong or any viscoelastic interfacial layer, are unable to stabilize emulsions efficiently in three-component TPP systems. The volume of the emulsion made with these proteins decrease to 0.7 percent of its initial volume in 1 minute. On the other hand, the emulsion containing bovine serum albumin was more stable: its volume was approximately 12 percent of the volume of the whole system after 1 min, but it decreased gradually to 2 percent, and after centrifuging (10 min at 672 g), it was just 1.5 percent. As it has been already discussed, bovine serum albumin readily absorbs at the interface, but the interfacial dilatational parameters show week lateral interactions, thus low short-term stability. However, the supposed multiple adsorption layers and conformational changes of the protein observed during longer period of time via the interfacial shear viscoelasticity contribute to the long-term stability. The same parameters for β-lactoglobulin are 25 percent after 1 minute, and a gradual decrease to approximately 7 percent after an hour, which decreases to its half after centrifuging. β-lactoglobulin not only adsorbs at a high rate, but also its conformational changes and

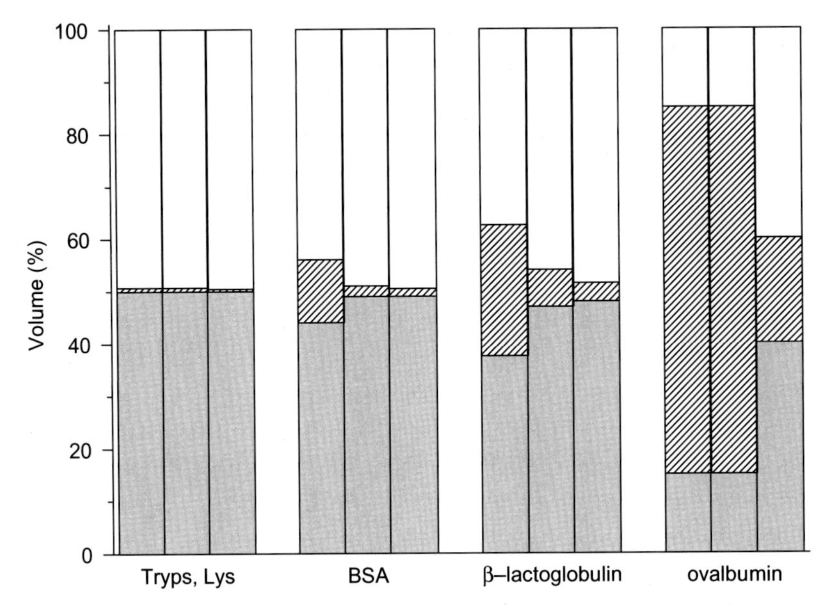

Fig. 6.8 ***Volume of the lower (gray), middle (crosshatched) and upper phases (white) for various proteins in TPP.*** The three columns correspond to state at the formation, after 1 h and following centrifugation.

the formation of a highly viscoelastic protein interfacial layer of coagulated molecules are very fast. Thus, both the long and the short-term stability of the emulsion are enhanced by β-lactoglobulin. The most stable emulsion was formed using ovalbumin, which decreases the interfacial tension more slowly; nevertheless, the development of coagulated protein film with high viscoelastic parameters is significantly high in this case. Therefore, creaming slowed down after 1 minute, and the volume of the emulsion did not change significantly after that, filling 70 percent of the total volume of the system. Even after centrifuging, its volume remained 20 percent.

Besides coalescence, another possible mechanism to destabilize emulsion is the increasing droplet size by isotherm dissolution. As both phases of the TPP systems contain all three components, namely *tert*-butanol, ammonium sulfate, and water, disproportionation through Ostwald ripening has to be considered. There could be a tendency to reach a thermodynamic equilibrium through diffusion of all components between the droplets and the continuous phase, increasing the larger droplets, and decreasing the smaller ones, hence accelerating the creaming/sedimentation. However, if the interfacial protein film is rigid, that could prevent Ostwald ripening as it would act against the shrinkage of droplets. That is supported by the fact that the two proteins that showed the highest dilatational moduli, i.e. ovalbumin and β-lactoglobulin, produced the most stable emulsions.

Microscopic images of the middle layer revealed that emulsion droplets covered with ovalbumin aggregated considerably even 1 min after the emulsification (Fig. 6.9A). That suggests that strong bonds formed between the droplets, protein molecules after a conformational change may form a bridge between the emulsion droplets. The droplets covered

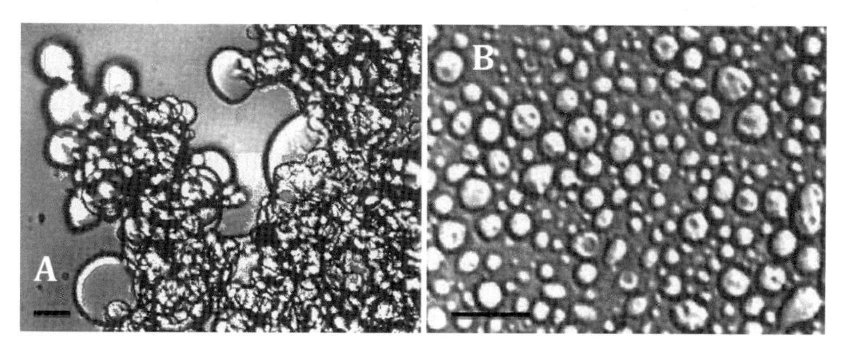

Fig. 6.9 ***Microscopic images of the emulsion containing ovalbumin (A) and bovine serum albumin (B).*** The image was taken 5 (A) and 1 min (B) after emulsification. Bars correspond to 100 μm. The system was the $\gamma = 2$ mN × m^{-1} one with protein concentration of 0.05 g × dm^{-3}. (Borbás, 2003b).

with bovine serum albumin were visibly smaller (Fig. 6.9B) and less deformed. They are more mobile for a longer period of time, arriving slowly into the middle phase by creaming.

By centrifuging the TPP systems, the middle phase becomes more concentrated and compressed. However, its emulsion character is still present, even if the droplets become deformed. Microscopic images were taken of a cross section of the middle phases after emulsification and centrifugation (Fig. 6.10). The smaller viscoelastic parameters of bovine serum albumin suggest lower resistance to deformations. Indeed, the higher distortions in the shape of the protein "skeleton" of the emulsion gel containing bovine serum albumin were observed. In agreement with that, the ovalbumin-covered droplets underwent a lower degree of coalescence and resisted deformations even during the strong mechanical effect of centrifugation, as shown in Fig. 6.10B, due to the higher dilatational elasticity and viscosity of the interfacial protein layer.

Based on the facts given above, it can be deduced that the middle phase formed during the TPP process is an emulsion; moreover, it is an emulsion gel. This finding is further supported by the composition analysis of the middle phase carried out by classical analytical methods coupled with near-infrared spectroscopy (Borbás et al., 2001) and also by the rheological study of the middle phase formation in situ (Borbás et al., 2003a; 2003c).

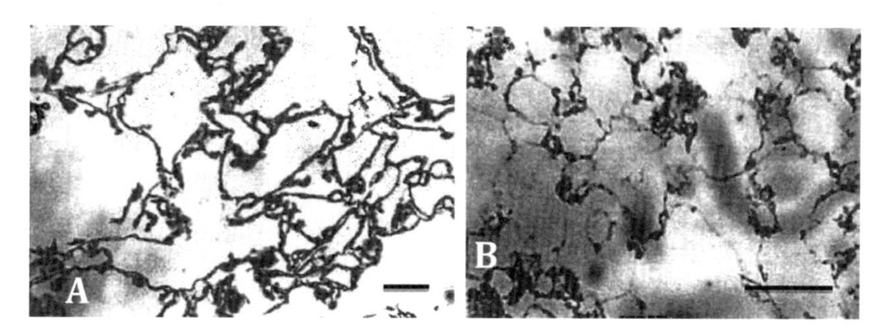

Fig. 6.10 ***Microscopic images of cross sections of paraffin embedded middle phases produced in the presence of bovine serum albumin (A) and ovalbumin (B) in TPP system.*** The bars correspond to 200 μm. (Borbás, 2003b).

The results of the analysis show that the composition of the middle phase falls into the heterogeneous region of the phase diagram with two immiscible phases, clearly indicating that the middle phase is not homogeneous, it is not a precipitate, rather an emulsion. This finding supports that the emulsion droplets covered with protein layers cream and aggregate into a gel, capturing the dispersed phase inside the gel. The cross-linking occurs between the proteins attached to the dispersed phase. The fluid part of the gels can be removed from the middle phase by intense centrifuging, which is beneficial for the recycling of the partitioning materials, and reducing the material demand of the TPP process for possible industrial or other large-scale use.

Moreover, the viscoelastic behavior of the middle phase also provides evidence for the middle layer being an emulsion gel. Young-moduli determined by uniaxial compression were found to be in the range of 1000–4000 $N \times m^{-2}$ for bovine serum albumin and ovalbumin containing TPP systems (Borbás et al., 2001; 2003a; 2003c; Borbás, 2003b). The elastic behavior of the middle phase suggests the existence of strong cohesion forces among the emulsion droplets. Therefore, it is expected that if a protein has high interfacial dilatational viscoelastic parameters, it can stabilize the emulsion properly, even resist the mechanical effects occurring and applied during the TPP process. In that case the middle phase is well separated and easily formed producing good yield.

6.3 Conclusion

Three phase partitioning (TPP) is found to be a simple, but powerful method for protein purification even from crude extracts. However, the detailed knowledge of the mechanism is highly desirable for the full understanding of the driving forces and optimal conditions of the efficient use of TPP. The previous findings are summarized into a concise scheme of the partitioning process (see Borbás, 2003b, and Table 6.3.)

The protein is dissolved in an aqueous medium (e.g. a crude extract), and ammonium sulfate is added to the macromolecular colloidal solution, then the system is

Table 6.3 The main processes occurring during TPP and the optimum protein properties for the efficient partitioning.

Steps of TPP

Processes occurring (with methods for detection)	Required property of the protein
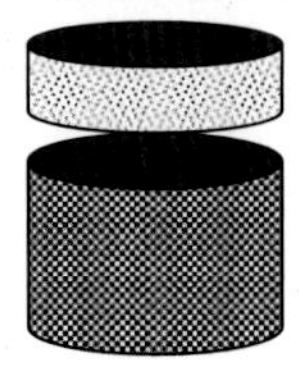	
Protein dissolved in the lower phase, or in ammonium sulfate solution: decrease in solubility, partial dehydration of proteins, binding of sulfate ions to the protein	• solubility of the protein in a high ionic strength solution

Steps of TPP

Processes occurring (with methods for detection)	Required property of the protein
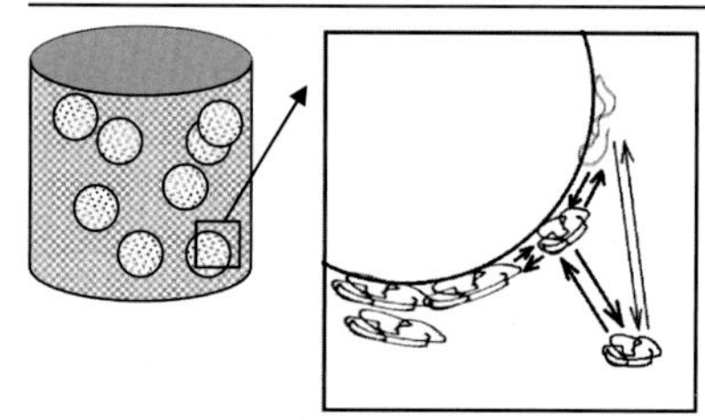	
Addition of tert-butanol, and the emulsification of the immiscible phases of TPP containing the protein • increase in the interfacial area • quick adsorption of protein (measurement of interfacial tension) • adsorbed protein layer hinders coalescence • further adsorption, formation of multiple protein layers (interfacial shear rheological measurements) • interactions between protein molecules (interfacial shear and dilatational rheological measurements) • conformational changes, unfolding, coagulation of the protein molecules • formation of viscoelastic and/or mechanically strong protein layers (interfacial shear and dilatational rheological measurements)	• high affinity towards the interface • fast protein adsorption • protein unfolds fast, quickly and reversibly changes its conformation • formation a strong viscoelastic interfacial layer • selection of a suitable pH and partitioning system
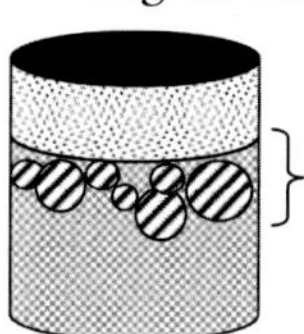	
Let the emulsion to settle, centrifuging of the system • creaming/sedimentation of the emulsion • possible coalescence of emulsion droplets, Ostwald ripening (microscopic observations) • flocculation, aggregation of droplets (microscopic observations), protein crosslinks between the droplets • formation of the emulsion gel, cohesive structure stabilized by the protein adsorbed at the interface of the emulsion droplets (rheological measurements, analysis of the composition of the middle phase)	• emulsion stabilization by protein • interactions to produce network
Removal of the middle phase from the system *Dissolution, regaining of the protein* *Further purification of the extracted protein* *Probable recycling of the materials used*	• reversible conformational changes

emulsified using appropriate amount of *tert*-butanol. The sudden and huge increase in the interfacial area enables for the protein molecules to adsorb at the interface. The adsorption is generally fast. The adsorbed protein layer or possible aggregates of protein at the interface of the emulsion droplets hinder coalescence. Further adsorption is followed by the building of multiple protein layer. Interactions may evolve between protein molecules due to conformational changes and unfolding. These processes lead to the formation of viscoelastic and mechanically strong protein layers, resisting shear and dilatational strain. When the emulsion is left at rest, a middle layer is formed as a result of creaming/sedimentation, which becomes cohesive. Coalescence of emulsion droplets and Ostwald ripening are actually slowed down or blocked by the adsorbed viscoelastic protein layer. The protein-covered emulsion droplets flocculate due the interaction between the proteins, which may even form crosslinks. Hence, a cohesive system, an emulsion gel is formed, stabilized and held together by a skeleton like interfacial protein layer. The emulsion gel is compressed by centrifugation, and it is ready for removal as a middle phase in TPP. Thus, the protein became enriched in the middle phase, while the other components of the protein-containing solution were selectively dissolved based on their polarity in the liquid phases of the TPP systems.

Based on these findings the proteins should possess certain properties in order to be optimally extracted by TPP. Firstly, they should be soluble in high ionic strength ammonium sulfate solution, and under these circumstances, it should have high affinity towards the interface of the two-liquid system. Secondly, it should form a preferably strong, viscoelastic film resisting dilatational and shear stress, by fast adsorption and/or quick unfolding and reversible conformational changes. Lateral interactions should be developed at the interface between the adsorbed particles in a short period of time in order to be able to stabilize the emulsion made of the immiscible liquids of the TPP systems. These can be promoted by choosing the appropriate composition of the partitioning system and the pH. Temperature may also affect the efficacy of extraction.

References

Anderson, C., Odegaard, B., Wilson, T., Lovrien, R.E., 1981. Am. Chem. Soc. Biol. Chem. Div. Meeting. Abstr., 62.

Benjamins, J., 2000. Ph.D. Thesis. University of Wageningen.

Borbás, R., Turza, S., Szamos, J., Kiss, É., 2001. Analysis of protein gels formed by interfacial partitioning. Colloid Polym. Sci. 279, 705–713.

Borbás, R., Kiss, É., Murray, B.S., 2003. Interfacial rheology and interfacial gelation partitioning. In: Dickinson, E., van Vliet, T. (Eds.), Food Colloids, Biopolymers and Materials. Roy. Soc. Chem., London, pp. 368–376.

Borbás, R., 2003. PhD Thesis. Eötvös Loránd University, Budapest, Hungary.

Borbás, R., Murray, B.S., Kiss, É., 2003c. Interfacial shear rheological behaviour of proteins in three phase partitioning systems. Colloids Surf. A 213, 93–103.

Bos, M.A., van Vliet, T., 2001. Interfacial rheological properties of adsorbed protein layers and surfactants: a review. Adv. Colloid Interf. Sci. 91 (3), 437–471.

Calabrese, V., Courtenay, J.C., Edler, K.J., Scott, J.L., 2018. Pickering emulsions stabilized by naturally derived or biodegradable particles. Current Opin. in Green and Sustainable Chem. 12, 83–90.

Cao, Ch, Lei, J., Zhang, L., Du, F.-.P., 2014. Equilibrium and dynamic interfacial properties of protein/ionic-liquid-type surfactant solutions at the decane/water interface. Langmuir 30, 13744–13753.
Chew, K.W., Ling, T.C., Show, P.L., 2018. Recent developments and applications of three-phase partitioning for the recovery of proteins. Separation & Purif. Rev. 48 (1), 52–64. doi:10.1080/15422119.2018.1427596.
Dennison, C., Lovrien, R., 1997. Three phase partitioning: concentration and purification of proteins. Protein Express. Purif. 11 (2), 149–161.
Dennison, C., 2011. Three-phase partitioning. In: Tschesche, H. (Ed.), Methods in Protein Biochemistry. Walter de Gruyter, Berlin, Germany, pp. 1–12.
Dickinson, E., Stainsby, G., 1982. Colloids in Food. Applied Science Publishers, London, pp. 287.
Dickinson, E., Matsamura, Y., 1994. Proteins and liquid interfaces: the role of the molten globule state. Colloids Surf. B 3 (1–2), 1–17.
Dickinson, E., 1998. Proteins at interfaces and in emulsions. Stability, rheology and interactions. J. Chem. Soc. Far. Trans. 94 (12), 1657–1669.
Gagaoua, M., Boucherba, N., Bouanane-Darenfed, A., Ziane, F., Nait-Rabah, S., Hafid, K., Boudechicha, H-R., 2014. Three-phase partitioning as an efficient method for the purification and recovery of ficin from Mediterranean fig (Ficus carica L.) latex. Separ. Purif. Techn. 132, 461–467.
Gagaoua, M., 2018. Aqueous methods for extraction/recovery of macromolecules from microorganisms of atypical environments: a focus on three phase partitioning. Methods in Microbiol. 45, 203–242.
Hoeben, A.M., van Hee, P., van der Lans, R., Kwant, G., derWielen, L., 2009. Design of a Counter-Current Interfacial Partitioning Process for the Separation of Ampicillin and Phenylglycine. Ind. & Eng. Chem. Res. 48 (16), 7753–7766. doi:10.1021/ie800680c.
Ketnawa, S., Rungraeng, N., Rawdkuen, S., 2017. Phase partitioning for enzyme separation: an overview and recent applications. Int. Food Res. J. 24 (1), 1–24.
Kiss, É., Szamos, J., Tamás, B., Borbás, R., 1998. Interfacial behavior of proteins in three-phase partitioning using salt-containing water/*tert*-butanol systems. Colloids Surf. A 142, 295–302.
Kiss, É., Borbás, R., 2003. Protein adsorption at liquid/liquid interface with low interfacial tension. Colloids Surf. B 31, 169–176.
Kulkarni, V.M., Rathod, V.K., 2014. Extraction of mangiferin from Mangifera indica leaves using three phase partitioning coupled with ultrasound. Ind. Crop. Prod. 52, 292–297.
Langevin, D., 2000. Influence of interfacial rheology on foam and emulsion properties. Adv. Colloid Interf. Sci. 88, 209–222.
Lovrein, R., Goldensoph, C., Anderson, P.C., Odegaard, B., 1987. Three phase partitioning (TPP) via *t*-butanol: enzymes separations from crudes. In: Burgess, R. (Ed.), Protein Purification: Micro to Macro. A.R. Liss Inc, New York, pp. 131–148.
Lucassen-Reynders, E.H., Benjamins, J., 1999. Dilational rheology of proteins adsorbed at fluid interfaces. In: Dickinson, E., RodriguezPatino, J.M. (Eds.), Food emulsions and foams: Interfaces, interactions and stability. Royal Society of Chemistry, London, pp. 195–206.
Makievski, A.V., Miller, R., Fainerman, V.B., Krägel, J., Wüstneck, R., 1999. Food Emulsions and Foams, Interfaces, Interactions and Stability E. Dickinson, J.M. Rodriguez Patino (Eds.). Royal Society of Chemistry, London, p. 269.
Moelbert, S., Normand, B., De Los Rios, P., 2004. Kosmotropes and chaotropes: modelling preferential exclusion, binding and aggregate stability. Biophys. Chem. 112 (1), 45–57. doi:10.1016/j.bpc.2004.06.012.
Mondal, K., Jain, S., Teotia, S., Gupta, M.N., 2006. Emerging options in protein bioseparation. Biotechn. Ann. Rev. 12, 1–29.
Murray, B.S., Dickinson, E., 1996. Interfacial rheology and the dynamic properties of adsorbed films of food proteins and surfactans. Food Sci. Techn. Int. 2 (3), 131–145.
Murray, B.S., 1998. Intefacial rheology of mixed food protein and surfactant adsorption layers with respect to emulsion and foam stability. In: Möbius, D., Miller, R. (Eds.), Proteins at Liquid Interfaces. Elsevier, Amsterdam, pp. 179–220.
Murray, B.S., 2019. Pickering emulsions for food and drinks. Curr. Opin. Food Sci. 27, 57–63.
Niehaus Jr., W.G., Dilts Jr., R.P., 1982. Purification and characterisation of mannitol dehidrogenase from *Aspergillus parasiticus*. J. Bacteriology 151, 243–250.

Péron, A., Cagna, A., Valade, M., Marchal, R., Manjean, A., Robillard, B., Aguié-Béghin, V., Douillard, R., 2000. Characterisation by drop tensiometry and by ellipsometry of the adsorption layer formed at the air/champagne wine interface. Adv. Colloid Interf. Sci. 88, 19–36.

Petkov, J.T., Gurkov, T.D., Campbell, B.E., Borwankar, R.P., 2000. Dilatational and shear elasticity of gel-like protein layers on air/water interface. Langmuir 16, 3703–3711.

Pezennec, S., Gauthier, F., Alonso, C., Graner, F., Croguennec, T., Brule, G., Renault, A., 2000. The protein net electric charge determines the surface rheological properties of ovalbumin adsorbed at the air-water interface. Food Hydrocoll 14, 463–472.

Pike, R.N., Dennison, C., 1989. Protein fractionation by three phase partitioning (TPP) in aqueous/t-butanol mixtures. Biotechn. Bioeng. 33, 221–228.

Singh, R.K., Gourinath, S., Sharma, S., Roy, I., Gupta, M.N., Betzel, C., Srinivasan, A., Singh, T.P., 2001. Enhancement of enzyme activity through three-phase partitioning: crystal structure of a modified serine protease at 1.5 angstrom resolution. Protein. Eng. 14, 307–313.

Szamos, J., Jánosi, A., Tamás, B., Kiss, É., 1998. A novel partitioning method as a possible tool for investigating meat. I. Z. Lebensm. Unters. Forsh. A 206, 208–212.

Tripp, B.C., Magda, J.J., Andrade, J.D., 1995. Adsorption of globular proteins at the air/water interface as measured via dynamic surface tension: concentration dependence, mass-transfer considerations, and adsorption kinetics. J. Colloid Interf. Sci. 173, 16–27.

Ward, A.F.H., Tordai, L., 1946. Time-dependence of boundary tensions of solutions; I. The role of diffusion in time-effects. J. Chem. Phys. 14, 453–461.

Yan, J.K., Wang, Y.Y., Qiu, W.Y., Ma, H., Wang, Z.B., Wu, J.Y., 2017. Three-phase partitioning as an elegant and versatile platform applied to nonchromatographic bioseparation processes. Crit. Rev. Food Sci. Nutr. 58 (14), 2416–2431. doi:10.1080/10408398.2017.1327418.

Yan, J.K., Wang, Y.Y., Qiu, W.Y., Wang, Z.B., Ma, H., 2018. Ultrasound synergized with three-phase partitioning for extraction and separation of Corbicula fluminea polysaccharides and possible relevant mechanisms. Ultrasonics Sonochem 40 (A), 128–134.

CHAPTER 7

Three-phase partitioning (TPP) of proteases from parasites, plants, tissue and bacteria for enhanced activity

Lauren E-A Eyssen, James Philip Dean Goldring, Theresa Helen Taillefer Coetzer
Biochemistry, School of Life Sciences, University of KwaZulu-Natal (Pietermaritzburg campus), Scottsville, South Africa

Chapter outline

7.1 Why we are interested in proteases

Proteases, also known as proteolytic enzymes or peptidases, are ubiquitous peptide bond hydrolyzing enzymes present in all forms of life (Bond, 2019; Lopez-Otin and Bond, 2008). Seven classes of proteases are distinguished (six hydrolazes and one lyase) based on the catalytic mechanism used for peptide bond cleavage. Six different nucleophiles are used for hydrolysis, i.e. the thiol of cysteine (cysteine proteases), or the hydroxyl of serine (serine proteases) or threonine (threonine proteases), or an activated water molecule bound to aspartate (aspartic proteases), glutamate (glutamic proteases) or a metal ion (mostly zinc, but also cobalt, copper or manganese) in metalloproteases. The seventh class, asparagine peptide lyases, cleave peptide bonds by cyclisation of an asparagine residue to a succinimide. Proteases of the different catalytic classes are grouped into families based on their amino acid sequence similarities and families are assembled into

Three Phase Partitioning
DOI: https://doi.org/10.1016/B978-0-12-824418-0.00008-4

clans based on three-dimensional structure similarities (Rawlings et al., 2017; Rawlings and Bateman, 2021).

These enzymes are not only responsible for protein catabolism but control many biological processes, including apoptosis, autophagy, blood coagulation, cell cycle progression, complement activation, inflammation, and wound healing in higher organisms. Consequently, aberrations in proteolysis underlie several pathological conditions such as cancer, neurodegenerative disorders, cardiovascular disease and inflammatory conditions (Turk, 2006). Several protease inhibitors have therefore been developed as therapeutic compounds such as angiotensin-converting enzyme (ACE) inhibitors to treat hypertension and the human immune deficiency virus (HIV) protease inhibitors that interrupt the virus life cycle in acquired immune deficiency (AIDS) patients. Proteases play vital roles in bacterial biochemistry and physiology as well as pathogenicity (Potempa and Pike, 2009), making them targets for developing new alternative antimicrobial compounds (Culp and Wright, 2017). Parasite-derived proteases play a role in establishing, sustaining and aggravating infection. This is accomplished by hydrolyzing components of the host extracellular matrix to enter cells or penetrate tissue and by evading and modulating the host immune response (McKerrow et al., 2006). Hence proteases are attractive parasite drug and diagnostic targets. In plants, proteases play a role in virtually every stage of the life cycle from mobilizing storage proteins during seed germination to plant immune responses and finally senescence and programmed cell death (Popovič and Brzin, 2007; Schaller, 2004; Thomas and van der Hoorn, 2018). Plant proteases are used in the food industry for cheese making and as meat tenderisers. Based on their use in traditional medicine, plant proteases have been explored in the pharmaceutical industry to treat cancer and digestive disorders and for immunomodulation and wound healing (Gonzalez-Rabade et al., 2011; Salas et al., 2008).

Proteases from microbial, plant and animal sources make up 60 percent of the global enzyme market where they find application in the leather, food and beverage industry, detergent and household soap market and the pharmaceutical sector. In 2019, the global protease market size was US$ 612 million, and it is expected to reach US$ 851.3 million by the end of 2026, with a compound annual growth rate of 4.8 percent during 2021–26 (360ResearchReports, 2020). In addition to purifying proteases to study their role in homeostasis, health and disease, determine their substrate specificity, catalytic mechanism and structure, industrial applications also require a simple and effective protease purification method. Three-phase partitioning (TPP) is a protein purification method that can be effected on a micro- or macro scale to extract, purify and concentrate proteins, and in the case of enzymes, such as proteases, enhance their activity (Dennison, 2012; Dennison and Lovrien, 1997; Gagaoua, 2018; Pike and Dennison, 1989b). Proteases have been purified using TPP from a large and diverse number of sources including parasite lysates, heterologous protein expression systems, plant and mammalian tissue and bacterial culture media (Fig. 7.1; Table 7.1).

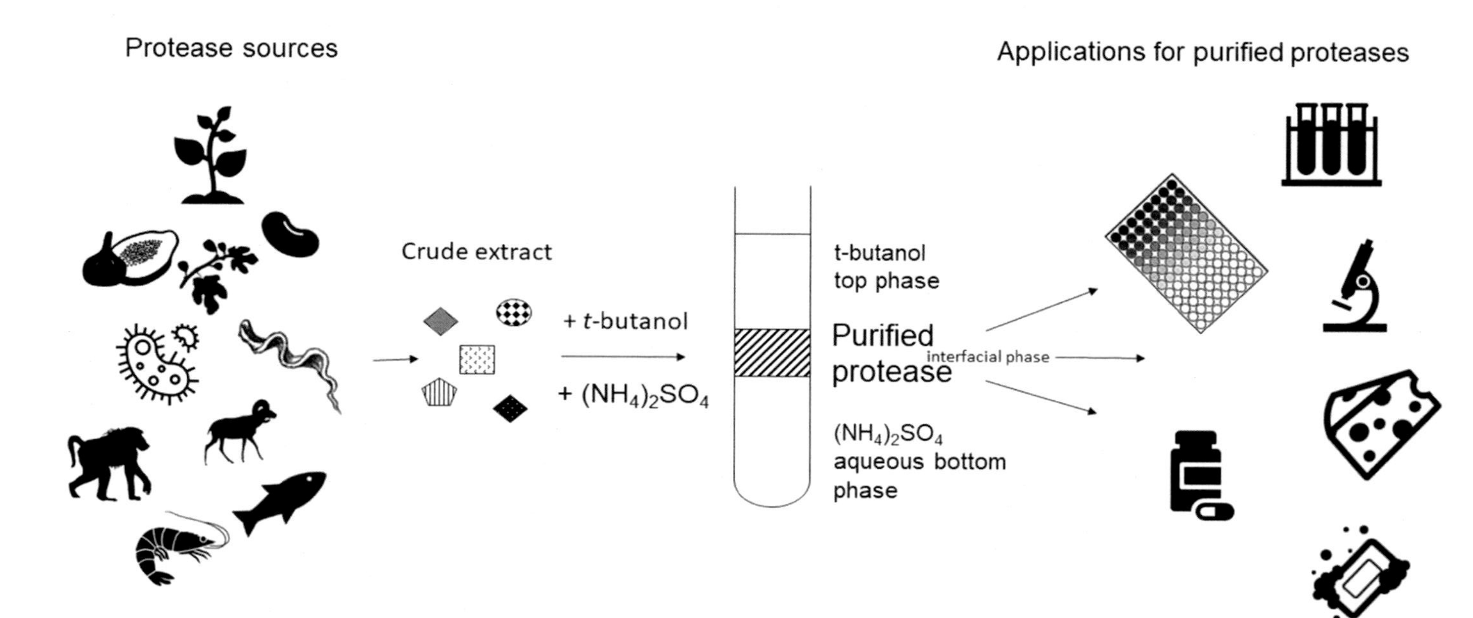

Fig. 7.1 ***Summary of three-phase partitioning for the isolation of proteases from different sources for various applications.*** Samples from different sources are homogenized or extracts prepared and ammonium sulfate/*t*-butanol added. After equilibration, samples are centrifuged to produce a top *t*-butanol, middle interfacial and bottom aqueous ammonium sulfate-containing phase. The isolated proteases find application in several fields such as the biomedical, biotechnology, laundry soap and food industries.

Table 7.1 Proteases and protease inhibitors isolated from different sources using three-phase partitioning.

Source	Protease and Catalytic Class or Inhibitor Family	Mol. Size kDa	pI[a]	pH	Crude extract:t-butanol	Percent (w/v) NH_2SO_4	Purification fold	Yield (percent)	Interfacial (I) or aqueous (A) phase	Reference
Trypanosoma brucei lysates	Trypanopain-Tb (*Tbb*CATL, cathepsin L-like) Cysteine C1	30	4.5	5.5	1:0.5	10–25	6	25	I	Troeberg et al. (1996)
Trypanosoma brucei lysates	Oligopeptidase B (*Tbb*OPB) Serine S9	80	5.6	5.5	1:0.5	10–25	15	63	I	Troeberg et al. (1996)
Trypanosoma brucei infected rat plasma	Pyroglutamyl peptidase type I (*Tbb*PGP) Cysteine C15	26	5.6	8	1:0.5	10–30	1.5	92	I	Morty et al. (2006)
Trypanosoma congolense lysates	Oligopeptidase B (*Tco*OPB) Serine S9	80	5.6	5.5	1:0.5	10–25	18	66	I	Morty et al. (1999)
Pichia pastoris GS115 recombinant expression supernatant	*Trypanosoma congolense* cathepsin L-like (*Tco*CATL) active site variants (CP_{SYN}; CP_{SHN}) Cysteine C1	28	4.7	4.2	1:0.5	30	11 (CP_{SYN}) 2.8 (CP_{SHN})	85.6 (CP_{SYN}) 80.1[b] (CP_{SHN})	I	Pillay et al. (2010)
Trypanosoma vivax lysates	Native vivapain (*Tvi*CATL, cathepsin L-like) Cysteine C1	44	5.1	7.4	1:0.5	10–20	1.5	71.6	I	Eyssen et al. (2018)

Pichia pastoris GS115 recombinant expression supernatant	Recombinant *Trypanosoma vivax Tvi*CATL catalytic domain Cysteine C1	28 and 33	5.1	4.2	1:0.5	40	21.68	95.04	I	Vather (2010)
Carica papaya fruit peels	Papain-like cysteine proteases Cysteine C1	30–60[c]	8.75 to 11.7	7	1:0.5 (steps 1 and 2)[d]	20 (step 1) 55 (step 2)	15.8 (step 1) 10.1 (step 2) 23.8 (step 2)	253.5 (step 1) 89.4 (step 2) 165.2 (step 2)	A (step 1) I (step 2) A (step 2)	Chaiwut et al. (2010)
Calotropis procera latex (Family Apocynaceae)	Procerain Cysteine C1	~39	9.1	7	1:0.5	60	6.92	132	I	Rawdkuen et al. (2010)
Ficus carica (fig)	Fic(a)in Cysteine C1	23.4	8.9	7	1:0.75	40	6.04	167	I	Gagaoua et al. (2014)
Zingiber officinale Roscoe rhizomes (ginger)	Zingibain Cysteine C1	33.8	4.38	7	1:1	50	14.91	215	A	Gagaoua et al. (2015)
Cucumis melo var. reticulatus (melon)	Cucumisin (subtilisin) Serine S8	68.4	8.7	8	1:1.25	60	4.61	156	I	Gagaoua et al. (2017)
Wrighta tinctoria (Family Apocynaceae)	Milk-clotting proteases Serine	95, 91, 83	3.9, 5.5, 5.4	5.5	1:1	60	22.6	89.5	I	Rajagopalan and Sukumaran (2018)
Litopenaeus vannamei (Pacific white shrimp) hepatopancreas	Alkaline proteases	36, 26	–	8	1:1	30	2.6	76	I	Senphan and Benjakul (2014)

(continued)

Table 7.1 (Cont'd)

Source	Protease and Catalytic Class or Inhibitor Family	Mol. Size kDa	pI[a]	pH	Crude extract:t-butanol	Percent (w/v) NH_2SO_4	Purification fold	Yield (percent)	Interfacial (I) or aqueous (A) phase	Reference
Pangasianodon gigas (giant catfish) viscera	Trypsin-like caseinolytic protease	~25	–	8	1:0.5	50	5	163	I	Rawdkuen et al. (2012)
Pangasianodon gigas (giant catfish) viscera	Alkaline proteases	39, 28, 24	–	8	1:0.5	50[e]	5.4	219	I	Ketnawa et al. (2014)
Cichla ocellaris (peacock bass) intestinal viscera	Collagenolytic protease Serine	60, 20, 10	–	7.5	1:0.5	30	5.07	238.2	I	de Melo Oliveira et al., (2020)
Bos taurus (bovine) spleen	Cathepsin D Aspartic A1	44	5.43	3.7	1:0.5	20–35	83.3	25.8	I	Jacobs et al. (1989)
Ovis aries (sheep) liver	Cathepsin L Cysteine C1	26	5.9	4.2	1:0.5	20–30	17	39	I	Pike and Dennison (1989a)
Papio ursinus (baboon) liver	Cathepsin L Cysteine C1	26	*ca.* 6	4.2	1:0.5	20–30	17.4	27	I	Coetzer et al. (1995)
Bacillus pumilus MP27 culture supernatant	Alkaline protease (subtilisin-like) Serine S8	53	–	8	1:1	20	16.65	187.5	I	Baweja et al. (2017)
Bacillus sp. ICTF2 culture supernatant	Alkaline protease (subtilisin-like) Serine S8	–	–	10	1:1	35	12	176	I	Kaur et al. (2019)
Bacillus natto NRRL-3666 fermentation broth	Nattokinase (subtilisin clan of serine NAT)	28	8.6	8	1:1.5	30	5.6	129.5	I	Garg and Thorat (2014)

Bacillus sphaericus MTCC 3672 microfiltration clarified fermentation broth	Fibrinolytic protease	55–70	6–7	9	1:0.5	80	7.98 (UATPP 16.15)	45 (UATPP 65)	A	Avhad et al. (2014)
Serratia marcescens NRRL B 23,112 cell-free fermentation broth	Serratiopeptidase (serralysin) Metalloprotease	50	4.61	7	1:1	30	4.2 (UATPP 9.4)	83 (UATPP 96)	I	Pakhale and Bhagwat (2016)
Glycine max (soybean) flour extract	Soybean trypsin inhibitor Inhibitor family I3	21	4.6	7	1:1	30	1.4 (66 or 73)[f]	91 (12 or 13)[f]	A (I)	Roy and Gupta (2002)
Eleusine coracana (ragi/finger millet) seeds	Bifunctional trypsin-like protease/amylase inhibitor Inhibitor family I6	14	6.5	7	1:1 (1:2)[d]	30 (60)[d]	8.7 (step 1)[d] 15.9 (step 2)	80.8 (step 1)[d] 31.5 (step 2)	A (step 1)[d] I (step 2)	Saxena et al. (2007)
Navy-red kidney- and adzuki beans	Trypsin inhibitors	*ca.*120, *ca.*60, *ca.*13	–	–	1:1	30	5.6[g]; 4.8[h] 14.1[g]; 9.5[h] 7.3[g]; 23.2[h]	315[g]; 160[h] 441[g]; 398[h] 228[g]; 28[h]	A[g];I[h] A[g];I[h] A[g];I[h]	Wati et al. (2009)

[a]pI values determined using Expasy compute MW/pI tool (https://web.expasy.org/compute_pi/) and UniProt Knowledgebase accession numbers for amino acid sequences (https://www.uniprot.org/);
[b]Fold-purification calculation in Table 1 of Ref. corrected;
[c]Sizes of proteins in papaya peel extracts showing proteolytic activity on non-reducing SDS-PAGE (papain, chymopapain, caricain and glycyl endopeptidase have similar sizes of ca. 23.4 kDa calculated from amino acid sequences at https://www.brenda-enzymes.org/);
[d]Aqueous phase from first TPP (step 1) re-extracted using TPP (step 2);
[e]Sodium citrate in place of ammonium sulfate;
[f]Second TPP on aqueous phase with 0.18 mM $ZnSO_4$ or 0.20 mM $CuSO_4$) in place of NH_4SO_4;
[g]Values for aqueous phase and,
[h]interfacial phase for nave, red kidney and adzuki beans respectively.

This review focuses on using TPP for the purification of proteases from various sources for a range of applications (Fig. 7.1). The conditions that need to be optimized for TPP will be discussed and illustrated with specific examples. An overview will be given of different proteases isolated using TPP from parasites, plants, bacteria, mammalian tissue and recombinant expression systems, either as a single-step procedure or preceding chromatography steps for various end-uses. The reported effects that TPP has on protease structure will be discussed in the context of a molecular basis for the observed enhanced specific activity of these hydrolytic enzymes.

7.2 Three-phase partitioning as a protease purification tool

Three-phase partitioning operates on salting out of proteins from a mixture of tertiary butanol (*t*-butanol, *t*-BuOH or 2-methylpropan-2-ol) and water. Tertiary butanol is completely miscible with water, but when sufficient ammonium sulfate is added, two phases are formed: a lower aqueous phase and an upper *t*-butanol phase. When protein-containing samples, contained in an aqueous buffer, are mixed with *t*-butanol, and adequate ammonium sulfate is added, protein is separated into the third (interfacial/middle) phase between the upper *t*-butanol and lower salt-containing aqueous phases (Dennison and Lovrien, 1997; Lovrien et al., 1987). A particular advantage of using TPP to isolate proteases from plant and animal material is that low molecular weight pigments and lipids partition into the *t*-butanol phase (Popovič et al., 2002; Rather and Gupta, 2013). The respective properties of *t*-butanol as a non-ionic, and ammonium sulfate as an ionic kosmotrope at the top of the Hoffmeister series, allow them to operate synergistically in an aqueous environment to stabilize protein structure (Dennison and Lovrien, 1997; Dobrzycki et al., 2019). Sulfate (SO_4^{2-}) remains a divalent anion down to pH 2.1 (pK of HSO_4^-). It would bind to discrete positively charged patches on proteins at pH values below their pI, i.e. around the slightly acidic to neutral pH at which TPP would most likely be conducted. With the addition of sulfate, protein conformation tightens because the hydrated divalent anion squeezes water out of the interior of the protein and the protein becomes a sulfate salt that represents a more stable conformation. It is thought that *t*-butanol does not denature protein structure because its branched structure prevents it from reaching the interior of globular proteins. Instead, it adsorbs to the protein surface which reduces the density, and hence the protein floats on top of the aqueous phase and below the organic *t*-butanol phase (Dennison, 2012; Roy et al., 2005). The interfacial phase has also been described as an emulsion gel comprising protein that envelops emulsion droplets by adsorbing to their surfaces (Borbás et al., 2001). The main parameters that need to be optimized before TPP can be used to isolate a protease from a crude mixture of proteins and other biomolecules are: 1) choice of organic solvent and its ratio to the crude extract; 2) choice of salt and its concentration; 3) the pH and 4) the temperature at which TPP is conducted.

7.3 Conditions considered during optimization of TPP

7.3.1 Choice of organic solvent and its ratio to crude extract

Tertiary butanol was originally used as the co-solvent in TPP because of its unique properties such as stabilizing protein structure (Dennison and Lovrien, 1997). However, although a number of studies reported comparing *t*-butanol to other organic solvents in TPP, *t*-butanol consistently resulted in the highest percentage enzyme activity or recovery (Table 7.1). During optimization of TPP for the isolation of cucumisin from *Cucumis melo*, *n*-propanol, *n*-butanol or isopropanol resulted in 50 percent and petroleum ether and ethanol in 20 percent enzyme activity recovery (Gagaoua et al., 2017), relative to that when using *t*-butanol. Likewise, for the isolation of alkaline proteases from the viscera of farmed catfish, *t*-butanol was superior to *n*-propanol, *n*-butanol or isopropanol in percent relative enzyme activity recovery (Ketnawa et al., 2014). Tertiary butanol also performed slightly better than *n*-pentanol as the TPP co-solvent for the isolation of the extracellular metalloprotease from *Serratia marcescens*, resulting in 81 percent (2.8-fold purification) compared to 75 percent recovery and 2.3-fold purification (Pakhale and Bhagwat, 2016). Dennison and Lovrien (1997) predicted that the levelling co-solvents, C1 and C2 alcohols such as methanol and ethanol, that tend to mask the differences between proteins, would not be as effective in promoting enzyme stability and activity as the differentiating co-solvents, C3-C5 alcohols, that accentuate the differences between proteins.

The ratio of the volume of the crude protein-containing extract to the volume of *t*-butanol is often considered. Although earlier studies indicated that twenty to thirty percent by volume *t*-butanol, representing a ratio of crude extract to *t*-butanol of 1:0.2 to 1:0.5, is optimal for many proteins (Dennison and Lovrien, 1997; Pike and Dennison, 1989b), there are exceptions. A comparison of different ratios of crude papaya peel extract to *t*-butanol with 40 percent (w/v) ammonium sulfate demonstrated that a 1:0.5 ratio partitioned most of the protease activity into the aqueous phase (Chaiwut et al., 2010). This ratio of crude extract to *t*-butanol was used in a two-step TPP method using 20 percent (w/v) ammonium sulfate in the first step that gave 253 percent recovery of protease activity in the aqueous phase. Re-extraction of the latter phase with the same amount of *t*-butanol and 55 percent (w/v) ammonium sulfate partitioned a substantial amount of protease into the interphase (89 percent recovery) with 165 percent recovery in the aqueous phase. Zingibain, the cysteine protease from a *Zingiber officinale* rhizome extract was also mostly partitioned into the aqueous phase using a 1:1 ratio of extract: *t*-butanol ratio and 50 percent (w/v) ammonium sulfate (Gagaoua et al., 2015). However, a two-step TPP was not compared for possible partitioning of zingibain into the interfacial phase. As shown in Table 7.1, the addition of a *t*-butanol volume equal to 30 percent of the crude extract (1:0.5 ratio) often results in the best yield and purification-fold for numerous proteases from different sources. Three phases were formed when test proteins were used with varying proportions of ammonium sulfate and *t*-butanol (Pike and Dennison, 1989b). However, the least protein was extracted into the interfacial phase at the lowest *t*-butanol levels. It is therefore prudent to determine the least-cost proportions of ammonium sulfate and *t*-butanol for industrial applications.

7.3.2 Choice of salt and concentration of salt used for TPP

A comparison of a number of different sulfates and chlorides used in TPP for the isolation of cucumisin from *Cucumis melo* fruit latex (Gagaoua et al., 2017) showed that ammonium sulfate was much more effective in terms of the percentage relative cucumisin activity extracted than potassium, sodium or magnesium sulfate. Using sodium and potassium chloride in TPP resulted in only one third enzyme activity. These results may be explained in terms of the relative positions of the cations and anions in the Hofmeister series (Okur et al., 2017) that places NH_4^+ ahead of K^+, Na^+ and Mg^{2+} regarding their effects on the protein solute (increased surface tension, harder to make a cavity, decrease in solubility of hydrocarbons, salting out, decreasing protein denaturation and increasing protein stability). The sulfate anion has the equivalent position as the ammonium cation, followed by $H_2PO_4^{2-}$, acetate, citrate and chlorine. Surprisingly, sodium citrate gave slightly better recovery of alkaline protease activity from farmed catfish viscera (143 percent and 4.65-fold purification) in comparison to that with ammonium sulfate (134 percent and 3-fold purification), followed by potassium phosphate in TPP (Ketnawa et al., 2014). In contrast to the one-step purification of cucumisin using TPP that gave a single protein band on reducing SDS-PAGE as well as in a casein-substrate gel, several alkaline proteases co-purified under the optimized conditions for TPP on catfish viscera. An earlier study on TPP isolation of alkaline proteases from giant catfish (*Pangasianodon gigas*) also resulted in co-purification of several protease activities using 50 percent (w/v) ammonium sulfate as the electrolyte (Rawdkuen et al., 2012).

In order to optimize of the amount of ammonium sulfate used in TPP, a constant ratio of crude *Calotropis procera* (apple of Sodom, family Apocynaceae) stem latex extract to *t*-butanol of 1:0.5 was combined with different amounts of ammonium sulfate. Although 50 percent ammonium sulfate gave the highest protease activity recovery (182 percent) in the interfacial phase (middle phase), protease purification-fold was only 0.95. *Re*-extraction of the aqueous phase, obtained with 30 percent ammonium sulfate, by increasing the ammonium sulfate to 60 percent (w/v) in the presence of the same amount of *t*-butanol, resulted in 132 percent protease activity recovery and a 7-fold purification (Rawdkuen et al., 2010). When individual standard proteins in solution were used to optimize the amount of ammonium sulfate required in TPP, it was shown that there is an inverse relationship between the amount of ammonium sulfate required to precipitate 90 percent of the protein and its molecular weight (Pike and Dennison, 1989b). However, for the isolation of proteases from crude extracts, the concentration of ammonium sulfate needs to be optimized in each case using suitable activity and protein quantitation assays. This general observation had the following exceptions: proteases of

60, 20 and 10 kDa co-purified from a shrimp (*Litopenaeus vannamei*) hepatopancreas extract using 30 percent ammonium sulfate and equal volumes of extract and *t*-butanol (Senphan and Benjakul, 2014), while 23 kDa ficin was isolated from *Ficus carica* using 40 percent (w/v) ammonium sulfate and 1:0.75 latex extract to *t*-butanol ratio.

7.3.3 Effect of pH and pI on TPP

The pH at which TPP is conducted has a profound effect on protein precipitation. At pH-values below the protein's isoelectric point (pI), proteins carry a net positive charge, thus facilitating divalent sulfate anion binding to effect salting out. The concomitant macromolecular tightening and conformational contraction explain the strong pH dependence of the other five mechanisms of sulfate salting out of proteins: ionic strength effects, kosmotropy, cavity surface tension enhancement, osmotic stressor (dehydration) and protein exclusion-crowding agent (Dennison and Lovrien, 1997). When TPP was carried out at different pH values on individual standard proteins, the greatest amount of protein was partitioned into the interfacial phase at pH-values below the pI where the protein would have a positive charge. In contrast, the partitioned protein was more soluble when TPP was conducted close to its pI (Pike and Dennison, 1989b). When a protease with a known pI was isolated using TPP at a pH below the pI, such as cathepsin L, pI ~ 5, from liver homogenates at pH 4.2, the cysteine protease was extracted into the interfacial phase (Coetzer et al., 1995; Pike and Dennison, 1989a; Pike et al., 1992). This was also the case during the isolation of ficin, pI 8.9, from the Mediterranean fig (*Ficus carica*) using TPP at pH 7 (Gagaoua et al., 2014) and using a pH close to the pI to isolate a subtilase (pI 4.6) from *Phaseolus vulgaris* L. (bean) leaves (Popovič et al., 2002). However, when TPP was conducted at pH 6 to isolate papain (pI 8.9) from *Carica papaya* latex, the protease partitioned into the bottom aqueous phase from where it was concentrated using ultrafiltration, producing a homogenous 23 kDa preparation in a single step (Hafid et al., 2020). It has also been predicted that should TPP be conducted at a pH much higher than the pI, the protein would have a net negative charge and be partitioned into the lower aqueous phase. This was the case when the cysteine protease zingibain, pI 4.38, was isolated from *Zingiber officinale* Roscoe rhizomes using TPP at pH 7 (Gagaoua et al., 2015). The 33.4 kDa zingibain partitioned exclusively into the aqueous phase and showed a single band of activity on a casein substrate gel while contaminating smaller proteins were extracted into the interfacial phase and were reported to have no proteolytic activity. Fibrinolytic proteases, pI 6–7, isolated from *Bacillus sphaericus* MTCC 3672 fermentation broth also partitioned into the bottom aqueous phase during TPP conducted at pH 9 (Avhad et al., 2014).

When TPP conditions are optimized for the isolation of proteases from starting material such as plant or animal tissue homogenates, plant latex, bacterial lysates or fermentation broth, that comprise mixtures of proteases with different (and unknown) pI-values, it is important to consider a range of pH values. To this end, either the extract is diluted with a buffer of the required pH and of sufficiently high concentration in mM to maintain the desired pH, or the pH of the extract is adjusted using a 500 mM NaOH or HCl solution. Following the optimisation of the concentration of ammonium sulfate and ratio of crude extract to *t*-butanol required for the isolation of a protease using TPP, the effect of pH over a range of pH-values (typically pH 3 to pH 8) on the percentage protease recovery and fold-purification is evaluated. If the pI of the protease to be isolated is known, it is recommended that a range of pH values of 2–4 units below the pI are investigated first. Should a protease be sensitive to a very low pH or higher pH, the addition of ammonium sulfate or *t*-butanol before pH adjustment would provide protection against denaturation (Dennison and Lovrien, 1997).

7.3.4 Effect of temperature on TPP

A distinct advantage of TPP is that it can be conducted at room temperature (ca. 25 °C) as opposed to using C1 and C2 alcohols and acetone for protein precipitation that needs to be done at very low temperatures to protect proteins against denaturation by these organic solvents (Dennison, 2003). Furthermore, *t*-butanol exerts its kosmotropic and crowding effects at room temperature and slightly above, in contrast to the C1 and C2 alcohols (Dennison and Lovrien, 1997).

In those studies where TPP was conducted at a range of temperatures, it was consistently found that around ambient temperature is optimal for the recovery and fold-purification of proteases from crude extracts (Gagaoua et al., 2015, 2017; Kaur et al., 2019). Although the fold-purification of alkaline proteases from farmed giant catfish viscera was higher when TPP was conducted at 37 and 45 °C than at 25 °C or ambient temperature (27–33 °C), the percentage recovery of protease at the latter temperatures was higher (Ketnawa et al., 2014). The authors elected to conduct TPP at 25 °C since they argued that the apparent increase in purification fold could be due to a combination of the removal of contaminating proteins and activation of the protease during TPP (see later). They obtained a 5.4-fold purification and 219 percent recovery of alkaline proteases using the optimized conditions comprising 50 percent (w/v) sodium citrate with a 1:0.5 (v/v) ratio of crude extract to *t*-butanol at pH 8.0 and 25 °C that showed proteolytic activity on casein substrate gels at 120, 39, 24, and below 18.2 kDa. Conducting TPP at 37 °C for the isolation of nattokinase (subtilisin NAT) from a *Bacillus natto* NRRL-3666 fermentation supernatant also resulted in a small increase in purification-fold compared to 25 °C (Garg and Thorat, 2014). In practice, *t*-butanol at 30 °C (above its melting point of 25 °C) is commonly added to the crude extract obtained by centrifugation at 4 °C in a ratio of crude extract:*t*-butanol of 1:0.5 and incubated at room temperature for an hour to effect partitioning.

Depending on the intended application for the isolated or homogenously purified protease, TPP is used as a one/two-step isolation procedure by itself, or it is followed by a number of chromatography steps. In the latter instances TPP provides an additional advantage in that the interfacial phase precipitate contains very little ammonium sulfate, obviating a need to dialyse before conducting ion exchange chromatography (Dennison and Lovrien, 1997; Pike and Dennison, 1989a). In a few instances, as detailed below, a single TPP step was sufficient to isolate a protease or protease inhibitor to homogeneity as evidenced by SDS-PAGE. In these cases TPP was conducted directly on a crude extract or was preceded by precipitation of proteins in the homogenate with 60 percent (w/v) ammonium sulfate, followed by extensive dialysis of the redissolved precipitate before TPP was conducted.

7.3.5 Parasite proteases as drug and diagnostic targets

Enzymatic characterisation of protease drug targets underpins the development of chemotherapeutic agents for parasitic diseases (Ferreira and Andricopulo, 2017; Giordani et al., 2016). To this end, catalytic site variants of the cathepsin L-like protease of the protozoan parasite, *Trypanosoma congolense*, that causes nagana in cattle, were recombinantly expressed in *Pichia pastoris* (Pillay et al., 2010). Three-phase partitioning provided an efficient upstream purification and concentrating step for proteases from the large volume of yeast expression supernatant. Conducting TPP at a pH below the pI of the protease led to the partitioning of the respective catalytic site variants into the interfacial phase, thereby providing the concentrating effect. In addition to >80 percent enzyme recovery, 11 and 2.8-fold purification were obtained for the two proteases. Further purification to electrophoretic homogeneity, required for enzyme kinetics studies or the production of antibodies, was achieved with anion and molecular exclusion chromatography. In earlier studies on proteases of *T. brucei* brucei, the model trypanosome that infects animals but that is not pathogenic, TPP was used as the first isolation step to partition a cysteine and serine protease from trypanosome lysates into the interfacial phase, away from a large number of trypanosome proteins (Troeberg et al., 1996). In a first step, TPP was conducted using 10 percent (w/v) ammonium sulfate to partition contaminants into the interfacial phase that was discarded. The amount of ammonium sulfate was then increased to 25 percent (w/v) to partition the proteases into the interfacial phase (indicated as 10–25 percent (w/v) ammonium sulfate in Table 7.1). Further purification of this fraction using different types of chromatography allowed for separation and purification of the cathepsin L-like cysteine protease (trypanopain or *Tbb*CATL) and the serine protease oligopeptidase B (*Tbb*OPB) as a prequel to their characterisation and studies on their interaction with a range of inhibitors in the context of identifying drug leads (Morty et al., 2000; Troeberg et al., 1999). In the context of developing a diagnostic test for animal African trypanosomiasis (nagana), the native cathepsin L-like protease (*Tvi*CATL or vivapain) was isolated from *T. vivax* lysates and the recombinant catalytic domain from *P. pastoris* expression supernatant using TPP (Eyssen et al., 2018; Vather, 2010). Once again TPP served as a useful concentrating step for the isolation of a protease from the large volume of the heterologous expression system supernatant and only an additional molecular exclusion chromatography step was required to remove high molecular mass contaminants. The purified recombinant *Tvi*CATL catalytic domain was used in an indirect antibody detection enzyme-linked immunosorbent assay (ELISA) to discriminate between *T. vivax*-infected and non-infected cattle sera. For the isolation of proteases from trypanosome lysates or plasma from trypanosome-infected rats (Table 7.1), TPP was first conducted using 10 percent (w/v) ammonium sulfate in the presence of 30 percent *t*-butanol (lysate or plasma:*t*-butanol ratio of 1:0.5). This step partitioned contaminating proteins into the interfacial phase, while the protease activity remained in the aqueous phase. A further increase of the amount of ammonium sulfate to 20–30 percent partitioned the protease activity into the interfacial phase that was subjected to further purification using chromatography steps that included cation exchange and molecular exclusion chromatography for the native form and only the latter for the recombinant form.

7.3.6 Plant proteases for milk-clotting in cheese making

Plant proteases are attractive alternatives to calf abomasum derived rennet for milk-clotting during cheese making, both from a supply and consumer (such as cultural, ethical or religious concerns or Lacto-vegetarian diet) point of view. The main protease in rennet is the aspartic protease chymosin (EC 3.4.23.4) that hydrolyzes κ-casein to initiate the clotting process (Banks and Horne, 2003). However, several plant cysteine and serine proteases were shown to have the same substrate specificity and are useful milk-clotting proteases (Ben Amira et al., 2017; Gagaoua and Hafid, 2016; Shah et al., 2014). Cucumisin, a subtilisin clan serine protease isolated from *Cucumis melo* juice or sarcocarp, has been used as an effective milk-clotting protease. Three-phase partitioning provided a much-simplified method for the isolation of cucumisin compared to ammonium sulfate precipitation, followed in most instances by cation exchange chromatography, and in some cases also one or more molecular exclusion chromatography steps (see Table 3 in Gagaoua et al. (2017). The percent recovery of cucumisin using TPP was almost three-fold higher than that of the best of the reported classical methods, and this one-step isolation method resulted in a homogenous 68 kDa protease that showed effective milk-clotting properties. Zingibain, a papain-like cysteine protease from *Zingiber officinale* (ginger, Family Zingiberaceae) rhizomes with milk-clotting activity, was purified using a single-step TPP method (Gagaoua et al., 2015). Several lengthier isolation methods that included acetone fractionation, ammonium sulfate precipitation and ion-exchange chromatography resulted in half or less of the percentage zingibain recovery. Apart from one study that did not show electrophoretic homogeneity of the isolated protease, the purification-fold of the 33.8 kDa TPP-purified zingibain was 1.5 to 20-fold higher. Although three serine proteases (95, 91 and 83 kDa) co-purified in a one-step TPP isolation procedure from *Wrighta tinctoria* (Family Apocynaceae) stem tissue, this preparation was an effective milk-clotting enzyme. It was particularly active in hydrolyzing the κ-casein subunit that is a critical step in forming the curd during cheese making (Rajagopalan and Sukumaran, 2018).

7.3.7 Bacterial proteases for laundry detergents and anti-inflammatory agents

Proteases, mainly those active and stable over a broad temperature and pH range, are essential constituents of laundry detergents. Several microorganisms such as *Bacillus spp* produce subtilisin family serine proteases with these properties that require isolation from large volumes of fermentation broth. Three-phase partitioning provided a cost-effective and straightforward purification method that could be used directly on crude bacterial culture supernatants (Table 7.1). One-step TPP purification procedures, that could be done at ambient temperature, were optimized for the isolation of alkaline protease activity from a *Bacillus* sp. ICTF2 strain, giving a 15-fold purification and 176 percent yield from the aqueous phase (Kaur et al., 2019) and from *B. pumilus* MP27 with a 16.7-fold purification and 187.5 percent yield from the interfacial phase (Baweja et al., 2017). Prior to TPP, the *B. pumilus* culture supernatant was concentrated five-fold by ultrafiltration using a 30 kDa cut-off membrane. The isolated alkaline proteases were optimally active at 40 °C and were shown to retain on average 75 percent activity in the presence of several commercial detergents, surfactants and oxidizing agents.

Proteases with fibrinolytic activity were isolated from bacteria using ultrasound-assisted three-phase partitioning (UATPP) for possible use as analgesics and anti-inflammatory drugs. The extracellular metalloprotease, serratiopeptidase (serralysin) was isolated from a cell-free *Serratia* marcescens NRRLB-23,112 broth (Pakhale and Bhagwat, 2016) and a fibrinolytic protease from *Bacillus sphaericus* MTCC 3672 (Avhad et al., 2014). A comparison of TPP and UATPP for the isolation of the 50 kDa serratiopeptidase showed that in addition to more than double the purification fold (9.4 compared to 4.2) and a higher yield (96 compared to 83 percent), UATPP isolation was completed in five minutes, compared to an hour for TPP. Additionally, an 80 kDa contaminating protein band remaining after TPP was removed in the UATPP step.

7.3.8 Fish and bacterial proteases in food biotechnology

The demand for proteolytic enzymes in the pharmaceutical and food biotechnology industries sparked an interest in isolating proteases from the by-products from fish processing as well as bacterial fermentation broth using a scalable, rapid and non-chromatographic method such as TPP (Yan et al., 2018). Casein substrate electrophoresis gels showed that the four proteolytic activities of 120, 39, 34 and 18 kDa were enriched in the interphase following TPP on a farmed giant catfish (*Pangasianodon gigas*) viscera extract (Ketnawa et al., 2014; Rawdkuen et al., 2012). The extracted alkaline protease preparation hydrolyzed soy, egg white and whey proteins in a concentration-dependent manner at pH 8 and 37 °C. It showed a broader substrate specificity and specific activity than commercial bovine trypsin. The catfish alkaline proteases show potential for producing protein hydrolysates, including those from legumes that contain trypsin inhibitors.

Subtilisin NAT (also called nattokinase) is a profibrinolytic enzyme that enhances blood clot lysis by tissue-type plasminogen activator by cleaving and inactivating plasminogen activator inhibitor (Urano et al., 2001). Three-phase partitioning proved to be a more effective purification method for nattokinase from a *Bacillus natto* NRRL-3666 fermentation supernatant than either ammonium sulfate precipitation or ultrafiltration (Garg and Thorat, 2014). The 28 kDa protease was purified to homogeneity in a single step at a pH below its pI. The protease partitioned into the interfacial phase giving a 5.6-fold purification and yield of 129 percent.

7.3.9 Protease inhibitors

Three-phase partitioning was also applied to the isolation of endogenous protease inhibitors from soybean flour and finger millet (ragi) seeds (Table 7.1). Protease inhibitors find application as affinity ligands for the purification of proteases and in biotechnology for the inhibition of unwanted proteolysis. Protease inhibitors are classified into families of homologous inhibitors based on the sequence similarity to that of the type inhibitor of the family (Rawlings et al., 2017). A 21 kDa soybean trypsin inhibitor (SBTI) was isolated from a soybean flour extract using metal-assisted TPP that is based on the affinity of His-, Cys-and Trp-residues in proteins for divalent cations, Cu^{2+}, Zn^{2+}, and Ni^{2+} (Roy and Gupta, 2002). Only a four-fold purification was initially achieved when TPP was conducted with 50 percent (w/v) ammonium sulfate and equal volumes of the extract and *t*-butanol. Since 91 percent of the inhibitor activity remained in the aqueous phase after conducting TPP with 30 percent (w/v) ammonium sulfate, this fraction was extensively dialyzed against water (pH adjusted to 7) to remove the salt before metal-assisted TPP was conducted. This method involved adding different concentrations of either $ZnSO_4$ or $CuSO_4$, followed by either one or two volumes of *t*-butanol. Soybean trypsin inhibitor partitioned into the interface with both metals and two volumes of *t*-butanol leading to a 12-fold purification (66 percent yield) with 0.18 mM $ZnSO_4$ and 13-fold purification (73 percent yield) with 0.20 mM $CuSO_4$. When Cu^{2+}or Zn^{2+}, was added to the aqueous phase, followed by two volumes of *t*-butanol, the metal cations pushed SBTI, that is rich in His-residues, out of solution to precipitate. A second round of TPP conducted with either Cu^{2+}- or Zn^{2+}-sulfate on the aqueous phase of the first TPP step purified 21 kDa SBTI to homogeneity as shown by reducing SDS-PAGE

Three-phase partitioning was used to isolate a bifunctional trypsin-like protease/amylase inhibitor from a finger millet (ragi) seed extract that was heat-treated to inactivate endogenous α-amylase activity (Saxena et al., 2007). The inhibitor remained in the aqueous phase after the first round of TPP using 30 percent (w/v) ammonium sulfate and an equal volume of extract to *t*-butanol. Conditions were optimized for the second TPP step on the aqueous phase that was first extensively dialyzed to remove the salt. The bifunctional inhibitor precipitated in the interfacial phase during the second TPP step using 60 percent ammonium sulfate and an aqueous phase to *t*-butanol ratio of 1:2. A 16-fold purification of the trypsin- and a 20-fold purification of the amylase inhibitory activity was obtained and yields of 31.46 and 39.5 percent respectively. By comparison, standard ammonium sulfate precipitation only gave 2.4- and 3.3-fold purification of the trypsin and amylase inhibitory activity respectively. This was borne out by the presence of multiple co-purifying proteins in the ammonium sulfate fraction analyzed by reducing SDS-PAGE in contrast to the 14 kDa single inhibitor protein band.

7.4 Effect of TPP on protease structure and activity

A particularly useful characteristic of TPP-isolated enzymes, including proteases, is their apparent enhanced activity (Lovrien et al., 1987; Roy and Gupta, 2004; Shah and Gupta, 2007). The effect of TPP on proteinase K structure was determined following the observation that the specific activity of the protease was increased by 210 percent after TPP (Singh et al., 2001). Purified proteinase K, a serine protease, was subjected to TPP (30 percent, w/v, ammonium sulfate in 50 mM acetate buffer, pH 6 to *t*-butanol ratio of 1:2) at 25 °C before crystallization and collection of X-ray diffraction data. Whereas the overall structure of the TPP-treated protease was similar to that of the untreated protease in an aqueous medium and the hydrogen-bonding in the catalytic triad was the same, there were notable differences in the substrate-binding site. A number of the side chains, of the amino acid residues that form the substrate-binding pocket, adopted several conformations because of the two-fold higher temperature (B) factor of the TPP-treated protease than that of the original protease. This suggests that the TPP-treated protease has a more flexible structure, and this is responsible for the enhanced activity. Similar findings were reported for TPP-treated alpha-chymotrypsin that showed a higher catalytic efficiency (2.7-fold increase in k_{cat}/K_m) and substrate conversion rate than the untreated protease (Roy and Gupta, 2004). The TPP-treated protease also showed higher activity in low water containing solvents that suggested a more flexible structure. A later study also showed that TPP-treated alpha-chymotrypsin undergoes minor structural changes that led to aggregation but no change in tertiary structure (Rather et al., 2012). However, the peptidolytic activity of alpha-chymotrypsin was enhanced three-fold in the aggregates. The authors ascribed the enhanced activity to an increase in flexibility of the protease after TPP treatment.

In a study to explore the effect of TPP on the structure of several enzymes, including the serine proteases proteinase K, alpha-chymotrypsin and subtilisin Carlsberg, a number of physicochemical properties were examined (Rather and Gupta, 2013). Following TPP, the percentage hydrolytic activity recovered from the interfacial phase for each of these proteases was 186, 182 and 386. Although in general all the tested enzymes, including the proteases, were more thermolabile after TPP, the effect on secondary structure was not uniform. The far-UV circular dichroism (CD) spectra showed that while proteinase K had an increase in α-helix and a decrease in β-sheet content after TPP, subtilisin Carlsberg experienced the opposite effects and alpha chymotrypsin showed very little change. There was, therefore not a consistent effect of TPP on protease structure, and it is appreciated that the interaction of both ammonium sulfate and *t*-butanol with proteins is complex and depend on the individual protein structures. It was also found that there was not much change in fluorescence emission spectra, surface hydrophobicity and hydrodynamic radii of the proteases after TPP treatment. These observations are indicative of minimal structural changes, no significant unfolding and possibly a slight tightening of the structures of alpha-chymotrypsin and subtilisin Carlsberg that showed very small decreases in hydrodynamic radii. The latter could be as a result of the shrinking effect of sulfate anion binding to cationic surface patches during TPP (Dennison and Lovrien, 1997). The increased flexibility of the substrate-binding site of proteinase K seemed to have been too subtle to affect the hydrodynamic radius of the protease. The authors concluded that the small localised differences in structure are just enough to make the structures of enzymes, including proteases, sufficiently less rigid to result in the observed higher specific activity recorded for TPP-isolated or -treated enzymes. The lack of significant conformational changes brought about by TPP emphasises its efficacy either upstream or downstream in protease isolation procedures where it is essential to maintain structural integrity.

7.5 Conclusions

The tour presented in this Chapter through the proteolytic activities isolated from a diverse range of organisms using TPP illustrates the universal applicability of this non-chromatographic protein isolation technique. Whether there is a need to isolate proteases from parasite lysates, or crude extracts of plant or animal tissue, or the large volumes of yeast or bacterial recombinant expression media or fermentation broth, conditions can easily be optimized to partition the protease of interest into the interfacial phase and thereby concentrate the protease. There are, however, also examples where contaminants were extracted into the interfacial phase, leaving the homogeneous protease preparation in the aqueous phase from where it was recovered by ultrafiltration. The conditions that need to be optimized is the percentage (w/v) of ammonium sulfate and the ratio of crude extract:t-butanol to effect optimal precipitation, the pH of the crude extract and the temperature at which TPP is conducted. Apart from a few cases such as when TPP was used to isolate the archetypal cysteine protease, papain, from *Carica papaya* latex (Hafid et al., 2020) or when metal-assisted TPP was used to isolate soybean trypsin inhibitor from soybean flour (Roy and Gupta, 2002), further chromatography steps are often required to purify proteases and protease inhibitors to homogeneity. The latter is mostly necessary for enzymatic characterization of proteases for biomedical applications such as diagnostics. Nevertheless, in each instance where TPP was used as an upstream step in the purification protocol, the high yield and fold purification underpinned the success of further downstream purification steps. An additional advantage is that several proteases showed enhanced activity after TPP, as reflected in the purification parameters. This may be ascribed to the more flexible structure that is the result of substrate binding pocket residues adopting different conformations as was shown for TPP-treated proteinase K (Singh et al., 2001). Three-phase partitioning is a cost-effective isolation method that is sufficiently gentle on protease structure to result in a stable preparation suitable for use in the target application such as biomedicine, biotechnology or the food and detergent industries.

References

ResearchReports 2020. Global Protease Market Size, Manufacturers, Supply Chain, Sales Channel and Clients, 2020-2026. Available at: https://www.360researchreports.com/global-protease-market-16192695 (accessed 25 September 2020).

Avhad, D.N., Niphadkar, S.S., Rathod, V.K., 2014. Ultrasound assisted three phase partitioning of a fibrinolytic enzyme. Ultrason. Sonochem. 21, 628–633.

Banks, J.M., Horne, D.S., 2003. CHEESES | Chemistry of Gel Formation. In: Caballero, B. (Ed.), Encyclopedia of Food Sciences and Nutrition (Second Edition). Academic Press, Oxford, pp. 1056–1062.

Baweja, M., Singh, P.K., Sadaf, A., Tiwari, R., Nain, L., Khare, S.K., Shukla, P., 2017. Cost effective characterization process and molecular dynamic simulation of detergent compatible alkaline protease from Bacillus pumilus strain MP27. Process Biochem. 58, 199–203.

Ben Amira, A., Besbes, S., Attia, H., Blecker, C., 2017. Milk-clotting properties of plant rennets and their enzymatic, rheological, and sensory role in cheese making: a review. Int. J. Food Prop. 20, S76–S93.

Bond, J.S., 2019. Proteases: history, discovery, and roles in health and disease. J. Biol. Chem. 294, 1643–1651.

Borbás, R., Turza, S., Szamos, J., Kiss, É., 2001. Analysis of protein gels formed by interfacial partitioning. Colloid Polym. Sci. 279, 705–713.

Chaiwut, P., Pintathong, P., Rawdkuen, S., 2010. Extraction and three-phase partitioning behavior of proteases from papaya peels. Process Biochem. 45, 1172–1175.

Coetzer, T.H., Dennehy, K.M., Pike, R.N., Dennison, C., 1995. Baboon (Papio ursinus) cathepsin L: purification, characterization and comparison with human and sheep cathepsin L. Comp. Biochem. Physiol. B. Biochem. Mol. Biol. 112, 429–439.

Culp, E., Wright, G.D., 2017. Bacterial proteases, untapped antimicrobial drug targets. J. Antibiot. 70, 366–377.

de Melo Oliveira, V., Carneiro da Cunha, M.N., Dias de Assis, C.R., Matias da Silva Batista, J., Nascimento, T.P., dos Santos, J.F., de Albuquerque Lima, C., de Araújo Viana Marques, D., de Souza Bezerra, R., Figueiredo Porto, A.L., 2020. Separation and partial purification of collagenolytic protease from peacock bass *(Cichla ocellaris)* using different protocol: Precipitation and partitioning approaches. Biocatal. Agric. Biotechnol. 24, 101509.

Dennison, C., 2003. Concentration of the extract. In: A Guide to Protein isolation, 2nd ed. Kluwer Academic Publishers, Dordrecht, The Netherlands, pp. 81–84.

Dennison, C., 2012. Three-phase partitioning. In: Tschesche, H. (Ed.), Methods in Protein Biochemistry. Walter de Gruyter GmbH & Co, Berlin/Boston.

Dennison, C., Lovrien, R., 1997. Three phase partitioning: concentration and purification of proteins. Protein Expr. Purif. 11, 149–161.

Dobrzycki, Ł., Socha, P., Ciesielski, A., Boese, R., Cyrański, M.K., 2019. Formation of crystalline hydrates by nonionic chaotropes and kosmotropes: case of piperidine. Cryst. Growth Des. 19, 1005–1020.

Eyssen, L.E., Vather, P., Jackson, L., Ximba, P., Biteau, N., Baltz, T., Boulangé, A., Büscher, P., Coetzer, T.H.T., 2018. Recombinant and native TviCATL from Trypanosoma vivax: enzymatic characterisation and evaluation as a diagnostic target for animal African trypanosomosis. Mol. Biochem. Parasitol. 223, 50–54.

Ferreira, L.G., Andricopulo, A.D., 2017. Targeting cysteine proteases in trypanosomatid disease drug discovery. Pharmacol. Ther. 180, 49–61.

Gagaoua, M., 2018. Chapter 8 - Aqueous methods for extraction/recovery of macromolecules from micro-organisms of atypical environments: a focus on three phase partitioning. In: Gurtler, V., Trevors, J.T. (Eds.), Methods in Microbiology. Academic Press, pp. 203–242.

Gagaoua, M., Boucherba, N., Bouanane-Darenfed, A., Ziane, F., Nait-Rabah, S., Hafid, K., Boudechicha, H.-R., 2014. Three-phase partitioning as an efficient method for the purification and recovery of ficin from Mediterranean fig (Ficus carica L.) latex. Sep. Purif. Technol. 132, 461–467.

Gagaoua, M., Hafid, K., 2016. Three phase partitioning system, an emerging non-chromatographic tool for proteolytic enzymes recovery and purification. Biosens. J. 5, 1000134.

Gagaoua, M., Hoggas, N., Hafid, K., 2015. Three phase partitioning of zingibain, a milk-clotting enzyme from Zingiber officinale Roscoe rhizomes. Int. J. Biol. Macromol. 73, 245–252.

Gagaoua, M., Ziane, F., Nait Rabah, S., Boucherba, N., Ait Kaki El-Hadef El-Okki, A., Bouanane-Darenfed, A., Hafid, K., 2017. Three phase partitioning, a scalable method for the purification and recovery of cucumisin, a milk-clotting enzyme, from the juice of Cucumis melo var. reticulatus. Int. J. Biol. Macromol. 102, 515–525.

Garg, R., Thorat, B.N., 2014. Nattokinase purification by three phase partitioning and impact of t-butanol on freeze drying. Sep. Purif. Technol. 131, 19–26.

Giordani, F., Morrison, L.J., Rowan, T.G., De Koning, H.P., Barrett, M.P., 2016. The animal trypanosomiases and their chemotherapy: a review. Parasitology 143, 1862–1889.

Gonzalez-Rabade, N., Badillo-Corona, J.A., Aranda-Barradas, J.S., Oliver-.S.alvador Mdel, C, 2011. Production of plant proteases in vivo and in vitro–a review. Biotechnol. Adv. 29, 983–996.

Hafid, K., John, J., Sayah, T.M., Domínguez, R., Becila, S., Lamri, M., Dib, A.L., Lorenzo, J.M., Gagaoua, M., 2020. One-step recovery of latex papain from Carica papaya using three phase partitioning and its use as milk-clotting and meat-tenderizing agent. Int. J. Biol. Macromol. 146, 798–810.

Jacobs, G.R., Pike, R.N., Dennison, C., 1989. Isolation of cathepsin D using three-phase partitioning in t-butanol/water/ammonium sulfate. Anal. Biochem. 180, 169–171.

Kaur, N., Gat, Y., Panghal, A., 2019. Cost-effective purification and characterization of an industrially important alkaline protease from a newly isolated strain of Bacillus sp. ICTF2. Ind. Biotechnol. 15, 20–24.

Ketnawa, S., Benjakul, S., Martínez-Alvarez, O., Rawdkuen, S., 2014. Three-phase partitioning and proteins hydrolysis patterns of alkaline proteases derived from fish viscera. Sep. Purif. Technol. 132, 174–181.

Lopez-Otin, C., Bond, J.S., 2008. Proteases: multifunctional enzymes in life and disease. J. Biol. Chem. 283, 30433–30437.

Lovrien, R., Goldensoph, C., Anderson, P.C., Odegaard, B., 1987. Three phase partitioning (TPP) via t-butanol: enzyme separation from crudes. In: Burgess, R. (Ed.), Protein Purification: Micro to Macro. A. R. Liss, New York, pp. 131–148.

McKerrow, J.H., Caffrey, C., Kelly, B., Loke, P., Sajid, M., 2006. Proteases in parasitic diseases. Annu. Rev. Pathol. 1, 497–536.

Morty, R.E., Authié, E., Troeberg, L., Lonsdale-Eccles, J.D., Coetzer, T.H., 1999. Purification and characterisation of a trypsin-like serine oligopeptidase from *Trypanosoma congolense*. Mol. Biochem. Parasitol. 102, 145–155.

Morty, R.E., Bulau, P., Pelle, R., Wilk, S., Abe, K., 2006. Pyroglutamyl peptidase type I from *Trypanosoma brucei*: a new virulence factor from African trypanosomes that de-blocks regulatory peptides in the plasma of infected hosts. Biochem. J. 394, 635–645.

Morty, R.E., Troeberg, L., Powers, J.C., Ono, S., Lonsdale-Eccles, J.D., Coetzer, T.H., 2000. Characterisation of the antitrypanosomal activity of peptidyl alpha-aminoalkyl phosphonate diphenyl esters. Biochem. Pharmacol. 60, 1497–1504.

Okur, H.I., Hladilkova, J., Rembert, K.B., Cho, Y., Heyda, J., Dzubiella, J., Cremer, P.S., Jungwirth, P., 2017. Beyond the Hofmeister Series: ion-Specific effects on proteins and their biological functions. J. Phys. Chem. B. 121, 1997–2014.

Pakhale, S.V., Bhagwat, S.S., 2016. Purification of serratiopeptidase from Serratia marcescens NRRL B 23112 using ultrasound assisted three phase partitioning. Ultrason. Sonochem. 31, 532–538.

Pike, R., Dennison, C., 1989a. A high yield method for the isolation of sheep's liver cathepsin L. Prep. Biochem. 19, 231–245.

Pike, R.N., Coetzer, T.H., Dennison, C., 1992. Proteolytically active complexes of cathepsin L and a cysteine proteinase inhibitor; purification and demonstration of their formation in vitro. Arch. Biochem. Biophys. 294, 623–629.
Pike, R.N., Dennison, C., 1989b. Protein fractionation by three phase partitioning (TPP) in aqueous/t-butanol mixtures. Biotechnol. Bioeng. 33, 221–228.
Pillay, D., Boulangé, A.F., Coetzer, T.H., 2010. Expression, purification and characterisation of two variant cysteine peptidases from Trypanosoma congolense with active site substitutions. Protein Expr. Purif. 74, 264–271.
Popovič, T., Brzin, J., 2007. Purification and characterization of two cysteine proteinases from potato leaves and the mode of their inhibition with endogenous inhibitors. Croat. Chem. Acta 80, 45–52.
Popovič, T., Puizdar, V., Brzin, J., 2002. A novel subtilase from common bean leaves. FEBS Lett. 530, 163–168.
Potempa, J., Pike, R.N., 2009. Corruption of innate immunity by bacterial proteases. J. Innate Immun. 1, 70–87.
Rajagopalan, A., Sukumaran, B.O., 2018. Three phase partitioning to concentrate milk clotting proteases from Wrightia tinctoria R. Br and its characterization. Int. J. Biol. Macromol. 118, 279–288.
Rather, G.M., Gupta, M.N., 2013. Three phase partitioning leads to subtle structural changes in proteins. Int. J. Biol. Macromol. 60, 134–140.
Rather, G.M., Mukherjee, J., Halling, P.J., Gupta, M.N., 2012. Activation of alpha chymotrypsin by three phase partitioning is accompanied by aggregation. PLoS One 7, e49241.
Rawdkuen, S., Chaiwut, P., Pintathong, P., Benjakul, S., 2010. Three-phase partitioning of protease from Calotropis procera latex. Biochem. Eng. J. 50, 145–149.
Rawdkuen, S., Vanabun, A., Benjakul, S., 2012. Recovery of proteases from the viscera of farmed giant catfish (Pangasianodon gigas) by three-phase partitioning. Process Biochem. 47, 2566–2569.
Rawlings, N.D., Barrett, A.J., Thomas, P.D., Huang, X., Bateman, A., Finn, R.D., 2017. The MEROPS database of proteolytic enzymes, their substrates and inhibitors in 2017 and a comparison with peptidases in the PANTHER database. Nucleic. Acids. Res. 46, D624–D632.
Rawlings, N.D., Bateman, A., 2021. How to use the MEROPS database and website to help understand peptidase specificity. Protein Sci. 30, 83–92.
Roy, I., Gupta, M.N., 2002. Three-phase affinity partitioning of proteins. Anal. Biochem. 300, 11–14.
Roy, I., Gupta, M.N., 2004. α-Chymotrypsin shows higher activity in water as well as organic solvents after three phase partitioning. Biocatal. Biotransformation 22, 261–268.
Roy, I., Sharma, A., Gupta, M.N., 2005. Recovery of biological activity in reversibly inactivated proteins by three phase partitioning. Enzyme Microb. Technol. 37, 113–120.
Salas, C.E., Gomes, M.T., Hernandez, M., Lopes, M.T., 2008. Plant cysteine proteinases: evaluation of the pharmacological activity. Phytochemistry 69, 2263–2269.
Saxena, L., Iyer, B.K., Ananthanarayan, L., 2007. Three phase partitioning as a novel method for purification of ragi (Eleusine coracana) bifunctional amylase/protease inhibitor. Process Biochem. 42, 491–495.
Schaller, A., 2004. A cut above the rest: the regulatory function of plant proteases. Planta 220, 183–197.
Senphan, T., Benjakul, S., 2014. Use of the combined phase partitioning systems for recovery of proteases from hepatopancreas of Pacific white shrimp. Sep. Purif. Technol. 129, 57–63.
Shah, M.A., Mir, S.A., Paray, M.A., 2014. Plant proteases as milk-clotting enzymes in cheesemaking: a review. Dairy Sci. Technol. 94, 5–16.
Shah, S., Gupta, M.N., 2007. Obtaining high transesterification activity for subtilisin in ionic liquids. Biochim. Biophys. Acta. Gen. Subj. 1770, 94–98.
Singh, R.K., Gourinath, S., Sharma, S., Roy, I., Gupta, M.N., Betzel, C., Srinivasan, A., Singh, T.P., 2001. Enhancement of enzyme activity through three-phase partitioning: crystal structure of a modified serine proteinase at 1.5 Å resolution. Protein. Eng. 14, 307–313.
Thomas, E.L., van der Hoorn, R.A.L., 2018. Ten Prominent Host Proteases in Plant-Pathogen Interactions. Int. J. Mol. Sci., 19.
Troeberg, L., Morty, R.E., Pike, R.N., Lonsdale-Eccles, J.D., Palmer, J.T., McKerrow, J.H., Coetzer, T.H., 1999. Cysteine proteinase inhibitors kill cultured bloodstream forms of Trypanosoma brucei brucei. Exp. Parasitol. 91, 349–355.

Troeberg, L., Pike, R.N., Morty, R.E., Berry, R.K., Coetzer, T.H.T., Lonsdale-Eccles, J.D., 1996. Proteases from Trypanosoma brucei brucei. Purification, characterisation and interactions with host regulatory molecules. Eur. J. Biochem. 238, 728–736.

Turk, B., 2006. Targeting proteases: successes, failures and future prospects. Nat. Rev. Drug Discov. 5, 785–799.

Urano, T., Ihara, H., Umemura, K., Suzuki, Y., Oike, M., Akita, S., Tsukamoto, Y., Suzuki, I., Takada, A., 2001. The profibrinolytic enzyme subtilisin NAT purified from Bacillus subtilis cleaves and inactivates plasminogen activator inhibitor type 1. J. Biol. Chem. 276, 24690–24696.

Vather, P., 2010. Vivapain: a cysteine peptidase from Trypanosoma vivax. University of KwaZulu-Natal, South Africa MSc.

Wati, R.K., Theppakorn, T., Benjakul, S., Rawdkuen, S., 2009. Three-phase partitioning of trypsin inhibitor from legume seeds. Process Biochem. 44, 1307–1314.

Yan, J.-.K., Wang, Y.-.Y., Qiu, W.-.Y., Ma, H., Wang, Z.-.B., Wu, J.-.Y., 2018. Three-phase partitioning as an elegant and versatile platform applied to nonchromatographic bioseparation processes. Crit. Rev. Food Sci. Nutr. 58, 2416–2431.

CHAPTER 8

Three phase partitioning of plant peroxidases

Yonca Avci Duman
Faculty of Arts and Sciences, Department of Chemistry, Kocaeli University, Umuttepe Campus, İzmit-Kocaeli, Turkey

Chapter outline

8.1 Peroxidases

Peroxidases exist in almost all living organisms. These are heme proteins; have molecular weights ranging from 30 to 150 kDa. Peroxidases (EC 1.11.1.X) belong to the class of oxidoreductases that contain ferriprotoporphyrin IX as the prosthetic group that catalyse the reduction of peroxides, such as hydrogen peroxide (H_2O_2) as well as the oxidation of a variety of both phenolic and nonphenolic compounds (e.g. catechol, pyrogallol, guaiacol, aromatic amines, indoles, sulfonates, etc.) in the presence of hydrogen peroxide (O'Brien, 2000; Hamid, 2009; Chanwun et al., 2013). Expressly, peroxidase activity occurs by oxidation of other co-substrates such as ferricyanides and ascorbate which are converted into harmless components (Chandra et al., 2011).

Three Phase Partitioning
DOI: https://doi.org/10.1016/B978-0-12-824418-0.00007-2

The catalytic reaction mechanism carried out by the peroxidase consists of three steps. In the first step, the porphyrin in the enzyme is oxidized by hydrogen peroxide to produce Compound I (Cpd I) which is an unstable intermediate product (Eq (8.1)). Later, Cpd II and free radical forms by reducing CpI with an appropriate electron donor (Eq (8.2)). Finally, the enzyme regenerates to the initial state and forms another radical by the reducing Cpd II with the second substrate (Eq (8.3)) (Kharatmol and Pandit, 2012).

$$Enz\left(Por - Fe^{III}\right) + H_2O_2 \rightarrow Cpd\ I\left(Por^{+} - Fe^{IV} = O\right) + H_2O \tag{8.1}$$

$$Cpd\ I\left(Por^{+} - Fe^{IV} = O\right) + AH_2 \rightarrow Cpd\ II\left(Por - Fe^{IV} - OH\right) + AH \tag{8.2}$$

$$Cpd\ II\left(Por - Fe^{IV} - OH\right) + AH_2 \rightarrow Enz\left(Por - Fe^{III}\right) + AH \tag{8.3}$$

8.2 Sources and functions of peroxidases

The peroxidases have been classified into heme containing and non-heme peroxidases (Pandey et al., 2017; Passardi et al., 2007a). As per PeroxiBase database, >80 percent of known peroxidase genes code for heme- peroxidases. Thiol peroxidase, alkylhydroperoxidase, NADH peroxidase are non-heme peroxidases and they account only a small proportion. Heme peroxidases have been subdivided into two superfamilies of peroxidase-cyclooxygenase superfamily (PCOXS, mammalian peroxidases) and the peroxidase-catalase superfamily (PCATS, non-mammalian peroxidases) (Passardi et al., 2007b; Zamocky and Obinger, 2010).

8.2.1 Mammalian peroxidases

Mammalian peroxidases differ from plant peroxidases with respect to the nature of prosthetic group, size, and amino acid sequence. Mammalian peroxidases have amino acids in the range of 576 to 738 and heme is covalently linked to the protein On the other hand, plant peroxidases have about 300 amino acids and the heme group is attached to the protein via non-covalent bonds only. Mammalian peroxidases protect the thyroid hormones by degrading peroxide. Plant peroxidases protect against biotic or abiotic stress and are also involved in the lignin and auxin metabolism. (O'Brien, 2000).

Important mammalian peroxidases include myeloperoxidase (MPO), eosinophil peroxidase (EPO), lactoperoxidase (LPO), thyroid peroxidase (TPO), Dual oxidase 1, and Dual oxidase 2 (DUOX1 and DUOX2). (Klebanof, 2005; Cheng et al., 2008). NOX, named for NADPH oxidase or DUOX for dual oxidase is the most reactive of a family of oxidases. The DUOX enzymes also contain a peroxidase domain. DUOX1 and 2 were first identified in the thyroid gland and named THOX1-2. THOX proteins are co-located with thyroid peroxidase in the apical membrane of human thyroid cells (De Deken et al., 2002), and these two enzymes catalyze thyroid hormone synthesis. Mutations in DUOX2 (THOX2) are known to either decrease or even completely stop the synthesis of thyroid hormone, thereby result in hypothyroidism. DUOX1 and DUOX2 are found in the salivary glands, rectum, trachea, and bronchi, and it has been suggested that these enzymes serve as sources of hydrogen peroxide for an LPO-mediated antimicrobial system on mucosal surfaces (Klebanoff, 2005; Moreno et al., 2002; Geiszt et al., 2003).

Myeloperoxidase (MPO) is an oxireductase (EC 1.11.1.7). This enzyme was first found by Agner who called it verdoperoxidase but now it is known as MPO. It is found mostly in neutrophils but it also occurs in monocytes though in lesser amounts. This heme containing enzyme in the presence of peroxide and halide ions, is effective in killing various microorganisms. It also catalyzes the formation of reactive oxygen intermediates, including hypochlorous acid (HOCl). The MPO/HOCl together play an important role in microbial disinfection by neutrophils. In addition, MPO has been known to be involved in tissue damage and the resulting inflammation in various inflammatory diseases. These results indicate MPO as a useful target for designing drugs for the treatment of inflammations (Tobler and Koeffler, 1991; Aratani, 2018).

Eosinophil peroxidase (EPO) is a heme containing major cationic protein of the cytoplasmic granules in human eosinophiles. It has been shown to be involved in the pathogenesis of cancer, asthma, and allergic inflammatory dysfunctions (Percopo et al., 2019; Ye et al., 2019).

Lactoperoxidase (LPO or salivary peroxidase) exists in many human exocrine secretions such as tears, milk, saliva, and vaginal fluid. As such it acts as a first defense against microorganisms that might enter the human body (Davies et al., 2008; Hillegass et al., 1990). Mammalian peroxidases display high sequence homology with the plant peroxidases (Jantschk et al., 2002). The human LPO has 712 amino acids and a very large number of these amino acids are also present in MPO. Main activity of lacto/salivary peroxidase is seen on the enamel of human teeth. Thyroid peroxidase catalyses the iodination of Tyr residues to form mono- and diiodotyrosines that plays an essential role in the thyroid function. Moreover, the thyroglobulin protein binds with the monoiodotyrosine to result in two important human hormones thyroxine and triiodothyronine (Davies et al., 2008).

8.2.2 Microbial peroxidases

Peroxidases from bacteria (*Bacillus sphaericus, Bacillus subtilis, Pseudomonas* sp., *Citrobacter* sp.), Cyanobacteria (*Anabaena* sp.), fungi (*Candida krusei, Coprinopsis cinerea, Phanerochaete chrysosporium*), actinomycetes (*Streptomyces* sp., *Thermobifida fusca*), and yeast are used in various industrial applications such as; textile dye degradation, paper-pulp industry, animal feed, degradation of pollutants from water, dye decolorization, sewage treatment, and as biosensors (Bansal, and Kanwar, 2013). Lignin peroxidase and manganese peroxidase are the fungal extracellular peroxidases that act on lignin polymers and play a critical role in the biodegradation of accumulated aromatic compounds. Besides, intracellular bacterial peroxidases are important defensive agents in protection against H_2O_2-induced cell damage (Koga et al., 1999). White-rot fungi are able to degrade lignin with the action of oxidative enzymes; primarily peroxidases. Brown-rot fungi can also degrade lignocellulose; during which lignin is modified (Granja-Travez et al., 2020). The use of the fungal enzymes at a large scale is constrained by challenges associated with their expression.

With the growing industrial applications there is an increasing interest in identifying lignolytic bacteria which contain lignin degrading enzymes. Up to now, a number of bacteria have been described to be lignin decomposers (Bugg et al., 2011; Ahmad et al., 2011). These include actinobacteria and proteobacteria which have an exclusive class of dye-decolorizing peroxidases (DyPs, EC1.11.1.19) (Brown et al., 2011, 2012). These enzymes are as good as the fungal oxidases for lignin degradation, but they are much simpler to obtain as their expression in active forms does not need post translational modification.

8.2.3 Plant peroxidases

The superfamily of plant peroxidases, which are heme-proteins can be further subdivided into three classes according to their function and cellular localization (Welinder et al. 1992). Class I peroxidases consist of both prokaryotes and eukaryotes and they are intracellular. Cytochrome c peroxidase (CCP; EC 1.11.1.5), ascorbate peroxidase (APX; EC 1.11.1.11) and catalase peroxidase (CP; EC 1.11.1.6) are in this class. They are important in combating oxidative stress by detoxification of ROS (H_2O_2) (Edreva et al., 1998; Skulachev, 1998; Passardi et al., 2007c; Shigeoka, 2002). The cytochrome c peroxidase (CCP)reduces hydrogen peroxide to water by taking reducing equivalents from cytochrome c. Ascorbate peroxidases (APx) are associated with hydrogen peroxide detoxification by using ascorbate as reducing agents, besides, in photo-defence of the chloroplasts and cytosol in higher plants (Sharp et al., 2003; Pandey et al., 2017). The catalase-peroxidases (CPs) are bi-functional and have been mostly reported in bacteria. CPs protect bacteria from oxidative stress by decomposing hydrogen peroxide and generating molecular oxygen (Bernroitner et al., 2009). Ascorbate peroxidases and cytochrome c-peroxidases appear to have evolved similarly (Smulevich et al., 2006; Yamada et al., 2002). Structurally, the class I peroxidases lack disulfide bonds, do not require calcium for their function and an endoplasmic reticulum signal sequence is absent. Lignin-decompositing peroxidases mostly belong to class II of non-mammalian heme-peroxidases, which are secreted by fungal peroxidases. These peroxidases are classified in four subgroups: (a) lignin peroxidases (LiP), (b) manganese peroxidases (MnP), (c) versatile peroxidases (VP), and (d) dyedecolorizing peroxidases (DyP). These are heme-containing and having a wide range substrate specificity in the presence of H_2O_2 towards phenolic, nonphenolic lignin substrates, and organic compounds (Hofrichter et al., 2010; Mnich et al., 2020). Classical secretory plant peroxidases (EC 1.11.1.7; Class III peroxidases; donor: H_2O_2

oxidoreductases) are also heme-containing enzymes and belong to the peroxidase-catalase superfamily (Lüthje et al., 2018; Zámocký et al., 2015). Soybean peroxidase (SBP), peanut peroxidase (PNP), horseradish peroxidases (HRP), etc. are members of Class III peroxidase (Hiraga et al., 2001; Caio and Dunand, 2009). These are involved in wide range of physiological processes such as lignification and suberization (El Mansouri et al., 1999), defense against pathogen infection (Vance et al., 1980; Fry, 1986), wounded healing (Larrimini and Rothstein, 1987; Hiraga et al., 2000; Roberts et al.; 2000) reactive oxygen species (ROS) (Mittler,2002; Asada, 1999; Pelligrineschi et al., 1995) auxin catabolism (Normanly, 1997), seed germination (Gaspar et al., 1982) fruit ripening (Chaubey and Malhotra, 2002) stress tolerance (Larkindale and Huag, 2004, 2004).

8.3 Biotechnological applications of plant peroxidases

Reduction of peroxides with electron-donating/accepting substrates makes peroxidases important in biotechnological applications. Some of the biotechnological applications of peroxidases have been reported as environmental, medical, agriculture, analytical, industries etc. (Sergeyeva et al., 1999; Adams, 1997; Jia et al., 2002). A number of the important applications of peroxidases are described below.

8.3.1 Pulp and paper industry

Lignin is a phenolic heteropolymer and causes problems in the use of lignocellulosic rich plant biomass for the pulp and paper industry. Therefore, the lignin has to be removed before the fabrication of well quality paper. The chemical process as a conventional approach for delignification produces various environmental pollutants. But, the enzymatic degradation of lignin as a biotechnological process constitutes a better option in this regard. The most significant lignin-degrading enzymes MnP, laccase, and LiP have been successfully used for bio pulping, biobleaching in the paper industry (Aehle, 2007; Regalado et al., 2004; Hatakka et al., 2003).

8.3.2 Bioremediation of phenolic compounds

Aromatic compounds such as phenols and its derivatives are frequently are useful in fiber bonding, wood composites, textiles, foundry resins, coatings, and abrasives. These compounds are known to be toxic as well as hazardous carcinogens and a major class of pollutants in wastewater from a number of food and chemical industries. The use of these toxic phenols is highly restricted in many countries and these must be removed from wastewater before their release in the environments (Karam and Nicell, 1997; Nicell et al., 1993). Peroxidases generate free-radicals from various aromatic pollutants and lead to their polymerization. This is potentially useful in bioremediation and wastewater treatment (Tatsumi et al., 1996; Bhunia et al., 2001; Wagner and Nicell; 2001; Wagner and Nicell, 2002; Klibanov et al., 1983; Bewtra et al., 1995). The use of horseradish peroxidase (HRP) for the bioremediation of wastewater containing phenols, cresols, and chlorinated phenols has been reported. The potential applications of soybean and turnip peroxidases for this have also been reported (Caza et al., 1999; Kinsley and Nicell, 2000; Kennedy et al., 2002).

8.3.3 Decolourization of industrial dyes

Dyes are synthetic aromatic compounds which are extensively used in paper printing, color photography, textile dyeing, and as an additive in petroleum products. Many of the synthetic dyes are biologically non-degradable and cause environmental problems (Forgiarini and de Souza et al., 2007; Spadaro et al., 1992). Although conventional methods such as chemical oxidation, reverse osmosis, and adsorption, can be used for their; these methods do have some disadvantages such as being expensive processes, difficulties of application, need for high energy necessity, and possible use/production of toxic substances. Accordingly, low in cost, energy consumption and more environment-friendly alternative methods like biotechnological degradation of dyes are now favored (An et al., 2002). Synthetic azo dyes such as orange II can be effectively decolourized by the Lignin peroxidases (LiPs) and manganese peroxidases (MnPs) (Chivukula et al., 1995). Remazol is another important azo dyes and HRP was reportedly successful in bioremediation of Remazol as well as phenol and substituted phenols even at the industrial level (Tatsumi et al., 1996; Bhunia et al., 2001).

8.3.4 Biosensor

Peroxidase- electrodes have been widely used for hydrogen peroxide and organic hydroperoxides analysis (Jia et al., 2002). Detection of glucose, alcohols, glutamate, choline with biosensors containing peroxidases co-immobilized with a hydrogen peroxide producing enzyme was reported (Ruzgas et al., 1996). HRP biosensors are used in several methods involving amperometric immunosensor, mass balance, potentiometric methods, photovoltaic spectroscopy, optical and chemiluminescent methods, etc. (Liu et al., 1997; Martin et al., 2002; Ghindilis et al., 1996; Wang et al., 2001; Choi et al., 2001; Rubtsova et al., 1998). Some other plant peroxidases like from potato, tobacco, peanut, soybean, etc. have been also used in the design of such biosensors. Because of its easy availability, high specific activity, and excellent electrochemical characteristics the sweet potato peroxidase may be advantageous for this application (Lindgren et al., 2000).

8.3.5 Analysis and diagnostic kits

Peroxidases are part of versatile diagnostic kits based on enzyme-conjugated antibody technology, because of their capability to produce stable chromogenic products (Krell, 1991). An example is the use of turnip root peroxidase in kits for detection of uric acid (Agostini et al., 2002). The mixture of cholesterol oxidase and cholesterol esterase, peroxidase is similarly useful in determining the levels of human serum cholesterol (Ragland et al., 2000; Malik and Pundir, 2002; Hirany et al., 1997). The amounts of

8-hydroxydeoxyguanosine and its analogues in urine have been estimated by using peroxidase based kits as these are useful in the detection of urinary and prostate cancers. In diabetes, hypoxia, and ischemia patients, the estimation of glucose and lactate has great importance in control of the pathologies. H_2O_2 responsive biosensors which use peroxidases have been useful in many such applications (Chiou et al., 2003).

8.3.6 Hair dyeing

Various enzymes like oxidases, peroxidases, etc. can be the best alternative to the harsh chemicals in hair dyeing applications. Conventional hair coloring dyes are synthesized with oxidative polymerization of dye precursors as phenols or aminophenols, etc. In this reaction, the H_2O_2 is used as initiator for the polymerization reaction but it decolorizes the natural hair pigment melanin and can cause hair damage (Pandey et al., 2007).

8.4 Three phase partitioning system

Enzymes, including peroxidases, have many industrial applications as described before. Unfortunately, purification cost of these valuable biocatalysts restricts their applications. Recent studies on enzyme purification have aimed at reducing of purification cost by reducing steps and process times. Three phase partitioning (TPP) is an important purification method. TPP uses *t*- butanol and ammonium sulfate to precipitate enzymes and proteins from aqueous solution. The simplicity of the method combined with its rapid result makes this a popular choice in large scale protein purification. TPP, a technique was firstly described by Lovrien's group (Tan and Lovrien, 1972). It has been used to purify several bio macromolecules in recent years. Conventional chromatographic purification methods are not simple and user friendly to be used for industrial applications. They are over costly, time-consuming and difficult to scale-up. Thus, a non-chromatographic approach, such as TPP, appears promising for large scale purification of peroxidases (Gagaoua and Hafid, 2016).

Fig. 8.1 illustrates steps in a classical purification protocol. Such protocols involve use of chromatography to obtain biomolecules in highly purified forms (Gote et al.; 2006; Bucar et al., 2013; Azmir et al., 2013; Kula et al., 1981). However, these chromatographic techniques have serious disadvantages like needing costly equipment requiring continuous maintenance, and time drain. Multiple steps require costly chemicals and involve high probability of contamination of the final desired product with some of the reagents. In light of this information, there is an increasing necessity in the research field for a time saving, simple, greener and economical method for obtaining large amounts of enzymes like peroxidase which is adequately pure for most of its applications (Duman and Kaya, 2013b).

An elegant non-chromatographic process, TPP, consists of mixing the crude protein extract with a salt (commonly ammonium sulfate, $(NH_4)_2SO_4$) and an organic solvent,

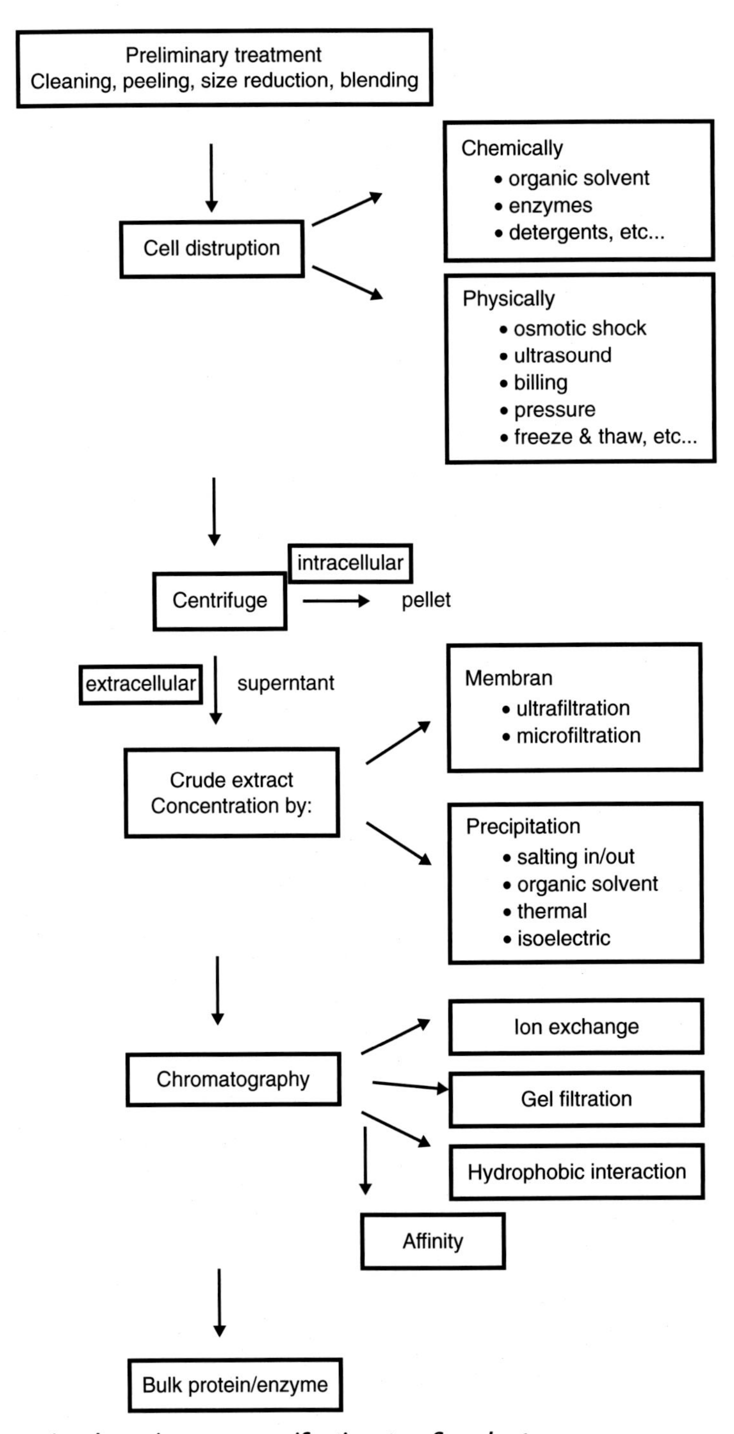

Fig. 8.1 ***Conventional protein-enzyme purification steps flow chart.***

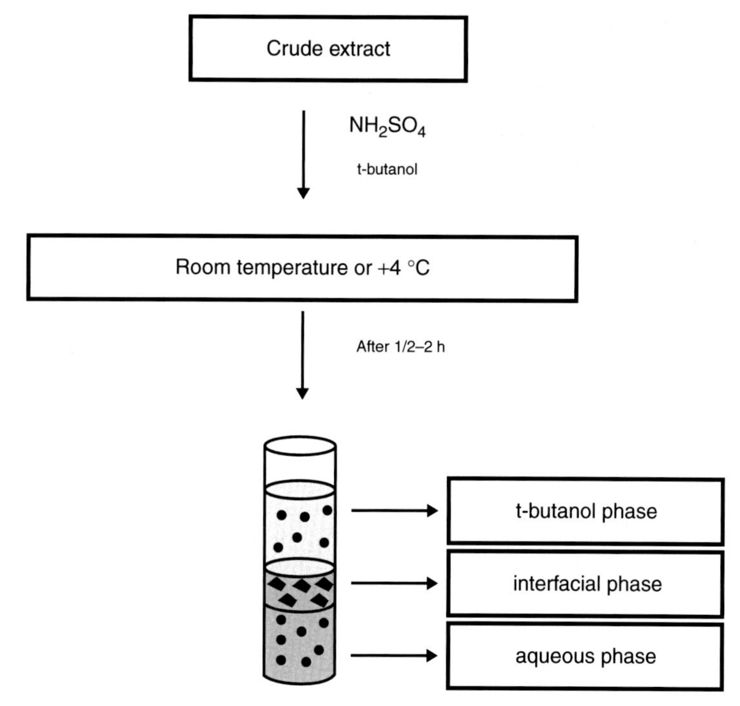

Fig. 8.2 ***Scheme for a typical TPP process.***

generally t-butanol (t-BuOH) which [when present in appropriate ratios] form three phases (Fig. 8.2).

The mixture of t-butanol and ammonium sulfate precipitates the protein out of the solution. During this process, the hydrophobic patches on the surface of the proteins bind with t-butanol to reduce the density of the proteins which makes them float above the denser aqueous salt phase. Approximately within an hour, an interfacial precipitate forms between the lower aqueous and upper organic phase that usually contains t-butanol (Dennison and Lovrien; 1997; Pike and Dennison, 1989). Use of TPP has been reported for the significant recovery and purification of a wide range of bioactive molecules, including proteins and enzymes (Dennison and Lovrien; 1997; Pike and Dennison, 1989; Szamos and Kiss, 1995; Roy and Gupta, 2002; Saxena et al., 2007; Wang et al., 2011). Separation and purification of proteins from the crude extracts by TPP is user and environment friendly because it is simple, quick, efficient, often a one-step process and does not need harmful chemicals. The synergetic effect of using both inorganic salt and t-butanol results in a number of phenomena operating simultaneously like i) salting out, ii) isoionic precipitation, iii) co-solvent precipitation, iv) osmolytic, and v) kosmotropic precipitation of proteins. TPP can be easily used on a large scale and can be applied directly with crude suspensions (Saxena et al., 2007; Yan et al., 2018). Phases are separated by centrifugation at 4 °C, the target biomolecules are selectively partitioned in the intermediate or bottom phases while contaminants such as unwanted proteins or carbohydrates to the other phases (Fig. 8.2). The upper organic phase containing non-polar compounds (like, lipids, etc.) is separated from the lower aqueous phase. (Gagaoua et al., 2014, 2015; Li, 2013; Rawdkuen, 2010; Rawdkuen et al., 2012). The upper organic phase is separated from the lower aqueous phase [containing polar molecules like proteins, carbohydrates etc.] by an interfacial protein precipitate (Chew et al., 2019; Duman et al., 2013a; Duman et al., 2013b; Duman et al., 2014; Gagaoua et al., 2014).

Although alcohols such as methanol, ethanol, 1-propanol, 2-propanol, and t-butanol are water-miscible, none of them dissolve the kosmotropic salts. Two liquid phases are formed (Chew et al., 2019). In biological and chemical processes, some ion effects in solution are specific to those ions. Several of these original ion effects follow the well-known Hofmeister series for anions and cations.

$$SO_4^{2-} > H_2PO_4^- > CH_3COO^- > Cl^- > NO_3$$

$$Mg^{+2} > Li^+ > Na^+ = K^+ > NH_4^+$$

The series also reflects their protein precipitation capacity (Nostro and Ninham, 2012). In the lyotropic series, before chlorine anions are kosmotropes (water-structure maker) whereas anions after chloride are chaotropic (water-structure breaker) (Yang, 1994; Yang et al., 1994). In the behavior of the kosmotropic effect of salt; as the ionic strength of the solution is increased, hydrophobic interactions will lead to proteins moving away from the water, causing them to cluster and collapse (salting-out). If a salt has a chaotropic effect due to low ionic strength, solubility will be increased by the hydrophilic interaction (salting in). The salting-out mechanism has been further discussed elsewhere was described (Duman, 2016).

8.4.1 Effect of salt on TPP

According to the Hoffmaster lyotropic series of ions $MgSO_4$ is the most preferred salt. However, ammonium sulfate, $(NH_4)_2SO_4$, is the most common salt used for protein salting-out in TPP because it has low cost, is readily available, stabilizes most of the proteins, and because of its high solubility even at low temperature. Moreover, $NH_4^+$and SO_4^{2-} are at the ends of their respective Hoffmeister series and have been shown to stabilize intermolecular interactions in macromolecules such as protein structures (Gagaoua and Hafid, 2016; Kiss et al., 1998; Roy and Gupta, 2002; Sharma and Gupta, 2001a; Szamos and Kiss, 1995). The salt has an important effect on the solubility of proteins, which makes the concentration of salt one of the major parameters in TPP. The ions of the salt act as counter-ions by binding to the anionic and cationic sites on the protein. Thus, at low concentration of the salts, protein solubility in water increases [salt-in phase]. However as the salt concentration increases, the hydration of the excess ions limits the availability of water molecules to the protein. Further beyond a point, the enough water molecules for keeping the protein in solution are not available any longer and salting out phase commences (Chew et al., 2019; Zhang, 2012).

8.4.2 Effect of t-butanol on TPP

The yield of the TPP process is affected by the concentrations of various alcohols such as n-propanol, n-butanol, isobutanol, and t-butanol. As mentioned above, t-butanol has been most frequently used in TPP by most of the workers. This increases the buoyancy of proteins by easily binding to them and thus supporting the floatation of proteins above the aqueous salt layer. t-Butanol has four carbons, is a non-ionic kosmotrope that is freely soluble in water. The added salt molecules results in its becoming a differentiating solvent. It does not cause denaturation of the partitioned enzyme. Its relatively larger size [as compared to lower alcohols] and spherical shape due to its branched structure hinders its diffusion into protein interiors. It also shows an important kosmotropic effect at 20–30 °C that improves the partitioning of enzyme and stabilizes the target protein (Dennison, 2011). Sulfate ion and t-butanol are together known to be excellent protein structure markers or kosmotropes (Dennison and Lovrien, 1997; Dennison, 2011; Moelbert et al., 2004).

8.4.3 Effect of temperature on TPP

Although the diffusion rate increases with increasing temperature, many enzymes are thermolabile even at moderate temperatures. The boiling point of the solvent also limits the temperature range which can be used in TPP. To avoid protein denaturation, TPP process is carried out at low temperatures generally in the range of +4 to 40 °C (Salvi and Yadav, 2020; Patil and Rathod, 2020; Kaur et al., 2020).

8.4.4 Effect of pH on TPP

The pH is considered to be a significant parameter in TPP. The ionization of amino acids involved in the protein is changed due to pH (Roy and Gupta, 2004; Sharma and Gupta, 2001; Roy et al., 2004). Thus partitioning of proteins is considerably dependent upon pH because of the electrostatic interactions among the phases and charged proteins (Duman and Kaya, 2013b). Proteins are reported to be more soluble during TPP when they are negatively charged (Chew et al., 2019). Different pH values will result in changes in the charge of the protein and this will also change the partitioning behavior.

8.5 TPP as an emerging technique for plant peroxidase purification

For the separation and purification of enzymes, the researchers aim to use environment-friendly, low cost and simple processes to achieve products with high purity. Three phase partitioning when compared to other techniques is an important method to reach this goal. Szamos and Hoschke (1992) reported a combination of two methods; ion-exchange chromatography and three-phase partitioning for the purification of HRP. This was purification of a peroxidase with a high fold purity using a few steps. TPP for HRP was studied using two ammonium sulfate saturation levels, first at 50 percent and then at 80 percent. Singh and Singh (2003) also reported purification and characterization of peroxidase from turnip. According to their study at large-scale, the turnip peroxidase purification by TPP was combined with metal affinity. This resulted in an 80.3-fold purification and a 20.3 percent yield of peroxidase. Narayan et al. (2008) studied extraction and purification of peroxidase from *Ipomoea palmata* by optimizing the parameters such as concentrations of t-butanol and ammonium sulfate and the temperature of extraction. TPP step was repeated twice. The ratio of the crude extract to t-butanol of 1:1 and 30 percent ammonium sulfate at 37 °C resulted in nearly 160 percent activity recovery and two-fold purification in the aqueous phase of the first cycle of TPP. At the second cycle of TPP, the purification fold increased to 18 times with about 81 percent activity recovery. The SDS analysis showed the molecular weight of peroxidase to be 20.1 kDa. Vetal and Rathod (2015) have shown the oranges peels' peroxidase purification by TPP for the first time. They have optimized the partitioning by using time, $(NH_3)_2SO_4$ saturation, pH, crude extract to t-butanol ratio, temperature, and speed of mixing. Fold purification and activity recovery with optimized parameters were reported as 18.20 and 93.96 percent, respectively. The optimization conditions were defined as; 50 percent ammonium sulfate saturation at pH 6 and temperature 30 °C with crude extracts to t-butanol ratio 1:1.5 (v/v), and 200 rpm for 80 min. The molecular weight of partially purified peroxidase was found 26 kDa by SDS PAGE analysis. The same researchers have evaluated the process with the ultrasound-assisted three-phase partitioning (UATPP) for the same enzyme. Optimization of $(NH_3)_2SO_4$ saturation, crude extract to t-butanol ratio, pH, temperature, ultrasonic frequency, ultrasonic power, and duty cycle were carried out. Optimization conditions were reported as, 50 percent $(NH_4)_2SO_4$ saturation, pH 6, 30 °C, enzyme extract to t-butanol ratio 1:1.5 (v/v), at 25 kHz frequency and 150 W ultrasonication power with 40 percent duty cycle, 6 min irradiation time. Under optimized conditions fold purification and activity recovery were reported as 24.28 and 91.84 percent, respectively. Researchers claimed that improved mass transfer with decreased operation time from 80 min to 6 min as compared to the conventional three-phase partitioning caused the higher purification fold of enzyme (Vetal and Rathod, 2016). Recently, Feltrin et al. (2017a) used TPP system and exclusion molecular chromatography to test and compare for purification of peroxidase from soybean meal. The experimental design methodology was used to elevate the enzyme extraction rate on mixing rate and time, pH, and extracting solvent volume. From this study, they found out 50 percent recovery and a 13.6 purification fold for 5 g soybean meal at 50 mL phosphate buffer 10 mmol L-1 pH 4.7, 60 min stirring rate at 100 rpm, an enzyme with a specific activity of 100 U mg-1. Their purified peroxidase had a more affinity to the substrate and reduced Deoxynivalenol levels by 80 percent; 20 percent more than the crude form of the enzyme. The same researchers have also investigated (Feltrin et al., 2017b) the recovery of peroxidase from rice bran with TPP for detoxification of deoxynivalenol (DON). In this case, they were able to recover 50 percent of soluble proteins with 5.7 purification fold at the interfacial phase by ammonium sulfate-acetone system. The ratio 1:2 for crude extract: acetone was the best condition with 30 percent ammonium sulfate saturation. Although in t-butanol is majorly used as an organic solvent in TPP; however Panadare and Rathod (2017) were the first researchers who employed dimethyl carbonate (DMC) in TPP as an alternative solvent to t-butanol. The molecular weight of DMC and t-butanol is not very different: 90.08 and 74.12, respectively. Similarly to t-butanol, DMC is also not expected to penetrate into the protein interior and hence does not denature proteins even when used in sufficient amounts so as to push the proteins into the middle layer during TPP. According to their result, extraction and purification of peroxidase can be applied efficiently in a single step with sodium citrate-DMC system using TPP process. They showed that optimization conditions were pH (7), temperature (40 C), salt concentration (20 percent), ratio of DMC: raw material (0.75:1), which resulted in peroxidase recovery and fold purification of 177 percent and 4.84, respectively. In a study by Yuzugullu et al. (2018), the purification and characterization of peroxidase from the medicinal plant, *Amsonia orientalis*, was described for the first time. Optimal purification parameters were found to be 20 percent (w/v) ammonium sulfate saturation at pH 6.0 and 25 °C with 1.0:1.0 (v/v) ratio of crude extract to t-butanol ratio for 30 min. According to their results, the activity recovery and fold purification for peroxidase was 162 percent and 12.5 at the bottom phase, respectively. They found the molecular mass of the enzyme as 59 kDa. K_m values of peroxidase were calculated as 1.88 and 2.0 mM for pyrogallol and hydrogen peroxide, respectively. Uckaya and Uckaya (2018) showed peroxidase purification from *Opuntia ficus indica* stem by a simple TPP system. Using this system, the intermediate phase was separated and optimized according to different process parameters as homogenate/t-butanol ratio (1:0.5), ammonium sulfate saturation (40 percent), pH (8), temperature (30 °C). The values of K_m (4.8 mM) and V_{max} (0.005 U/mL.min) with 2.25-fold purification and 106.64 percent recovery were reported. The molecular weight of the partitioned enzyme was found to be about 28 kDa. They demonstrated that this protocol using TPP was the better method as it did not require any column chromatography step.

8.6 Conclusion

Plant peroxidases have been applied in bioremediation of environmental pollutants such as phenolic compounds, delignification in paper and pulp industry, diagnostic kit development based on immunoassays, organic and polymer synthesis, and biosensor technology. Although having wide range industrial applications, unfortunately, the purification cost of these valuable biocatalysts restricts their uses in such areas. Three phase partitioning (TPP) is a good alternative for the purification and recovery for industrial important enzymes as a low cost, single-step process, and simple for purification and recovery. It has shown potential as an alternative to the conventional separation and purification methods for the recovery of proteins. This method can be well adapted to large-scale production by doing some optimizations. The recovery and fold purification of the protein depends on process parameters like pH, temperature, solvent and salt type, their concentrations and degree of mixing; characteristics of proteins of course also plays a role. Even so, the TPP technique is still an important method for the purification and recovery of biomolecules. Hence, the combination of other conventional steps with TPP system to improve the separation process appears promising. The applications of such an approach in the purification of enzymes like peroxidases especially for applications which do not require enzyme preparations of very high grade purity is a good illustration of what this technique can achieve.

References

Adams, J.B., 1997. Regeneration and the kinetics of peroxidase inactivation. Food Chem. 60, 201–206.

Aehle, W., 2007. Enzymes in Industry: Production and Applications. John Wiley & Sons.

Agostini, E., Hernández-Ruiz, J., Arnao, M.B., Milrad, S.R., Tigier, H.A., Acosta, M., 2002. A peroxidase isoenzyme secreted by turnip (Brassica napus) hairy-root cultures: inactivation by hydrogen peroxide and application in diagnostic kits. Biotechnol. Appl. Biochem. 35, 1–7.

Ahmad, M., Roberts, J.N., Hardiman, E.M., Singh, R., Eltis, L.D., Bugg, T.D., 2011. Identification of DypB from Rhodococcus jostii RHA1 as a lignin peroxidase. Biochemistry 50, 5096–5107.

An, S.Y., Min, S.K., Cha, I.H., Choi, Y.L., Cho, Y.S., Kim, C.H., Lee, Y.C., 2002. Decolorization of triphenylmethane and azo dyes by Citrobacter sp. Biotechnol. Lett. 24 (12), 1037–1040.

Aratani, Y., 2018. Myeloperoxidase: its role for host defense, inflammation, and neutrophil function. Arch. Biochem. Biophys. 640, 47–52.

Asada, K., 1999. The water-water cycle in chloroplast: scavenging of active oxygens and dissipation of excess photons. Annu. Rev. Plant Physiol. Plant Mol. Biol. 50, 601–639.

Azmir, J., Zaidul, I.S.M., Rahman, M.M., Sharif, K.M., Mohamed, A., Sahena, F., et al., 2013. Techniques for extraction of bioactive compounds from plant materials: a review. J. Food Eng. 117, 426–436.

Bansal, N., Kanwar, S.S., 2013. Peroxidase(s) in environment protection. Sci. World. J. 2013, 714639.

Bernroitner, M., Zamocky, M., Furtm€uller, P.G., Peschek, G.A., Obinger, C., 2009. Occurrence, phylogeny, structure, and function of catalases and peroxidases in cyanobacteria. J. Exp. Bot. 60, 423–440.

Bewtra, J.K., Biswas, N., Henderson, W.D., Nicell, J.A., 1995. Recent advances in treatment of selected hazardous wastes. Water Qual. Res. J. Canada 30, 115–126.

Bhunia, A., Durani, S., Wangikar, P.P., 2001. Horseradish peroxidase catalyzed degradation of industrially important dyes. Biotechnol. Bioeng. 72, 562–567.

Brown, M.E., Barros, T., Chang, M.C., 2012. Identification and characterization of a multifunctional dye peroxidase from a lignin-reactive bacterium. ACS Chem. Biol. 7, 2074–2081.

Brown, M.E., Walker, M.C., Nakashige, T.G., Iavarone, A.T., Chang, M.C., 2011. Discovery and characterization of heme enzymes from unsequenced bacteria: application to microbial lignin degradation. J. Am. Chem. Soc. 133, 18006–18009.

Bucar, F., Wube, A., Schmid, M., 2013. Natural product isolation–how to get from biological material to pure compounds. Nat. Prod. Rep. 30, 525–545.

Bugg, T.D., Ahmad, M., Hardiman, E.M., Singh, R., 2011. The emerging role for bacteria in lignin degradation and bio-product formation. Curr. Opin. Biotechnol. 22, 394–400.

Caza, N., Bewtra, J.K., Biswas, N., Taylor, K.E., 1999. Removal of phenolic compounds from synthetic wastewater using soybean peroxidase. Water Res. 33, 3012–3018.

Chandra, R., Abhishek, A., Sankhwar, M., 2011. Bacterial decolorization and detoxification of black liquor from rayon grade pulp manufacturing paper industry and detection of their metabolic products. Bioresour. Technol. 102, 6429–6436.
Chanwun, T., Muhamad, N., Chirapongsatonkul, N., Churngchow, N., 2013. Hevea brasiliensis cell suspension peroxidase: purification, characterization and application for dye decolorization. AMB Exp 3, 1–9.
Chaubey, A., Malhotra, B.D., 2002. Mediated biosensors. Biosens. Bioelectron. 17, 441–456.
Cheng, G., Salerno, J.C., Cao, Z., Pagano, P.J., Lambeth, J.D., 2008. Identification and characterization of VPO1, a new animal heme-containing peroxidase. Free Radic. Biol. Med. 45, 1682–1694.
Chew, K.W., Ling, T.C., Show, P.L., 2019. Recent developments and applications of three-phase partitioning for the recovery of proteins. Sep. Purif. Rev. 48, 52–64.
Chiou, C.C., Chang, P.Y., Chan, E.C., Wu, T.L., Tsao, K.C., Wu, J.T., 2003. Urinary 8-hydroxydeoxyguanosine and its analogs as DNA marker of oxidative stress: development of an ELISA and measurement in both bladder and prostate cancers. Clin. Chim. Acta 334, 87–94.
Chivukula, M., Spadaro, J.T., Renganathan, V., 1995. Lignin peroxidase-catalyzed oxidation of sulfonated azo dyes generates novel sulfophenyl hydroperoxides. Biochemistry 34, 7765–7772.
Choi, J.W., Lim, I.H., Kim, H.H., Min, J., Lee, W.H., 2001. Optical peroxide biosensor using the electrically controlled-release technique. Biosens. Bioelectron. 16, 141–146.
Cosio, C., Dunand, C., 2009. Specific functions of individual class III peroxidase genes. J. Exp. Bot. 60, 391–408.
Davies, M.J., Hawkins, C.L., Pattison, D.I., Rees, M.D., 2008. Mammalian heme peroxidases: from molecular mechanisms to health implications. Antioxid. Redox Signal. 10, 1199–1234.
De Deken, X., Wang, D., Dumont, J.E., Miot, F., 2002. Characterization of ThOX proteins as components of the thyroid $H_{(2)}O_{(2)}$-generating system. Exp. Cell Res. 273, 187–196.
Dennison, C., 2011. Three-phase partitioning. In: Tschesche, H. (Ed.), Three-phase partitioning. Methods in protein biochemistry, 1–5.
Dennison, C., Lovrien, R., 1997. Three phase partitioning: concentration and purification of proteins. Protein Expr. Purif. 11, 149–161.
Duman, Y.A., Kaya, E., 2013a. Three-phase partitioning as a rapid and easy method for the purification and recovery of catalase from sweet potato tubers (Solanum tuberosum). Appl. Biochem. Biotechnol. 170, 1119–1126.
Duman, Y.A., Kaya, E., 2013b. Purification, recovery, and characterization of chick pea (Cicer arietinum) β-galactosidase in single step by three phase partitioning as a rapid and easy technique. Protein Expr. Purif. 91, 155–160.
Duman, Y., 2016. Üçlü Faz Ayrımı (ÜFA) ile Geleneksel Enzim Saflaştırma Tekniğinin Karşılaştırılması; ÜFA ile Saflaştırılan β-Galaktosidazın Termodinamik Özellikleri. Iğdır Üniversitesi Fen Bilimleri Enstitüsü Dergisi 6, 107–117 In Turkish.
Duman, Y., Kaya, E., 2014. Purification and recovery of invertase from potato tubers (Solanum tuberosum) by three phase partitioning and determination of kinetic properties of purified enzyme. Turk. J. Biochem.y/Turk Biyokimya Dergisi 39, 443–448.
Edreva, A., Yordanov, I., Karkjieva, R., Georgieva, I.D., Gesheva, E., 1998. Involvement of peroxidases in heat shock responses of bean plants. Plant Peroxidase Newsletter 11, 27–34.
El Mansouri, I., Mercado, J.A., Santiago-Dómenech, N., Pliego-Alfaro, F., Valpuesta, V., Quesada, M.A., 1999. Biochemical and phenotypical characterization of transgenic tomato plants overexpressing a basic peroxidase. Physiol. Plant. 106, 355–362.
Feltrin, A.C.P., Fontes, M.R.V., Gracia, H.D.K., Badiale-Furlong, E., Garda-Buffon, J., 2017a. Peroxidase from soybean meal: obtention, purification and application in reduction of deoxynivalenol levels. Química Nova 40, 908–915.
Feltrin, A.C.P., Garcia, S.D.O., Caldas, S.S., Primel, E.G., Badiale-Furlong, E., Garda-Buffon, J., 2017b. Characterization and application of the enzyme peroxidase to the degradation of the mycotoxin DON. J. Environ. Sci. Health B 52, 777–783.
Forgiarini, E., de Souza, A.A.U., 2007. Toxicity of textile dyes and their degradation by the enzyme horseradish peroxidase (HRP). J. Hazard. Mater. 147, 1073–1078.
Fry, S.C., 1986. Cross-linking of matrix polymers in the growing cell walls of angiosperms. Annu. Rev. Plant Physiol. 37, 165–186.

Gagaoua, M., Boucherba, N., Bouanane-Darenfed, A., Ziane, F., Nait-Rabah, S., Hafid, K., Boudechicha, H.R., 2014. Three-phase partitioning as an efficient method for the purification and recovery of ficin from Mediterranean fig (Ficus carica L.) latex. Sep. Purif. Technol. 132, 461–467.

Gagaoua, M., Hafid, K., 2016. Three phase partitioning system, an emerging non-chromatographic tool for proteolytic enzymes recovery and purification. Biosens. J. 5, 135.

Gagaoua, M., Hoggas, N., Hafid, K., 2015. Three phase partitioning of a milk-clotting enzyme from Zingiber officinale Roscoe rhizomes. Int. J. Biol. Macromol. 73, 245–252.

Gaspar, T., Penel, C., Thorpe, T., Greppin, H., 1982. Peroxidases 1970-1980A Survey of Their Biochemical and Physiological Roles in Higher Plants. Universite de Geneve, Geneva.

Geiszt, M., Witta, J., Baffi, J., Lekstrom, K., Leto, T.L., 2003. Dual oxidases represent novel hydrogen peroxide sources supporting mucosal surface host defense. FASEB J. 17, 1502–1504.

Ghindilis, A.L., Atanasov, P., Wilkins, E., 1996. Potentiometric immunoelectrode for fast assay based on direct electron transfer catalyzed by peroxidase. Sens. Actuators B 34, 528–532.

Gote, M.M., Khan, M.I., Gokhale, D.V., Bastawde, K.B., Khire, J.M., 2006. Purification, characterization and substrate specificity of thermostable a-galactosidase from Bacillus stearothermophilus (NCIM5146). Process Biochem. 41, 1311–1317.

Granja-Travez, R.S., Persinoti, G.F., Squina, F.M., Bugg, T.D., 2020. Functional genomic analysis of bacterial lignin degraders: diversity in mechanisms of lignin oxidation and metabolism. Appl. Microbiol. Biotechnol. 104, 3305–3320.

Hamid, M., 2009. Potential applications of peroxidases. Food Chem. 115, 1177–1186.

Hatakka, A., Lundell, T., Hofrichter, M., Maijala, P., 2003. Manganese peroxidase and its role in the degradation of wood lignin In Applications of enzymes to ligninocellulosics. In: Mansfield, S.D., Saddler, J.N. (Eds.), Manganese peroxidase and its role in the degradation of wood lignin In Applications of enzymes to ligninocellulosics. ACS Symp. Ser. 855, 230–243.

Hillegass, L.M., Griswold, D.E., Brickson, B., Albrightson-Winslow, C., 1990. Assessment of myeloperoxidase activity in whole rat kidney. J. Pharmacol. Methods 24, 285–295.

Hiraga, S., Ito, H., Yamakawa, H., Ohtsubo, N., Seo, S., Mitsuhara, I., Ohashi, Y., 2000. An HR-induced tobacco peroxidase gene is responsive to spermine, but not to salicylate, methyl jasmonate, and ethephon. Mol. Plant Microbe Interact. 13, 210–216.

Hiraga, S., Sasaki, K., Ito, H., Ohashi, Y., Matsui, H., 2001. A large family of class III plant peroxidases. Plant Cell Physiol. 42, 462–468.

Hirany, S., Li, D., Jialal, I., 1997. A more valid measurement of low-density lipoprotein cholesterol in diabetic patients. Am. J. Med. 102, 48–53.

Hofrichter, M., Ullrich, R., Pecyna, M.J., Liers, C., Lundell, T., 2010. New and classic families of secreted fungal heme peroxidases. Appl. Microbiol. Biotechnol. 87, 871–897.

Jantschko, W., Furtmüller, P.G., Allegra, M., Livrea, M.A., Jakopitsch, C., Regelsberger, G., Obinger, C., 2002. Redox intermediates of plant and mammalian peroxidases: a comparative transient-kinetic study of their reactivity toward indole derivatives. Arch. Biochem. Biophys. 398, 12–22.

Jia, J., Wang, B., Wu, A., Cheng, G., Li, Z., Dong, S., 2002. A method to construct a third-generation horseradish peroxidase biosensor: self-assembling gold nanoparticles to three-dimensional sol− gel network. Anal. Chem. 74, 2217–2223.

Karam, J., Nicell, J.A., 1997. Potential application of enzymes in waste treatment. J. Chem. Technol. Biotechnol. 69, 141–153.

Kaur, G., Sharma, S., Das, N., 2020. Comparison of catalase activity in different organs of the potato (Solanum tuberosum L.) cultivars grown under field condition and purification by three-phase partitioning. Acta Physiol. Plant. 42, 10.

Kennedy, K., Alemany, K., Warith, M., 2002. Optimisation of soybean peroxidase treatment of 2, 4-dichlorophenol. Water Sa 28, 149–158.

Kharatmol, P.P., Pandit, A.B., 2012. Extraction, partialpurification and characterization of acidic peroxidasefrom cabbage leaves. J. Biochem. Technol. 4, 531–540.

Kinsley, C., Nicell, J.A., 2000. Treatment of aqueous phenol with soybean peroxidase in the presence of polyethylene glycol. Bioresour. Technol. 73, 139–146.

Kiss, E., Szamos, J., Tamas, B., Borbas, R., 1998. Interfacial behavior of proteins in three-phase partitioning using salt-containing water/tert-butanol systems. Colloids Surf. A Physicochem. Eng. Asp. 142, 295–302.

Klebanoff, S.J., 2005. Myeloperoxidase: friend and foe. J. Leukoc. Biol. 77, 598–625.
Klibanov, A.M., Tu, T.M., Scott, K.P., 1983. Peroxidase-catalyzed removal of phenols from coal-conversion waste waters. Science 221, 259–261.
Koga, S., Ogawa, J., Choi, Y.M., Shimizu, S., 1999. Novel bacterial peroxidase without catalase activity from Flavobacterium meningosepticum: purification and characterization. Biochim. Biophys. Acta 1435, 117–126.
Krell, H.W., 1991. Peroxidase: an important enzyme for diagnostic test kits. In: Lobarzewsky, J., Greppin, H., Penel, C., Gasparm, T. (Eds.), Biochemical, Molecular and Physiological Aspects of Plant Peroxidases. University M Curie, Lublind Poland and University of Geneva, Geneva Switzerland, pp. 469–478.
Kula, M.R., Kroner, K.H., Hustedt, H., Schutte, H., 1981. Technical aspects of extractive enzyme purification. Ann. N.Y. Acad. Sci. 369, 341–354.
Lagrimini, L.M., Rothstein, S., 1987. Tissue specificity of tobacco peroxidase isozymes and their induction by wounding and tobacco mosaic virus infection. Plant Physiol. 84, 438–442.
Larkindale, J., Huag, B., 2004. Thermotolerance and antioxidant systems in Agrosfis stolonifira: involvement of salicylic acid, abscisic acid, calciurn, hydrogen peroxide and ethylenc. J. Plant Physiol. 161, 405–413.
Li, Z., 2013. Simultaneously concentrating and pretreating of microalgae Chlorella spp. by three-phase partitioning. Bioresour. Technol. 149, 286–291.
Lindgren, A., Ruzgas, T., Gorton, L., Csöregi, E., Ardila, G.B., Sakharov, I.Y., Gazaryan, I.G., 2000. Biosensors based on novel peroxidases with improved properties in direct and mediated electron transfer. Biosens. Bioelectron. 15, 491–497.
Liu, H., Ying, T., Sun, K., Li, H., Qi, D., 1997. Reagentless amperometric biosensors highly sensitive to hydrogen peroxide, glucose and lactose based on N-methyl phenazine methosulfate incorporated in a Nafion film as an electron transfer mediator between horseradish peroxidase and an electrode. Anal. Chim. Acta 344, 187–199.
Lo Nostro, P., Ninham, B.W., 2012. Hofmeister phenomena: an update on ion specificity in biology. Chem. Rev. 112, 2286–2322.
Lüthje, S., Martinez-Cortes, T., 2018. Membrane-bound class III peroxidases: unexpected enzymes with exciting functions. Int. J. Mol. Sci. 19, 2876.
Malik, V., Pundir, C., 2002. Determination of total cholesterol in serum by cholesterol esterase and cholesterol oxidase immobilized and coimmobilized on to arylamine glass. Biotechnol. Appl. Biochem. 35, 191–197.
Martin, S.P., Lynch, J.M., Reddy, S.M., 2002. Optimisation of the enzyme-based determination of hydrogen peroxide using the quartz crystal microbalance. Biosens. Bioelectron. 17, 735–739.
Mnich, E., Bjarnholt, N., Eudes, A., Harholt, J., Holland, C., Jørgensen, B., Mikkelsen, J.D., 2020. Phenolic cross-links: building and de-constructing the plant cell wall. Nat. Prod. Rep. 37, 919–961.
Moelbert, S., Normand, B., De Los Rios, P., 2004. Kosmotropes and chaotropes: modelling preferential exclusion, binding and aggregate stability. Biophys. Chem. 112, 45–57.
Moreno, J.C., Bikker, H., Kempers, M.J., van Trotsenburg, A.S., Baas, F., de Vijlder, J.J., Vulsma, T., Ris-Stalpers, C., 2002. Inactivating mutations in the gene for thyroid oxidase 2 (THOX2) and congenital hypothyroidism. N. Engl. J. Med. 347, 95–102.
Narayan, A.V., Madhusudhan, M.C., Raghavarao, K S M S, 2008. Extraction and purification of Ipomoea peroxidase employing three-phase partitioning. Appl. Biochem. Biotechnol. 151, 263–272.
Nicell, J.A., Bewtra, J.K., Biswas, N., Taylor, E., 1993. Reactor development for peroxidase catalyzed polymerization and precipitation of phenols from wastewater. Water Res. 27, 1629–1639.
Normanly, J., 1997. Auxin metabolism. Physiol. Plant. 100, 431–442.
O'Brien, P.J., 2000. Peroxidases. Chem. Biol. Interact. 129, 113–139.
Panadare, D.C., Rathod, V.K., 2017. Extraction of peroxidase from bitter gourd (Momordica charantia) by three phase partitioning with dimethyl carbonate (DMC) as organic phase. Process Biochem. 61, 195–201.
Pandey, V.P., Awasthi, M., Singh, S., Tiwari, S., Dwivedi, U.N., 2017. A comprehensive review on function and application of plant peroxidases. Biochem. Anal. Biochem. 6, 308.
Passardi, F., Bakalovic, N., Teixeira, F.K., Pinheiro-Margis, M., Penel, C., et al., 2007c. Prokaryotic origins of the peroxidase superfamily and organellar-mediated transmission to eukaryotes. Genomics 89, 567–579.
Passardi, F., Theiler, G., Zamocky, M., Cosio, C., Rouhier, N., et al., 2007a. PeroxiBase: the peroxidase database. Phytochemistry 68, 1605–1611.

Passardi, F., Zamocky, M., Favet, J., Jakopitsch, C., Penel, C., et al., 2007b. Phylogenetic distribution of catalase-peroxidases: are these patches of order in chaos? Gene 397, 101–113.

Patil, S.S., Rathod, V.K., 2020. Synergistic effect of ultrasound and three phase partitioning for the extraction of curcuminoids from Curcuma longa and its bioactivity profile. Process Biochem. 93, 85–93.

Pelligrineschi, A., Kis, M., Dix, I., Kavanagh, T.A., Dix, P.J., 1995. Expression of horseradish peroxidase in transgenic tobacco. Biochem. Soc. Trans. 23, 247–250.

Percopo, C.M., Krumholz, J.O., Fischer, E.R., Kraemer, L.S., Ma, M., Laky, K., Rosenberg, H.F., 2019. Impact of eosinophil-peroxidase (EPX) deficiency on eosinophil structure and function in mouse airways. J. Leukoc. Biol. 105, 151–161.

Pike, R.N., Dennison, C., 1989. Protein fractionation by three-phase partitioning in aqueous/t-butanol mixtures. Biotechnol. Bioeng. 33, 335–339.

Ragland, B.D., Konrad, R.J., Chaffin, C., Robinson, C.A., Hardy, R.W., 2000. Evaluation of a homogeneous direct LDL-cholesterol assay in diabetic patients: effect of glycemic control. Clin. Chem. 46, 1848–1851.

Rawdkuen, S., 2010. Three-phase partitioning of protease from Calotropis procera latex. Biochem. Eng. J. 50, 145–149.

Rawdkuen, S., Vanabun, A., Benjakul, S., 2012. Recovery of proteases from the viscera of farmed giant catfish (Pangasianodon gigas) by three-phase partitioning. Process Biochem. 47, 2566–2569.

Regalado, C., García-Almendárez, B.E., Duarte-Vãzquez, M.A., 2004. Biotechnological applications of peroxidases. Phytochem. Rev. 3, 243–256.

Mittler, R., 2002. Oxidative stress, antioxidants and stress tolerance. Trends Plant Sci. 7, 405–410.

Roy, I., Gupta, M.N., 2002. Three-phase affinity partitioning of proteins. Anal. Biochem. 300, 11–14.

Roy, I., Sharma, A., Gupta, M.N., 2004. Three-phase partitioning for simultaneous renaturation and partial purification of Aspergillus niger xylanase. Biochim. Biophys. Acta 1698, 107–110.

Rubtsova, M.Y., Kovba, G.V., Egorov, A.M., 1998. Chemiluminescent biosensors based on porous supports with immobilized peroxidase. Biosens. Bioelectron. 13, 75–85.

Ruzgas, T., Csöregi, E., Katakis, I., Kenausis, G., Gorton, L., 1996. Preliminary investigations of an amperometric oligosaccharide dehydrogenase-based electrode for the detection of glucose and some other low molecular weight saccharides. J. Mol. Recognit. 9, 480–484.

Salvi, H.M., Yadav, G.D., 2020. Extraction of epoxide hydrolase from Glycine max using microwave assisted three phase partitioning with dimethyl carbonate as green solvent. Food Bioprod. Proc. 124, 159–167.

Saxena, L., Iyer, B.K., Ananthanarayan, L., 2007. Three phase partitioning as a novel method for purification of ragi (Eleusine coracana) bifunctional amylase/protease inhibitor. Process Biochem. 42, 491–495.

Sergeyeva, T.A., Lavrik, N.V., Rachkov, A.E., Kazantseva, Z.I., Piletsky, S.A., El'skaya, A.V., 1999. Hydrogen peroxide–sensitive enzyme sensor based on phthalocyanine thin film. Anal. Chim. Acta 391, 289–297.

Sharma, A., Gupta, M.N., 2001a. Purification of pectinases by three-phase partitioning. Biotechnol. Lett. 23, 1625–1627.

Sharma, S., Gupta, M.N., 2001b. Purification of phospholipase D from Dacus carota by three-phase partitioning and its characterization. Protein Expr. Purif. 21, 310–316.

Sharp, K.H., Mewies, M., Moody, P.C.E., Raven, E.L., 2003. Crystal structure of the ascorbate peroxidase–ascorbate complex. Nat. Struct. Biol. 10, 303–307.

Shigeoka, S., Ishikawa, T., Tamoi, M., Miyagawa, Y., Takeda, T., et al., 2002. Regulation and function of ascorbate peroxidase isoenzymes. J. Exp. Bot. 53, 1305–1319.

Singh, N., Singh, J., 2003. A method for large scale purification of turnip peroxidase and its characterization. Prep. Biochem. Biotechnol. 33, 125–135.

Skulachev, V.P., 1998. Cytochrome C in the apoptotic and antioxidant cascades. FEBS Lett. 423, 275–280.

Smulevich, G., Jakopitsch, C., Droghetti, E., Obinger, C., 2006. Probing the structure and bifunctionality of catalase-peroxidase (KatG). J. Inorg. Biochem. 100, 568–585.

Spadaro, J.T., Gold, M.H., Renganathan, V., 1992. Degradation of azo dyes by the lignin-degrading fungus Phanerochaete chrysosporium. Appl. Env. Microbiol. 58, 2397–2401.

Szamos, J., Kiss, E., 1995. Three-phase partitioning of crude protein extracts. J. Colloid Interf. Sci. 170, 290–292.

Szamos, J., Hoschke, A., 1992. Purification of horseradish peroxidase by three-phase partitioning (TPP). Acta Alimentaria 21, 253–260.

Tan, K.H., Lovrien, R., 1972. Enzymology in aqueous-organic cosolvent binary mixtures. J. Biol. Chem. 247, 3278–3285.

Tatsumi, K., Wada, S., Ichikawa, H., 1996. Removal of chlorophenols from wastewater by immobilized horseradish peroxidase. Biotechnol. Bioeng. 51, 126–130.

Tobler A., Koeffler H.P., 1991. Myeloperoxidase: localization, structure, and function. In Blood Cell Biochemistry (Harris, J.R., Ed.), Plenum Press, New York, Volume 3, pp. 255-288.

Uckaya, F., Uckaya, M., 2018. A rapid and efficient method for purification of peroxidase from Opuntia-ficus indica stem with decolorization effect. Glob. J. Sci. Front. Res. 18, 1–9.

Vance, C.P., Kirk, T.K., Sherwood, R.T., 1980. Lignification as a mechanism of disease resistance. Annu. Rev. Phytopathol. 18, 259–288.

Vetal, M.D., Rathod, V.K., 2015. Three phase partitioning a novel technique for purification of peroxidase from orange peels (Citrus sinenses). Food Bioprod. Proc 94, 284–289.

Vetal, M.D., Rathod, V.K., 2016. Ultrasound assisted three phase partitioning of peroxidase from waste orange peels. Green Process. Synth. 5, 205–212.

Wagner, M., Nicell, J.A., 2001. Peroxidase-catalyzed removal of phenols from a petroleum refinery wastewater. Water Sci. Technol. 43, 253–260.

Wagner, M., Nicell, J.A., 2002. Detoxification of phenolic solutions with horseradish peroxidase and hydrogen peroxide. Water Res. 36, 4041–4052.

Wang, H.H., Chen, C.L., Jeng, T.L., Sung, J.M., 2011. Comparisons of a-amylase inhibitors from seeds of common bean mutants extracted through three phase partitioning. Food Chem. 128, 1066–1071.

Wang, X.Q., Li, L.S., Van der Meer, B.W., Jin, J., Tang, D., Hui, Z., Li, T.J., 2001. Comparison of photovoltaic behaviors for horseradish peroxidase and its mimicry by surface photovoltage spectroscopy. Biochim. Biophys. Acta 1544, 333–340.

Welinder, K.G., Mauro, J.M., Norskov-Lauritsen, L., 1992. Structure of plant and fungal peroxidases. Biochem. Soc. Trans. 20, 337–340.

Yamada, Y., Fujiwara, T., Sato, T., Igarashi, N., Tanaka, N., 2002. The 2.0 Å crystal structure of catalase-peroxidase from Haloarcula marismortui. Nat. Struct. Biol. 9, 691–695.

Yan, J.K., Wang, Y.Y., Qiu, W.Y., Ma, H., Wang, Z.B., Wu, J.Y., 2018. Three-phase partitioning as an elegant and versatile platform applied to nonchromatographic bioseparation processes. Crit. Rev. Food Sci. Nutr. 58, 2416–2431.

Yang, Y., 1994. Modelling cationic dyeing-Consideration of ionic and hydrophobic interactions in a modified Donnan approach. J. Soc. Dye. Colour. 110, 98–103.

Yang, Y., Li, S., Lan, T., 1994. Ion sorption by polyamide with consideration of ionic interaction and other physical interactions. J. Appl. Polym. Sci. 51, 81–87.

Ye, L., Wang, H., Li, H., Liu, H., Lv, T., Song, Y., Zhang, F., 2019. Eosinophil peroxidase over-expression predicts the clinical outcome of patients with primary lung adenocarcinoma. J. Cancer. 10, 1032–1038.

Yuzugullu Karakus, Y., Acemi, A., Işık, S., Duman, Y., 2018. Purification of peroxidase from Amsonia orientalis by three-phase partitioning and its biochemical characterization. Sep. Sci. Technol. 53, 756–766.

Zámocký, M., Obinger, C., 2010. Molecular phylogeny of heme peroxidases. In: Torresl, E., Ayala, M. (Eds.), Biocatalysis Based on Heme Peroxidases. Springer, Heidelberg, pp. 7–35.

Zámocký, M., Hofbauer, S., Schaffner, I., Gasselhuber, B., Nicolussi, A., Soudi, M., Pirker, K.F., Furtmüller, P.G., Obinger, C., 2015. Independent evolution of four heme peroxidase superfamilies. Arch. Biochem. Biophys. 574, 108–119.

Zhang, J., 2012. Protein-protein interactions in salt solutions. In: Cai, W., Hong, H. (Eds.), Protein-Protein Interactions - Computational and Experimental Tools. InTech, pp. 359–376.

CHAPTER 9

Macro-(affinity ligand) facilitated three phase partitioning

Converting TPP into an affinity based process

Ipsita Roy[a], Munishwar Nath Gupta[b]

[a]Department of Biotechnology, National Institute of Pharmaceutical Education and Research (NIPER), S.A.S. Nagar, Punjab, India

[b]Former Emeritus Professor, Department of Biochemical Engineering and Biotechnology, Indian Institute of Technology, New Delhi, India

Chapter outline

9.1 Introduction

Three phase partitioning (TPP) has been most extensively used for isolation, concentration and partial purification of proteins/enzymes. The applications in other sectors though are also attracting increasing attention as indicated by various chapters in this book. Macro-(affinity ligand) facilitated three phase partitioning (MLFTPP) is a version of TPP in which affinity interaction has been incorporated. This chapter describes this underexploited format of TPP. It also includes an account of some success stories of its applications for protein purification. The next chapter describes applications of MLFTPP for refolding/reactivation of proteins/enzymes from their inactive forms. Before describing the process of MLFTPP, some further background to the importance of affinity based methods of protein purification is provided.

9.1.1 Importance of affinity interactions in protein purification

Both downstream and upstream methods generally used in case of proteins have been described in Chapter 3 of this book. Over the years, it has been recognized

Three Phase Partitioning
DOI: https://doi.org/10.1016/B978-0-12-824418-0.00006-0

that downstream component contributes a very high percent to the overall production costs of enzymes (Ghosh and Mattiasson 1993; Labrou 2014; Sadana and Beelaram 1994). In this context, there are a number of important generalizations which must be kept in mind. Firstly, many applications of enzymes in industry actually do not require very high purity (Roy and Gupta 2000). Hence many industrial grade enzymes do not even have the purity level of a single band during electrophoretic analysis. Such enzyme preparations are generally used straightaway in various industries. Enzymes have been increasingly used for organic synthesis and/or enantiomeric resolution of organic compounds during the last several decades (Carrea and Riva 2000; Shah and Gupta 2007; Thayer 2006). Most such studies also tend to use industrial preparations as such. That works out generally fine. However, in some cases, contaminating enzymes may lead to side reactions. In such cases, a high resolution purification method such as an affinity based method may be required. Also, pharmaceutical proteins require extremely high purity and an affinity step in some form or another is generally used in their production.

The second important point is that more the number of steps in protein purification, higher is the cost (Spears 1993; Straathof 2011). One simple reason for this is that no method ever gives complete recovery of activity. So, losses increase in a non-linear fashion. For this reason, the earlier approach of using affinity based approaches towards the end of the protocol has been abandoned (Mondal et al. 2006; Mondal and Gupta 2006). The lowering cost of affinity ligands has contributed a lot to make this possible. While earlier, such affinity ligands were generally coenzymes, inhibitors or costly proteins like lectins or antibodies, in many cases these have been largely replaced by dyes, metal ions, synthetic peptides, etc. (Labrou 2002; Nascimento et al. 2012; Scopes 1987; Teng et al. 1999). It must be however realized that often it is at the cost of selectivity of the affinity ligand. A monoclonal antibody is definitely more selective towards its target antigen (Moser and Hage 2010). A biomimetic dye, although more expensive, has far better selectivity as compared to its normal textile dye counterpart (Lowe et al. 1992). However, judicious screening of even relatively inexpensive affinity ligands can lead to a choice of an affinity ligand which shows adequate selectivity towards the desired protein. To sum up, there is a trade-off involved and hence it is wise to keep in mind the desired purity of the protein for the intended application before setting out to design a protein purification protocol!

A related third feature of affinity based methods has been abolition of the distinction between upstream and downstream phases. The commonest example of this is the use of fusion tags while cloning proteins. The kits based upon "Ni^{2+} column" (immobilized metal ion affinity chromatography (IMAC) media) have become the most frequently used purification step by molecular biologists (Kuo and Chase 2011; Porath 1992). The beads linked with affinity ligands for breaking the cells would be another good example of integration of downstream and upstream processes.

The last but not the least is the development of protocols for carrying out affinity interactions in free solutions. Use of antibodies to precipitate corresponding antigens (or vice versa) (Handlogten et al. 2014) or precipitation of dehydrogenases with dimers of coenzymes (Larsson and Mosbach 1979) have been early examples of such approaches. These suffer from the disadvantage that one has to determine the exact ratio of the two for obtaining maximum precipitation of the desired protein as the precipitation curves often have a sharp maximum. Hence the use of smart polymer-affinity ligands has been very useful in this regard; MLFTPP also uses such "polymeric affinity ligands". Apart from precipitation, affinity interactions in free solutions can also be carried out in two-phase extractions (Kamihira et al. 1992; Teotia and Gupta 2004).

Aqueous two phase systems (ATPS) were made more selective by use of a polymer-affinity ligand rather than just the polymeric phase (Johansson 1984). One of the large scale processes for purification of formate dehydrogenase used affinity partitioning (Cordes and Kula 1994). Arnold's group used metal affinity for isolation of heme proteins (Wuenschell et al. 1990).

Like all other precipitation techniques, TPP is a single theoretical plate process, i.e. its selectivity to discriminate between various proteins in a crude mixture of proteins is rather limited. Hence, significant level of purification observed by us in many case was a surprise. However, in most of our early attempts, the starting materials were industrial preparations of enzymes which, to start with, had much higher levels of target enzymes as compared to any contaminating proteins. Nevertheless, at least in one case, it was shown by our group that fold purification obtained was better than the use of ammonium salt fractionation (Sharma et al. 2000a). In many respects, this works better than ATPS (Albertsson 1971; Walter et el. 1991; Wu et al. 2017). ATPS either leaves the protein in salt solutions (and hence it needs concentration) or in the polymeric phase (from where it needs to be separated).

As TPP does not involve any matrix/polymer, the affinity component cannot be built-in a simple way. An early attempt was made by our group to introduce metal affinity interactions in TPP (Roy and Gupta 2002a). The addition of low amounts (non-precipitating level) of copper and zinc ions to the dissolved interfacial precipitate before a second cycle of TPP resulted in 13-fold purification of soybean trypsin inhibitor with 75 percent recovery of the starting activity in the crude extract of soybean flour.

However, TPP of water soluble polymers opened up the possibility of converting TPP into a general affinity based process for separation of proteins/enzymes.

9.2 Water soluble polymers and smart polymers

The use of water soluble polymers in protein separations is quite old. Both polyethylene glycol (PEG) and dextran have been known to be used for fractionation of proteins by precipitation (Ingham 1984; Iverius and Laurent 1967; Roy and Gupta 2000; Simpson 2006). Later, polyacrylic acid has also been used (Sternberg and Hershberger 1974).

9.2.1 Carrageenans

Red seaweeds contain a family of linear sulfated polysaccharides which are collectively known as carrageenans (Pereira and van De Velde 2011). These are different from another major class of sulfated polyanions known as heparinoids. The interest in these increased considerably after World War II as the food processing industry expanded and the search for food grade additives which could modify/control rheology as well as texture of the processed food materials expanded (Kanmani and Rhim 2014). However, these naturally occurring polymers have a long history. There used to be a village on the south Irish coast called Carraghen. The name of these polysaccharides is obviously derived from that village. Its inhabitants were reported to make flans by boiling milk with *Chondrus crispus* (a red seaweed, also known as Irish moss). Apart from that, carrageenans are also produced from the genus *Eucheuma, Gigartina* and *Iridaca*. Early applications of this polysaccharide were in beer clarification, textile sizing and as a thickener. Subsequent applications and emerging markets include food applications (for improving clarity of wines and vinegar, creating milk based drinks, such as chocolate drinks, of desirable viscosity and preventing phase separation in them, improving mouth feel of ice creams and inhibiting ice crystal formations in them, thickening sauces and dressings and controlling their viscosity, preventing dehydration of frozen poultry and preventing fat separation in meats) (Guiseley et al. 1980, Kanmani and Rhim 2014). It is also useful in pet foods, canned meat/fish, toothpastes, air fresheners, breads and noodles. Non-food applications include sectors such as paper, textile printing, flame retardants, pesticide formulations and in oil completion fluids (van de Welde and Ruiter 2002). It is also valuable as a component of industrial slurries and suspensions in many fields. The commercial applications lead to its large demand and production at a large scale. This in turn makes this available as an inexpensive polysaccharide for biotechnological applications.

These polysaccharides are building materials of cell walls of red seaweeds. There are three main classes of this polymer which are called kappa, lambda and iota (Barbeyron et al. 2000; Necas and Bartosikova 2013). All are sulfated galactans in which the linear chain has alternating units of 3-linked-beta-D-galactopyranose and 4-linked-alpha-D-galactopyranose. The kappa, iota and lambda have one, two and three sulfate groups per two successive monosaccharides. Some structural variants in which hydroxyl groups are modified are invariably present; any algal extract therefore is generally a mixture of such molecules. The three main classes of the polysaccharide differ in their solubility properties (Necas and Bartosikova 2013). As gelling is responsible for most of the applications, this is a useful way to classify carrageenans. Kappa is the most common class and sodium salt of this polysaccharide is soluble in hot water. Sodium salt of iota class polymer is soluble in both cold and hot water; the corresponding salt of lambda class has limited solubility in cold water but dissolves completely in hot water.

Let us look at kappa class more closely as almost all applications related to TPP are described with this class. While the sodium salt is soluble, addition of some other ions to the solution leads to formation of gels (Michel et al. 1997). A more appropriate description (in our context) will be the gelatinous precipitate. In general, precipitates are more hydrated than gels. While using precipitation, there is a trade off involved. Compact precipitates with less water are to be preferred as extra water may retain contaminating proteins. On the other hand, real gels in practice are difficult to deal with and dissociation of the bound protein is more difficult. So, the best ion is potassium as calcium produces brittle gels with kappa-carrageenan (Michel et al. 1997). In fact, if the application is immobilization (of proteins), higher concentrations of potassium ions have been used to lower the pore size of gels so that the proteins do not leach out (Roy and Gupta 2003a). Proteins/enzymes immobilized in kappa-carrageenan gels have been used both for biotransformations and bioanalysis (van de Velde et al. 2002).

Our laboratory used kappa-carrageenan in affinity precipitation of pullulanase (Roy and Gupta 2003a) and yeast alcohol dehydrogenase (Mondal et al. 2003a). The polymer showed selectivity for pullulanase as such, attachment of the dye Cibacron blue as an affinity ligand was used to separate the dehydrogenase. In both cases, potassium ion was used to precipitate the macro-affinity ligand. This smart polymer which is easily available as a food grade product, is inexpensive and easier to link an affinity ligand, would also be useful in MLFTPP.

9.2.2 Alginates

A major advance in our understanding of this important class of polysaccharides came from the work reported from the University of Trondheim, Norway, in late 1980s. This anionic polymer is mostly isolated from some brown macro-algae which occur in sea although bacterial alginates have also been characterized (Hay et al. 2013; Marinho-Soriano et al. 2006; Sutherland 1994). This is a polydispersed linear polymer consisting of D-mannuronic acid and L-guluronic acid. These monomers can occur in blocks or alternately and these sequences do influence the properties of the molecule. To start with, the pKa values of -COOH groups in mannuronic acid and guluronic acid are 3.38 and 3.65, respectively. While the polymer occurs as mixed salts of sodium, magnesium, calcium, it is extracted as the soluble salt of sodium. Given those pKa values, it can be precipitated by lowering the pH to below 2.0. This is again a food grade marine product and its applications in ice creams and many other processed food products, textile industries, paper industries, etc. has resulted in its easy availability in large quantities as an inexpensive product (Qin et al. 2018). Alginates were also the first material to be used for whole cell immobilization (Moreno-García et al. 2018; Smidsrød and Skjåk-Bræk 1990). Alginates are commercially used in entrapping probiotic cells (Cook et al. 2014). Their biocompatibility has resulted in applications as scaffolds in tissue engineering (Bidarra et al. 2014; Dar et al. 2002; Yang et al. 2002; Wang et al. 2003). Alginate beads are also used in designing controlled drug release systems (Lopes et al. 2017). In many such applications, conversion to the insoluble form by addition of calcium ions (which easily replace sodium ions) has been exploited (Smidsrød and Skjåk-Bræk 1990). The relative solubility of salts of various ions have been reported and even explored in some applications. As the calcium salt can be easily solubilized with EDTA or even by using common buffers like phosphate or citrate, it is definitely a reversibly soluble-insoluble class of smart polymer. Earlier reports of use of calcium alginate beads as affinity material for pectinase attracted our attention (Kester and Visser 1990). Subsequently, it was found useful in precipitation of not only pectinase (Gupta et al. 1993) but for lipase (Sharma and Gupta 2001) and amylases (Sharma et al. 2000b) as well. Its affinity towards many of these enzymes was also exploited in two phase affinity extractions (Jain et al. 2006; Teotia and Gupta 2001, 2004; Teotia et al. 2001) and expanded bed chromatography (Roy et al. 2004; Roy and Gupta 2002b, 2002c). Alginate (as mentioned briefly in Chapter 2 of this book) was, in fact, the first water soluble polymer used to establish proof-of-the concept of MLFTPP by purification of xylanase.

The fortuitous affinity of alginate towards so many different classes of enzymes is interesting. It is worth speculating why alginate should have this broad affinity. The charges on the polymer and two different kinds of uronic acids in its sequence probably create suitable sets of points of interactions with some of these proteins. These interactions collectively seem to be adequate to confer the necessary selectivity. The applications of this polysaccharide in tissue engineering and crafting of drug delivery systems have also emerged over the years (Banks et al. 2019; Pandolfi et al. 2017). We have also found that esterification (Sharma et al. 2004) as well as microwave treatment (Mondal et al. 2004a) can alter/modulate its selectivity towards some enzymes. Perhaps, more work on this material can bring out its versatility as an affinity material even further.

9.2.3 Chitosan

Unlike the two polymers discussed above, chitosan as such does not occur naturally. However, chitin, poly(N-acetyl glucosamine), occurs in crab shells, fungal cell walls and insect exo-skeletons (Deshpande 1986). Chitin is an important renewable resource and more than a dozen volumes devoted to it (which have been published continually with updates on its applications) testify to diverse applications in many sectors (Muzzarelli 1977; van den Broek and Boeriu 2019). Its deacetylation is easily carried out and liberation of free amino groups produces a range of partially deacetylated (> 80 percent) soluble product called chitosan (Shin-ya et al. 2001; Vårum et al. 1991). Many of the applications of chitin actually involve its use in chitosan form.

While most of the commercially produced polysaccharides are anionic, chitosan (due to the presence of free amino groups) is cationic in nature. Both chitin and chitosan are food grade polymers. Chitin (and chitosan) are biodegradable and can be easily sterilized (one of the early applications of chitin was in making surgical threads!) (Nakajima et al. 1986).

Chitosan becomes soluble below pH 6.5 (although the exact value can be modulated by changing buffer composition, etc.). It is a pH sensitive smart polymer as its solubility can be changed in a reversible fashion easily by altering pH (Du et al. 2015; Teotia et al. 2004). As already mentioned, this polysaccharide has many commercial applications in many diverse areas. Chitin is also biologically relevant as it is part of the defense system of plants against many pathogens.

Chitinases are an important class of enzymes. Lysozyme is also a chitinase. Many chitinases also degrade chitosan although more specific chitosanases have also been reported. The chitooligomers, obtained by enzymatic and nonenzymatic methods, are quite useful products. The smaller ones were used by John Rupley in establishing the mechanism of action of lysozyme (Rupley and Gates 1967).

Lectins are defined as carbohydrate binding proteins of non-immune origin which agglutinate many cells in a facile fashion, notably the red blood cells (because of which they have also been called haemagglutinins). One common way of classifying lectins is by their specificity of binding to carbohydrates. Chitin binding lectins (for example, in potato, tomato and rice) also bind to chitosan and an early application of chitosan in affinity precipitation was in purification of wheat germ lectin (Tyagi et al. 1996). Subsequently, the lectins from potato and tomato were also obtained by affinity precipitation. Chitosan has also been used as a macro-(affinity ligand) by incorporating it in PEG phase in two-phase affinity extractions (Teotia et al. 2006).

9.2.4 Eudragits

Unlike the above polymers, eudragit polymers used by us in bioseparations are synthetic polymers. Decades back, Rohm GmbH, Darmstadt introduced Eupergit class of carriers for applications in biotechnology. Eupergit C series became very well known for immobilization of enzymes and affinity chromatography (Boller et al. 2002; Katchalski-Katzir and Kraemer 2000). There was another class of polymers from the same company which attracted less attention from biotechnologists. These were Eudragit class of polymers; two of these, Eudragit L and Eudragit S, are better known as enteric coatings resistant to gastric fluids (Barbosa et al. 2019). These had attracted our attention because of our Indo-Swedish collaboration on affinity precipitation. Both are co-polymers of methacrylic acid and methylmethacrylate. In case of Eudragit S-100, the ratio of ester to acid side chains is 2:1; in Eudragit L-100, corresponding ratio is 1:1. That makes the former more hydrophobic in nature. Both become insoluble when their carboxyl groups become protonated and can no longer form H-bonds with water. Above that pH, these are soluble in aqueous buffers. Both, hence, are pH-sensitive reversibly soluble-insoluble class of synthetic polymers. As these are enteric polymers, both are non-toxic and very attractive for any biotechnological applications. Their original application also ensured that these were not costly and hence any process developed using these will not suffer cost-wise because of their choices (Roy and Gupta 2003b).

Eudragit S-100 is soluble above pH 5.5 and its precipitation is more or less complete by pH 4.7. The corresponding value for Eudragit L-100, the one with more free carboxyl (and less ester) groups, is 4.0. As more enzymes tend to be more stable around pH 4.7; the S-form has been found to be somewhat more useful. Search for better precipitation conditions have led to efforts to develop other precipitation strategies for these polymers. Notably, Eudragit S-100 could be totally precipitated at high concentrations of both polymer and Ca^{2+} ions at 25 °C after about 20 h (Guoqiang et al. 1995). Another set of conditions was higher temperature of 65 °C and moderate concentrations of calcium ions (50 mM). The free carboxyl groups in Eudragit polymers provide ready-made sites for linking affinity ligands. The well-known carbodiimide coupling with or without NHS (N-hydroxy succinimide) can be employed (Hermanson 2013). Now, any modification at these carboxyl groups which are critically involved in solubility of these polymers is expected to affect their solubility behavior. The conjugate of Eudragit S-100 with the dye Cibacron blue precipitated at a more moderate temperature of 40 °C in the presence of Ca^{2+} ions. Guoquiang et al. (1994) used these precipitation conditions to devise their "thermoprecipitation" strategy to purify lactate dehydrogenase (LDH) and pyruvate dehydrogenase from porcine muscle. The bound enzymes were neatly dissociated sequentially with different concentrations of salt. A similar approach was used by the same group in purifying yeast alcohol dehydrogenase (Gupta et al. 1996). Limited recycling was possible for the conjugate to be reused. While the textile dye and the polymer may be inexpensive, the conjugation step adds to the cost.

As with a few polysaccharides, Eudragit S-100 as such was found to show some selectivity in binding enzymes. Using low pH, xylanase from *T. viride* was precipitated as its complex with the polymer (Gupta et al. 1994). The recovery of activity was 89 percent and SDS-PAGE analysis showed one major band of the enzyme along with a lower molecular weight impurity. The enzyme activity could be released from the polymer by solubilizing the polymer-enzyme precipitate at pH 7.0 in phosphate buffer containing 1 M NaCl. Xylanase has numerous industrial applications such as clarification of fruit juices and wines, valorization of agricultural and forestry waste to obtain low molecular weight carbohydrates, improvements in rheological properties of cereal flours (Collins et al. 2005; Coughlan and Hazlewood 1993; Wong et al. 1988). Its major application, however, has been to replace the use of chlorine in pre-bleaching of pulp in paper manufacturing industry. For this purpose, it is critical that xylanase is free of any cellulase activity which may damage the cellulose for its subsequent use in paper industry (Kirk and Jeffries 1996). The electrophoretic analysis indicated that the purified preparation was free of any cellulase (Gupta et al. 1994).

We have already mentioned that this affinity interaction was the one which was used to develop MLFTPP; the advantage being that any pre-clarification of the feed is not necessary while using MLFTPP. With similar aims, affinity precipitation has been integrated with two-phase extractions. Eudragit S-100-Cibacron dye could be incorporated in the top PEG phase of a PEG-dextran two phase system (Guoqiang et al. 1994). While without this macro-(affinity ligand), the enzyme LDH from porcine muscle extract partitioned into the lower phase along with many other impurities; in the two-phase affinity system, the enzyme partitioned into the top phase by binding to Eudragit-dye conjugate. The conjugate with bound enzyme could be neatly precipitated from the PEG phase by lowering the pH. Furthermore, the enzyme could be easily dissociated from the polymer by treatment with 0.5 M NaCl (Guoquiang et al. 1994).

In some cases however, the dissociation of the enzyme from its complex with this or a similar polymer may not be so straight forward. Such a situation was encountered in the case of D-LDH from L. *bulgaricus* (Guoquiang et al. 1993). Elution of the enzyme was made possible by using the surfactants Tween 20 or Triton X-100. Of course, this necessitates removal of the surfactant bound to the protein for many applications. An inexpensive hydrophobic zeolite Y has been found to work well for this purpose (Blum and Eriksson 1991). In fact, the same zeolite could also be used to remove the surfactant bound to the polymer as well which allowed recycling of the carrier. It is noteworthy that purification of this LDH this way obviated the need for conjugating any affinity ligand to polymer.

Another very impressive integration of affinity precipitation with two-phase extraction was reported by Kamihara et al. (1992). Protein A is a valuable affinity ligand for purification for monoclonal antibodies. In fact, affinity chromatography on Protein A columns is one of the two/three step protocols which are used for purification of monoclonals. Recombinant Protein A was purified by using a conjugate of Eudragit S-100 and immunoglobulin G in a two-phase affinity system. Protein A was recovered after precipitating the complex of the polymer with bound Protein A. One advantage of such integrations has been that actual separation utilized two-phase systems. Scale up of such systems has been carried out in many cases and that experience makes it facile to carry out such protocols at commercial levels. In such systems, affinity precipitation constitutes a part of the recovery/elution step.

9.3 Smart biocatalysts

As mentioned above, most of these smart (reversibly soluble-insoluble polymers) have suitable functionalities so that affinity ligands can be attached to create desired macro-(affinity ligands). Identical chemistries can be used to link proteins/enzymes to such polymers. There are sufficient examples in the literature which show that this does not seem to abolish their property of being reversibly soluble-insoluble. This has made it possible to design smart biocatalysts (Roy et al. 2003; Roy and Gupta 2003c, 2006a). Such smart biocatalysts represent a special class of immobilized enzymes. The purpose of this section is not to describe applications of these smart biocatalysts but glean some insights from that area which are equally useful in designing macro-(affinity ligands) for MLFTPP.

Immobilized enzymes often show enhanced stability (as compared to the free enzyme) but even more important is the acquired trait of reusability. Conventional immobilized enzymes (with immobilization on solid supports) often suffer from issues of intra-particle mass transfer and mass transfer in the reaction media (Cabral and Kennedy 1993; Cao 2005; Küchler et al. 2016; Sheldon and van Pelt 2013). Soluble supports, in the form of smart polymers, avoid these issues as reactions can be carried out in free solution and the biocatalyst recovered for use after the catalytic cycle by precipitating it with a suitable stimulus (Roy et al. 2013). Useful data on this was provided by an excellent study of Arasaratnam et al. (2000). Their results showed that accessibility to trypsin linked to Eudragit S-100 by soybean trypsin inhibitor was much better in the soluble form of the bioconjugate as compared to the situation when the bioconjugate was present as a precipitate. Examples of polymers which can work well even in non-aqueous media in a smart way have been described. Some illustrative examples of these smart biocatalysts are discussed below. The relevance of this discussion can be further appreciated by the fact that similar conjugation chemistries and design principles apply in obtaining such smart biocatalysts and macro-(affinity ligands) for MLFTPP. Incidentally, increase in operational stability is also desirable in both immobilization as well as MLFTPP. After all, that is what becomes a major factor in reusability of the bioconjugate. Greater the reusability, lower is the overall operational cost. This is because conjugation is invariably a fairly costly step.

Often, these smart biocatalysts have shown higher stability than the corresponding free enzyme (Sardar et al. 2000). In fact, use of such water soluble polymers was explored much before the concept of smart polymers came into existence. Lysozyme covalently linked to alginic acid via cynaogen bromide coupling showed remarkable reusability in hydrolysis of *M. lysodectikus* cells; the biocatalyst retained 75 percent activity even after seven cycles (Charles et al. 1974). One early design lesson was learnt during the attempt to use trypsin linked to a highly negatively charged synthetic polymer for hydrolysis of the positively charged substrate protein. Because of the repulsion between the substrate and the polymer, the proteolytic activity of the biocatalyst was found to be very low (van Leemputten and Horisberger 1976).

Fujimura's work with a methacrylate copolymer showed that carbodiimide coupling (a frequently used coupling method to form an amide bond) is much more efficient if carried out with the soluble form of the polymer rather than its precipitated form (Fujimura et al. 1987). Numerous studies in the areas of affinity chromatography and protein immobilization have repeatedly shown that non-covalent immobilization is less reliable than covalent method as the ligand/protein can slowly leach off the matrix during the application (Guisan 2006; Rehm et al. 2016; Sheldon and van Pelt 2013). If the application is in either food or medical domain, regulatory agencies insist on this check. In many studies, this is evaluated by suspending the bioconjugate in simple aqueous buffers. Results of such experiments can be misleading. It is always best to check the reusability by actually carrying out the reaction (irrespective of the method used for immobilization). This is more so if the substrate in the application is of industrial grade as the impurities in the substrate may inactivate the enzymes or just clog the matrix.

One non-covalent method which can lead to exceptional stability of linkages is affinity immobilization (Roy and Gupta 2006b; Saleemuddin 1999; Sassolas et al. 2020). For example, biotin-avidin pair, wherever involved, leads to highly stable linkages. Similarly, polyvalent interactions, such as in antigen-antibody pairs, involve avidity rather than a simple binding constant and result in a fairly high stability, good enough for all practical purposes. Sardar et al. (2008) have also described how multilayers of the enzyme pectinase and the polymer alginate (having an affinity for the enzyme) could be created. The idea behind such assemblies is to deposit high amount of enzyme activity on the available surface area. In principle, it should be possible to create such assembled macro-(affinity ligands) in the soluble form and try their use in MLFTPP.

One factor which can lower the activity of an immobilized protein is steric hindrance from the matrix. This factor also becomes important in affinity chromatography. The use of soluble polymers as matrices decreases this effect but it is still present to some extent. Experience with PEG as a water soluble polymer (PEG is a classical stealth polymer used in drug delivery and extensively used to conjugate drug molecules) revealed that such steric hindrance is further reduced if the ligand is linked to one end via end group activation (Hermanson 2013). There could be an occasional disappointment as observed in the case of PEG-chymotrypsin when the bioconjugate did not hydrolyze BSA as the substrate protein was excluded from the PEG phase (Chiu et al. 1993). It is also desirable that multiple PEG chains do not surround the protein as that again creates steric hindrance for macromolecular substrates. Here, thermosensitive oligomers of NIPAAm [poly(N-isopropyl acrylamide)] provide a better option. The conjugates of trypsin prepared with such oligomers actually yielded better than expected results (Ding et al. 1996; Raghava et al. 2006). As more oligomers per enzyme molecule were used, the esterase as well as amidase activity of trypsin increased. It would be interesting to prepare macro-(affinity ligand) with such oligomers and explore their use in MLFTPP.

Hoffman's group has also carried out site-specific conjugation with poly-NIPAAm polymers in case of streptavidin which is a better choice than avidin while exploiting affinity interactions with biotin for separation purposes because of its lower (but still high enough) binding constant towards biotin (Fong et al. 1999; Stayton et al. 2004).

Smart biocatalysts have been used for hydrolysis of starch, proteins and cellulose (Roy et al. 2003). In case of Eudragit L-100 conjugated to xylanase, fluorescence spectroscopy revealed conformational changes in the enzyme upon binding to the polymer (Sardar et al. 2000). It was also possible to correlate this with effectiveness factor (a measure of how much enzyme activity survives as a result of immobilization) with increasing enzyme to polymer ratio.

As MLFTPP involves t-butanol (although, as discussed elsewhere in this book, a number of alternatives have been tried in case of TPP), it may be relevant to mention that conjugates of water soluble polymers have also been used in nearly anhydrous media (Inada et al. 1995; Ito et al. 1999; Liu et al. 2011). PEG-enzyme conjugates actually become soluble in organic solvents like benzene (although later work with light scattering showed that the particles are of colloidal sizes rather than forming a true solution) (Khan et al. 1992). Enzymes in non-aqueous media display some useful and interesting properties (as discussed in Chapter 2). Ito's group had described an interesting smart biocatalyst (photoresponsive) which works well in some organic solvents (Ito et al. 1999).

The above perspective shows that as far as design of macro-(affinity ligands) for use in MLFTPP is concerned, there is a lot which remains unexplored.

9.4 MLFTPP

Many water soluble polymers share some common structural features with proteins. Both classes of macromolecules have side chains with charges; both have some non-polar components. This simple thought prompted us to investigate whether water soluble polymers would also form an interfacial precipitate when subjected to TPP. The choice of alginate among the early ones was dictated by more than one consideration (Sharma and Gupta 2002a). High purity alginate was not easily available. So, it was considered worthwhile to investigate if it could be purified by TPP. The specific binding of alginate to pectinase (Gupta et al. 1993), starch degrading enzymes (Teotia and Gupta 2002) and phospholipase D (Jain et al. 2006; Teotia and Gupta 2004) is known.

Three different commercial preparations of alginate were tried (Sharma and Gupta 2002a). In all the cases, optimization of the usual parameters of TPP resulted in precipitation of alginates in the interfacial layer. The most troublesome impurities in alginates are polyphenols (Dusseault et al. 2006; Martinsen et al. 1989). Fluorescence emission spectroscopy showed that repeated TPP cycles got rid of most of the polyphenolic content (Sharma and Gupta 2002a).

Alginate was not the only polysaccharide which can be precipitated by TPP. Starch and its modified forms (oxidized starch and cationic starch) were obtained with 90 percent recovery by TPP (Mondal et al. 2004b). Here the aim was not purification but to examine the structural consequences of subjecting starch to TPP. FT-IR spectroscopy indicated decrease in H-bonding. An interesting result was that TPP-treated starch was less biodegradable by wheat germ amylase (Mondal et al. 2004b). Starch had been tried as a biodegradable wrapping material in food industry (Abreu et al. 2015). The major constraint is that amylases are ubiquitous and hence starch is may be subjected to biodegradation during the intended application. Currently, some forms of modified starch are successfully used for this purpose (Masmoudi et al. 2016). However, the above result on the use of TPP in altering biodegradability of starch remains unexplored.

Having established that many water soluble polymers can be precipitated by TPP, it naturally led us to explore what happens if, say, xylanase is present along with Eudragit when the latter is subjected to TPP. It was found that xylanase (selectively) bound to Eudragit appeared as a precipitate after TPP was carried out (Sharma and Gupta 2002b). It could be dissociated from the polymer and purified.

The focus of using these twin techniques, TPP and MLFTPP, was on purifying industrial enzymes. The reason was that both techniques, being single plate separation processes, are not expected to provide high purity preparations of proteins required for therapeutic and other medical purposes (like scaffolds for tissue engineering). On the other hand, any inexpensive and quick method for significant improvement in specific activity should be welcome in case of industrial enzymes. That is where these techniques should be positioned among the purification options discussed in chapter 3.

Thus, with this thinking, MLFTPP was next tried for purification of glucoamylase and pullulanase (Mondal et al. 2003b). Degradation of starch to glucose requires a combination of enzymes to break both 1→4 and 1→6 bonds present in this polysaccharide. While glucoamylases hydrolyze the latter bond as well, that activity is rather slow. Pullulanase, on the other hand, is specific for breaking that branching point. This early application of MLFTPP for purification was also chosen to validate that the process works with a polymer other than Eudragit as well. In this case, alginate was a natural choice as its affinity for amylases was already known (Sharma et al. 2000b). Both alginate and an esterified preparation of alginate (commercially available with degree of esterification with propylene glycol in the range of 40–60 percent) were used to carry out MLFTPP of glucoamylase (Mondal et al. 2003b). In both cases, MLFTPP led to glucoamylase appearing along with the polymer from which it could be dissociated by a buffer containing 1 M maltose (Table 9.1). Requirement of such a high concentration of maltose (a known inhibitor of the enzyme) reflected the high binding constant between the polymer and the enzyme. Fold purification (19–20 times) was quite high but recovery of activity was significantly higher (about 83 percent) in case where esterified alginate was used to carry out MLFTPP (Mondal et al. 2003b). Pullulanase could be purified similarly with the help of esterified alginate. Fold purification was about 38 and the activity recovery was about 89 percent. In both cases, the enzymes purified with MLFTPP showed single bands when analyzed by SDS-PAGE. The starting commercial preparations of enzymes in both cases, of course, showed multiple bands/smear upon electrophoretic analysis.

Around the same time, MLFTPP using alginate was also attempted to purify alpha-amylases from much cruder preparations than commercially available industrial enzymes (Mondal et al. 2003c). The starting materials were porcine pancreatic extract, wheat germ extract and the enzyme excreted by an active culture of *B. amyloliquefaciens* (MTCC-610, obtained from MTCC, IMTech, Chandigarh, India). Here, again, the same alginate ester was used. In all three cases, the preparation obtained after the single step purification by MLFTPP was found to be electrophoretically pure. The fold purifications obtained with porcine, wheat germ and the microbial souces were 10, 55 and 5.5, respectively. The corresponding activity recoveries were 92, 77 and 74 percent, respectively (Mondal et al. 2003c). These sources were diverse, representing animal, plant and microbial sources.

Table 9.1 Purification of glucoamylase and pullulanase from esterified alginate (Reproduced from Mondal et al. 2003b, with permission from Elsevier). Crude glucoamylase preparation (containing 438 U) or crude pullulanase preparation (containing 441 U) was added to 1 ml of esterified alginate (final concentration 0.5 percent, w/v). This was followed by the addition of 20 percent (w/v) ammonium sulphate and t-butanol in a ratio of 1:1, v/v (aqueous to t-butanol ratio). Three phases were formed upon incubating this solution at 37 °C for 1 h The interfacial precipitate (consisting of alginate bound enzyme) formed between the lower aqueous and upper t-butanol phase was collected and washed as indicated. Enzyme activity in the interfacial precipitate was eluted with 1 M maltose at 4 °C for 4 h The activity initially added was taken as 100 percent in each case.

Steps	Activity (U)	Protein (mg)	Specific activity (U/mg)	Yield (percent)	Fold purification
Glucoamylase					
Crude	438	2.4	183	100	1
Aqueous phase	14	2.1	7	3	–
Washing (0.05 M acetate buffer, pH 4.5)	2	0.2	10	–	–
Elution (1M maltose in 0.05M acetate buffer, pH 4.5)	364	0.1	3640	83	20
Pullulanase					
Crude	441	3.8	116	100	1
Aqueous phase	17	0.6	28	4	–
Washing (0.05 M acetate buffer, pH 5.0)	2	0.4	5	–	–
Elution (1 M maltose in 0.05 M acetate buffer, pH 5.0)	392	0.09	4356	89	38

There were a few more valuable insights gained about MLFTPP as a process. One early worry was whether the free enzyme(s) would also precipitate as a result of competing TPP process. In all the three cases, under optimized conditions employed for MLFTPP, no significant precipitation of the free enzyme was observed in the absence of the polymer (Mondal et al. 2003c). This control indicated that direct TPP of the enzyme(s) did not play an important role. Secondly, the crude extracts without any pre-clarification could be used. The third lesson related to the process design. In case of wheat germ extract, the polymer bound enzyme remained in the aqueous phase. The aqueous phase was dialyzed and again subjected to TPP (that is, no fresh polymer was added, except much lower concentration of ammonium sulfate was needed in this step to obtain the enzyme activity bound to the polymer as the interfacial precipitate). The take home lesson was that each system behaves differently even if the target enzyme and the macro-(affinity ligand) being employed are same.

As pointed out earlier, MLFTPP is not expected to have a high resolution capability like chromatographic processes. However, a straight comparison between its result and the corresponding purification obtained with similar affinity chromatography has never been carried out. There are numerous protein purification protocols in which fold purification is increased by running a series of purification steps in sequence. Sharma et al.

(2003) looked at a model system of a commercial preparation Pectinex Ultra SP-L which has cellulase activity along with pectinase activity. This preparation was subjected to sequential MLFTPP steps. In the first step, alginate was used as the macro-(affinity ligand) to obtain pectinase purified 13 fold with 96 percent recovery of activity. The pectinase enzyme could be dissociated from the polymer by using 1 M NaCl in acetate buffer. The preparation left behind was subjected to another MLFTPP step but with chitosan as the macro-(affinity ligand) to mop up cellulase. Cellulase so obtained was purified 16 fold and the recovery (in terms of activity units) was 92 percent. In this case, dissociation of the enzyme-chitosan complex was carried out by 1 M sodium phosphate. Both preparations, when analyzed by SDS-PAGE, revealed single bands (Sharma et al. 2003).

Another noteworthy observation has been that in MLFTPP carried out so far, elution has required only inexpensive chemicals. The costliest eluting agent has been maltose. Given that these are batch processes, it should not be difficult to set up a recovery/recycle step if required. MLFTPP needs to be explored further with a variety of target enzymes and macro-(affinity ligands). The discussion on smart biocatalysts included the facile way specific macro-(affinity ligands) can be prepared. Many such polymer-affinity ligands have already been used in protein purification by affinity precipitation.

9.5 Conclusion/future perspectives

The water soluble polymers which have been used to carry out MLFTPP are very small in number. It may be noted that it may not be necessary for such polymers to be smart. If a water soluble polymer as such shows inherent affinity for a target protein/some other class of molecule, all that is required is that it precipitates during TPP. In other cases, it is possible that a conjugate of a water soluble polymer with the affinity ligand would precipitate even if the polymer as such does not! Furthermore, the crosslinking chemistry used for linking the polymer with the affinity ligand may influence its precipitation during TPP. A large number of crosslinking reagents of diverse nature are available (Gupta et al. 2020). All that is perhaps required is the presence of both hydrophilic and hydrophobic moieties in the polymer. A large number of bacterial polysaccharides are known (Sutherland 1994). Similarly, a large number of synthetic water soluble polymers are also known.

We initially thought that the tricky part of designing a MLFTPP protocol was to have precipitation conditions for the polymer or its conjugate under which the target protein does not precipitate. On further analysis, if the polymer or its conjugate with the affinity ligand has sufficiently high binding constant for the target protein, preincubation before addition of ammonium sulfate and t-butanol is anyway going to tie up almost all of the protein.

In affinity precipitation experiments, we observed that use of higher concentration of polymers invariably led to the formation of precipitates from which recovery of the target protein was not easy. We have not investigated whether the same constraint is there in MLFTPP as well. We have also not examined the release kinetics of the protein

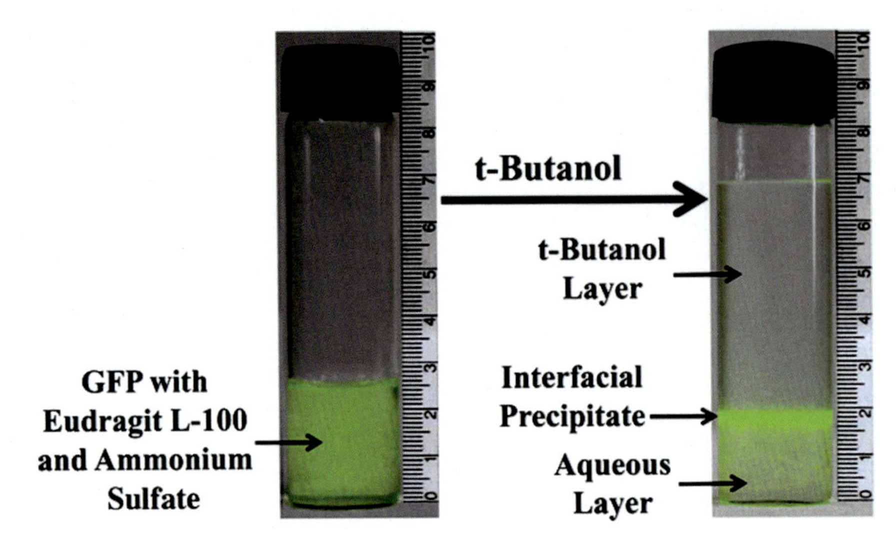

Fig. 9.1 ***MLFTPP of recombinant GFP with Eudragit L-100*** *(Reproduced from Gautam et al. 2012, with permission from Elsevier)*. First, GFP (2 mg ml^{-1}) was added with 0.5 percent (w/v) Eudragit L-100 and 30 percent (w/v) ammonium sulphate in 50 mM Tris–HCl buffer (pH 7.5). Later, the addition of t-butanol (1:2, v/v) to it and centrifugation at 3000 g for 5 min at 25 °C resulted in the formation of three phases.

in the absence of the dissociating reagents/conditions. It may be very interesting to examine these precipitates from the viewpoint of antigen formulations instead of the use of adjuvants to form local depot of antigens for eliciting immune responses.

Limited data is available to support the fact that as compared to TPP, the proteins recovered by MLFTPP are closer to the conformations of the 'native' structure (Fig. 9.1) (Gautam et al. 2012; Sharma et al. 2004). So, for subsequent structural characterization of conformations of new proteins, MLFTPP is to be preferred over TPP. In many cases, where a pre-fractionation step is desirable for a proteomic study, techniques like MLFTPP constitute a powerful option.

Industrial level centrifugation to separate cell debris, etc. is a complicated option; that has made membrane separations popular at the upstream stage of protein production (Eykamp 1997). Along with two-phase affinity extractions and expanded bed affinity chromatography, MLFTPP is an option to combine affinity step along with separation of the insoluble matter which generally was found to form the fourth solid phase at the bottom of the vessels. MLFTPP, in addition, separates many lipophilic compounds /pigments, etc. in the t-butanol phase (Sharma and Gupta 2002a). Many other chapters in this book describe separations of such compounds by TPP; an integrated process to obtain both protein and oil from soybean meal was mentioned in Chapter 2. MLFTPP offers the same possibilities with the added advantage of a built-in affinity step. For that reason, it is worth exploring for separation of other classes of both small and large molecular weight compounds/materials.

One question which may be asked is whether MLFTPP offers any advantage over the simpler TPP process. Limited data shows that in terms of fold purification or activity recovery, both seem to give similar results. The main advantage of MLFTPP over TPP may be that the former results in a protein structures closer to the native structure (Gautam et al. 2012; Sharma et al. 2004). The higher activity (Chapter 2 and Chapter 10) of the enzymes seen after TPP is not observed after MLFTPP. Is it because the enzyme in MLFTPP is actually bound to the polymeric ligand when exposed to ammonium sulfate and t-butanol? That may result in the polymer preventing any significant interactions with ammonium sulfate and/or t-butanol. That will also avoid structural consequences of the synergistic effects of these compounds on proteins which were so beautifully discussed by Dennison and Lovrein (1997).

In the next chapter, we look at another valuable but underexploited application of MLFTPP, namely in refolding/reactivation of proteins.

References

Abreu, A.S., Oliveira, M., de Sá, A., Rodrigues, R.M., Cerqueira, M.A., Vicente, A.A., Machado, AV, 2015. Antimicrobial nanostructured starch based films for packaging. Carbohydr. Polym. 129, 127–134.

Albertsson, P.A., 1971. Partitioning of Cell Particles and Macromolecules, 2nd edn. Wiley, New York.

Arasaratnam, V., Galaev, I.Y., Mattiasson, B., 2000. Reversibly soluble biocatalyst: optimization of trypsin coupling to Eudragit S-100 and biocatalyst activity in soluble and precipitated forms. Enzyme Microb. Technol. 27, 254–263.

Barbeyron, T., Michel, G., Potin, P., Henrissat, B., Kloareg, B., 2000. ι-Carrageenases constitute a novel family of glycoside hydrolases, unrelated to that of κ-carrageenases. J. Biol. Chem. 275, 35499–35505.

Banks, S.R., Enck, K., Wright, M., Opara, E.C., Welker, M.E., 2019. Chemical modification of alginate for controlled oral drug delivery. J. Agric. Food Chem. 67, 10481–10488.

Barbosa, J.A.C., Al-Kauraishi, M.M., Smith, A.M., Conway, B.R., Merchant, H.A., 2019. Achieving gastroresistance without coating: formulation of capsule shells from enteric polymers. Eur. J. Pharm. Biopharm. 144, 174–179.

Bidarra, S.J., Barrias, C.C., Granja, P.L., 2014. Injectable alginate hydrogels for cell delivery in tissue engineering. Acta Biomater. 10, 1646–1662.

Blum, Z., Eriksson, H., 1991. Utilization of zeolite Y in the removal of anionic, cationic and nonionic detergents during purification of proteins. Biotechnol. Tech. 5, 49–54.

Boller, T., Meier, C., Menzler, C., 2002. Eupergit oxirane acrylic beads: How to make enzymes fit for biocatalysis. Org Proc Res Dev 6, 509–519.

Cabral, J.M.S., Kennedy, J.F., 1993. Immobilization techniques for altering thermal stability of enzymes. In: Gupta, MN (Ed.), Thermostability of enzymes. Narosa Publishing House/Springer Verlag, New Delhi/Berlin, pp. 162–179.

Carrea, G., Riva, S., 2000. Properties and synthetic applications of enzymes in organic solvents. Angew. Chem. Int. Ed. Engl. 39, 2226–2254.

Cao, L., 2005. Immobilised enzymes: science or art? Curr. Opin. Chem. Biol. 9, 217–226.

Charles, M., Coughlin, R.W., Hasselberger, F.X., 1974. Soluble–insoluble enzyme catalysts. Biotechnol. Bioeng. 16, 1553–1556.

Chiu, H.C., Zalipsky, S., Kopecková, P., Kopecek, J., 1993. Enzymatic activity of chymotrypsin and its poly(ethylene glycol) conjugates toward low and high molecular weight substrates. Bioconjug. Chem. 4, 290–295.

Collins, T., Gerday, C., Feller, G., 2005. Xylanases, xylanase families and extremophilic xylanases. FEMS Microbiol. Rev. 29, 3–23.

Cook, M.T., Charalampopoulos, D., Khutoryanskiy, V.V., 2014. Microencapsulation of probiotic bacteria into alginate hydrogels. In: Connon, CJ, Hamley, IW (Eds.), Hydrogels in Cell-Based Therapies1st ed. Royal Society of Chemistry, London, UK., pp. 95–111.

Cordes, A., Kula, M.R., 1994. Large-scale purification of formate dehydrogenase. Methods Enzymol. 228, 600–608.

Coughlan, M.P., Hazlewood, G.P., 1993. Beta-1,4-D-xylan-degrading enzyme systems: biochemistry, molecular biology and applications. Biotechnol. Appl. Biochem. 17, 259–289.

Dar, A., Shachar, M., Leor, J., Cohen, S., 2002. Optimization of cardiac cell seeding and distribution in 3D porous alginate scaffolds. Biotechnol. Bioeng. 80, 305–312.

Dennison, C., Lovrien, R., 1997. Three phase partitioning: concentration and purification of proteins. Protein Expr. Purif. 11, 149–161.

Deshpande, M.V., 1986. Enzymatic degradation of chitin and its biological applications. J. Sci. Ind. Res. 45, 277–281.

Ding, Z., Chen, G., Hoffman, A.S., 1996. Synthesis and purification of thermally sensitive oligomer-enzyme conjugates of poly(N-isopropylacrylamide)-trypsin. Bioconjug. Chem. 7, 121–126.

Du, H., Liu, M., Yang, X., Zhai, G., 2015. The design of pH-sensitive chitosan-based formulations for gastrointestinal delivery. Drug Discov. Today 20, 1004–1011.

Dusseault, J., Tam, S.K., Menard, M., Polizu, S., Jourdan, G., Yahia, L., Halle, JP, 2006. Evaluation of alginate purification methods: effect on polyphenol, endotoxin, and protein contamination. J. Biomed. Mater. Res. 76A, 243–251.

Eykamp, W., 1997. Membrane separations in downstream processing. In: Goldberg, E. (Ed.), Handbook of Downstream Processing. Springer, Dordrecht, pp. 90–139.

Fong, R.B., Ding, Z., Long, C.J., Hoffman, A.S., Stayton, P.S., 1999. Thermoprecipitation of streptavidin via oligonucleotide-mediated self-assembly with poly(N-isopropylacrylamide). Bioconjug. Chem. 10, 720–725.

Fujimura, M., Mori, T., Tosa, T., 1987. Preparation and properties of soluble-insoluble immobilized proteases. Biotechnol. Bioeng. 29, 747–752.

Gautam, S., Dubey, P., Singh, P., Varadarajan, R., Gupta, M.N., 2012. Simultaneous refolding and purification of recombinant proteins by macro-(affinity ligand) facilitated three-phase partitioning. Anal. Biochem. 430, 56–64.

Ghosh, S., Mattiasson, B., 1993. Isolation and purification of proteins. In: Gupta, MN (Ed.), Thermostability of enzymes. Narosa Publishing House/Springer Verlag, New Delhi/Berlin, pp. 24–43.

Guisan, J.M., 2006. Immobilization of enzymes as the 21st century begins. In: Guisan, JM (Ed.), Immobilization of Enzymes and Cells2nd ed. Humana Press Inc, NJ, pp. 1–13.

Guiseley, K.B., Stanley, N.F., 1980. Whitehouse PA. Carrageenan. In: Davidson, RL (Ed.), Handbook of Water-Soluble Gums. McGraw-Hill Book Co., New York 5-1 to 5-30.

Guoqiang, D., Batra, R., Kaul, R., Gupta, M.N., Mattiasson, B., 1995. Alternative modes of precipitation of Eudragit S-100: a potential ligand carrier for affinity based separations. Bioseparation 5, 339–350.

Guoqiang, D., Kaul, R., Mattiasson, B., 1993. Purification of *Lactobacillus bulgaricus*D-lactate dehydrogenase by precipitation with an anionic polymer. Bioseparation 3, 333–341.

Guoqiang, D., Kaul, R., Mattiasson, B., 1994. Integration of aqueous two-phase extraction and affinity precipitation for the purification of lactate dehydrogenase. J. Chromatogr. A 668, 145–152.

Gupta, M.N., Guoqiang, D., Mattiasson, B., 1993a. Purification of endo-polygalacturonase by affinity precipitation using alginate. Biotechnol. Appl. Biochem. 18, 321–327.

Gupta, M.N., Guoqiang, D., Kaul, R., Mattiasson, B., 1994. Purification of xylanase from *Trichoderma viride* by precipitation with an anionic polymer Eudragit S 100. Biotechnol. Tech. 8, 117–122.

Gupta, M.N., Kaul, R., Guoqiang, D., Dissing, U., Mattiasson, B., 1996. Affinity precipitation of proteins. J. Mol. Recognit. 9, 356–359.

Gupta, M.N., Perwez, M., Sardar, M., 2020. Protein crosslinking: uses in chemistry, biology and biotechnology. Biocatal Biotransf 38, 178–201.

Handlogten, M.W., Stefanick, J.F., Deak, P.E., Bilgicer, B., 2014. Affinity-based precipitation via a bivalent peptidic hapten for the purification of monoclonal antibodies. Analyst 139, 4247–4255.

Hay, I.D., Ur Rehman, Z., Moradali, M.F., Wang, Y., Rehm, B.H., 2013. Microbial alginate production, modification and its applications. Microb. Biotechnol. (6), 637–650.

Hermanson, G.T., 2013. Bioconjugate techniques, 3rd edition. Academic Press, Waltham, USA.

Inada, Y., Furukawa, M., Sasaki, H., Kodeva, Y., Hiroto, M., Nishimura, H., Matsushima, A, 1995. Biomedical and biotechnological applications of PEG- and PM-modified proteins. Trends Biotechnol. 13, 86–91.

Ingham, K.C., 1984. Protein precipitation with polyethylene glycol. Methods Enzymol. 104, 351–356.

Ito, Y., Sugimura, N., Kwon, O.H., Imanishi, Y., 1999. Enzyme modification by polymers with solubilities that change in response to photoirradiation in organic media. Nat. Biotechnol. 17, 73–75.

Iverius, P.H., Laurent, T.C., 1967. Precipitation of some plasma proteins by the addition of dextran or polyethylene glycol. Biochim. Biophys. Acta 133, 371–373.

Jain, S., Mondal, K., Gupta, M.N., 2006. Applications of alginate in bioseparation of proteins. Artif. Cells Blood. Substit. Immobil. Biotechnol. 34, 127–144.

Johansson, G., 1984. Affinity partitioning. Methods Enzymol. 104, 356–364.

Kamihira, M., Kaul, R., Mattiasson, B., 1992. Purification of recombinant protein A by aqueous two-phase extraction integrated with affinity precipitation. Biotechnol. Bioeng. 40, 1381–1387.

Kanmani P., Rhim J.W., 2014, Physicochemical properties of gelatin/silver nanoparticle antimicrobial composite films. Food Chem. 148, 162–169.

Katchalski-Katzir, E., Krämer, D.M., 2000. Eupergit® C, a carrier for immobilization of enzymes of industrial potential. J. Mol. Catal. B. Enzymol. 10, 157–176.

Kester, H.C., Visser, J., 1990. Purification and characterization of polygalacturonases produced by the hyphal fungus *Aspergillus niger*. Biotechnol. Appl. Biochem. 12, 150–160.

Khan, S.A., Halling, P.J., Bosley, J.A., Clark, A.H., Peilow, A.D., Pelan, E.G., Rowlands, DW, 1992. Polyethylene glycol-modified subtilisin forms microparticulate suspensions in organic solvents. Enzyme Microb. Technol. 14, 96–100.

Kirk, T.K., Jeffries, T.W., 1996. Roles for microbial enzymes in pulp and paper processing. In: Jeffries, TW, Viikari, L (Eds.), Roles for microbial enzymes in pulp and paper processing. Enzymes for Pulp and Paper Processing 655, 2–24.

Küchler, A., Yoshimoto, M., Luginbühl, S., Mavelli, F., Walde, P., 2016. Enzymatic reactions in confined environments. Nature Nanotech 11, 409–420.

Kuo, W.H., Chase, H.A., 2011. Exploiting the interactions between poly-histidine fusion tags and immobilized metal ions. Biotechnol. Lett. 33, 1075–1084.

Labrou, N.E., 2002. Dye-ligand affinity adsorbents for enzyme purification. Mol. Biotechnol. 20, 77–84.

Labrou, N.E., 2014. Protein purification: an overview. Methods Mol. Biol. 1129, 3–10.

Larsson, P.O., Mosbach, K., 1979. Affinity precipitation of enzymes. FEBS Lett. 98, 333–338.

Liu, M., Xu, M., Loh, X.J., Abe, H., Tsumuraya, T., Fujii, I., Li, J, Son, TI, Ito, Y, 2011. PEGylated antibody in organic media. J. Biosci. Bioeng. 111, 564–568.

Lopes, M., Abrahim, B., Veiga, F., Seiça, R., Cabral, L.M., Arnaud, P., Andrade, JC., Ribeiro, AJ., 2017. Preparation methods and applications behind alginate-based particles. Expert Opin Drug Deliv 14, 769–782.

Lowe, C.R., Burton, S.J., Burton, N.P., Alderton, W.K., Pitts, J.M., Thomas, J.A., 1992. Designer dyes: 'biomimetic' ligands for the purification of pharmaceutical proteins by affinity chromatography. Trends Biotechnol. 10, 442–448.

Marinho-Soriano, E., Fonseca, P.C., Carneiro, M.A., Moreira, W.S., 2006. Seasonal variation in the chemical composition of two tropical seaweeds. Bioresour. Technol. 97, 2402–2406.

Martinsen, A., Skjåk-Braek, G., Smidsrød, O., 1989. Alginate as immobilization material: I. Correlation between chemical and physical properties of alginate gel beads. Biotechnol. Bioeng. 33, 79–89.

Masmoudi, F., Bessadok, A., Dammak, M., Jaziri, M., Ammar, E., 2016. Biodegradable packaging materials conception based on starch and polylactic acid (PLA) reinforced with cellulose. Environ Sci Pollut Res 23, 20904.

Mondal, K., Gupta, M.N., 2006. The affinity concept in bioseparation: evolving paradigms and expanding range of applications. Biomol. Eng. 23, 59–76.

Mondal, K., Gupta, M.N., Roy, I., 2006. Affinity-based strategies for protein purification. Anal. Chem. 78, 3499–3504.

Mondal, K., Mehta, P., Gupta, M.N., 2004a. Affinity precipitation of *Aspergillus niger* pectinase by microwave-treated alginate. Protein Expr. Purif. 33, 104–109.

Mondal, K., Roy, I., Gupta, M.N., 2003a. κ-Carrageenan as a carrier in affinity precipitation of yeast alcohol dehydrogenase. Protein Expr. Purif. 32, 151–160.

Mondal, K., Sharma, A., Gupta, M.N., 2003b. Macroaffinity ligand-facilitated three-phase partitioning for purification of glucoamylase and pullulanase using alginate. Protein Expr. Purif. 28, 190–195.
Mondal, K., Sharma, A., Gupta, M.N., 2004b. Three phase partitioning of starch and its structural consequences. Carbohydr. Polym. 56, 355–359.
Mondal, K., Sharma, A., Lata, G.M.N, 2003c. Macroaffinity ligand-facilitated three-phase partitioning (MLFTPP) of alpha-amylases using a modified alginate. Biotechnol. Prog. 19, 493–944.
Moreno-García, J., García-Martínez, T., Mauricio, J.C., Moreno, J., 2018. Yeast immobilization systems for alcoholic wine fermentations: actual trends and future perspectives. Front Microbiol 9, 241.
Michel, A.S., Mestdagh, M.M., Axelos, M.A., 1997. Physico-chemical properties of carrageenan gels in presence of various cations. Int. J. Biol. Macromol. 21, 195–200.
Moser, A.C., Hage, D.S., 2010. Immunoaffinity chromatography: an introduction to applications and recent developments. Bioanalysis 2, 769–790.
Muzzarelli, R.A.A., 1977. Chitin. Pergamon Press, London.
Nakajima, M., Atsumi, K., Kifune, K., Miura, K., Kanamaru, H., 1986. Chitin is an effective material for sutures. Jpn. J. Surg. 16, 418–424.
Nascimento, K.S., Cunha, A.I., Nascimento, K.S., Cavada, B.S., Azevedo, A.M., 2012. Aires-Barros MR. An overview of lectins purification strategies. J. Mol. Recognit. 25, 527–541.
Necas, J., Bartosikova, L., 2013. Carrageenan: a review. Vet Med 58, 187–205.
Pandolfi, V., Pereira, U., Dufresne, M., Legallais, C., 2017. Alginate-based cell microencapsulation for tissue engineering and regenerative medicine. Curr. Pharm. Des. 23, 3833–3844.
Pereira, L., Van De Velde, F., 2011. Portuguese carrageenophytes: carrageenan composition and geographic distribution of eight species (Gigartinales, Rhodophyta). Carbohydr. Polym. 84, 614–623.
Porath, J., 1992. Immobilized metal ion affinity chromatography. Protein Expr. Purif. 3, 263–281.
Qin, Y., Jiang, J., Zhao, L., Zhang, J., Wang, F., 2018. Applications of alginate as a functional food ingredient. In: Grumezescu, AM, Holban, AM (Eds.), Applications of alginate as a functional food ingredient. Handbook of food bioengineering, biopolymers for food design, 409–429.
Raghava, S., Mondal, K., Gupta, M.N., Pareek, P., Kuckling, D., 2006. Preparation and properties of thermoresponsive bioconjugates of trypsin. Artif. Cells Blood. Substit. Immobil. Biotechnol. 34, 323–336.
Rehm, F.B., Chen, S., Rehm, B.H., 2016. Enzyme engineering for in situ immobilization. Molecules 21, 1370.
Roy, I., Gupta, M.N., 2000. Current trends in affinity-based separation of proteins/enzymes. Curr. Sci. 78, 587–591.
Roy, I., Gupta, M.N., 2002a. Three-phase affinity partitioning of proteins. Anal. Biochem. 300, 11–14.
Roy, I., Gupta, M.N., 2002b. Unexpected affinity of polysaccharides and its application in separation of enzymes on fluidized beds. Sep. Sci. Technol. 37, 1591–1610.
Roy, I., Gupta, M.N., 2002c. Purification of a bacterial pullulanase on a fluidized bed of calcium alginate beads. J. Chromatogr. A 950, 1591–1610.
Roy, I., Gupta, M.N., 2003a. κ-Carrageenan as a new smart macroaffinity ligand for the purification of pullulanase. J. Chromatogr. A 998, 103–108.
Roy, I., Gupta, M.N., 2003b. Smart polymeric materials: emerging biochemical applications. Chem. Biol. 10, 1161–1171.
Roy, I., Gupta, M.N., 2003c. Repeated enzymatic hydrolysis of polygalacturonic acid, chitosan and chitin using a novel reversibly-soluble pectinase with the aid of κ-carrageenan. Biocatal Biotransf 21, 297–304.
Roy, I., Gupta, M.N., 2006. Immobilization of enzymes on smart polymers. In: Guisan, JM (Ed.), Immobilization of Enzymes and Cells2nd edition. Humana Press, Totowa, USA, pp. 87–96.
Roy, I., Gupta, M.N., 2006. Bioaffinity immobilization. In: Guisan, JM (Ed.), Immobilization of Enzymes and Cells2nd edition. Humana Press, Totowa, USA, pp. 107–116.
Roy, I., Jain, S., Teotia, S., Gupta, M.N., 2004. Evaluation of microbeads of calcium alginate as a fluidized bed medium for affinity chromatography of *Aspergillus niger* pectinase. Biotechnol. Prog. 20, 1490–1495.
Roy, I., Sharma, S., Gupta, M.N., 2003. Smart biocatalysts: design and application. Adv. Biochem. Eng. Biotechnol. 86, 159–189.
Rupley, J.A., Gates, V., 1967. Studies on the enzymic activity of lysozyme, II. The hydrolysis and transfer reactions of N-acetylglucosamine oligosaccharides. Proc. Natl Acad. Sci. USA 57, 496–510.
Sadana, A., Beelaram, A., 1994. Efficiency and economics of bioseparation: some case studies. Bioseparation 4, 221–235.

Saleemuddin, M., 1999. Bioaffinity based immobilization of enzymes. Adv Biochem Eng Bioeng 64, 203–226.
Sardar, M., Roy, I., Gupta, M.N., 2000. Simultaneous purification and immobilization of *Aspergillus niger* xylanase on the reversibly soluble polymer Eudragit™ L-100. Enzyme Microb. Technol. 27, 672–679.
Sardar, M., Varandani, D., Mehta, B., Gupta, M.N., 2008. Affinity directed assembly of multilayers of pectinase. Biocatal Biotransf 26, 313–320.
Sassolas, A., Hayat, A., Marty, J.L., 2020. Immobilization of enzymes on magnetic beads through affinity interactions. Methods Mol. Biol. 2100, 189–198.
Scopes, R., 1987. Protein Purification. Principles and Practice. Springer Verlag, New York.
Shah, S., Gupta, M.N., 2007. Kinetic resolution of (±)-1-phenylethanol in [Bmim][PF_6] using high activity preparations of lipases. Bioorg. Med. Chem. Lett. 17, 921–924.
Sharma, A., Gupta, M.N., 2002a. Three phase partitioning of carbohydrate polymers: separation and purification of alginates. Carbohydr. Polym. 48, 391–395.
Sharma, A., Gupta, M.N., 2002b. Macroaffinity ligand-facilitated three-phase partitioning (MLFTPP) for purification of xylanase. Biotechnol. Bioeng. 80, 228–232.
Sharma, S., Gupta, M.N., 2001. Alginate as a macroaffinity ligand and an additive for enhanced activity and thermostability of lipases. Biotechnol. Appl. Biochem. 33, 161–165.
Sharma, A., Mondal, K., Gupta, M.N., 2003. Separation of enzymes by sequential macroaffinity ligand-facilitated three-phase partitioning. J. Chromatogr. A 995, 127–134.
Sharma, A., Roy, I., Gupta, M.N., 2004. Affinity precipitation and macroaffinity ligand facilitated three-phase partitioning for refolding and simultaneous purification of urea-denatured pectinase. Biotechnol. Prog. 20, 1255–1258.
Sharma, A., Sharma, S., Gupta, M.N., 2000a. Purification of alkaline phosphatase from chicken intestine by three-phase partitioning and use of phenyl-sepharose 6B in the batch mode. Bioseparation 9, 155–161.
Sharma, A., Sharma, S., Gupta, M.N., 2000b. Purification of wheat germ amylase by precipitation. Protein Expr. Purif. 18, 111–114.
Sheldon, R.A., van Pelt, S., 2013. Enzyme immobilisation in biocatalysis: why, what and how. Chem. Soc. Rev. 42, 6223–6235.
Shin-ya, Y., Lee, M.Y., Hinode, H., Kajiuchi, T., 2001. Effects of N-acetylation degree on N-acetylated chitosan hydrolysis with commercially available and modified pectinases. Biochem. Eng. J. 7, 85–88.
Simpson, R.J., 2006. Precipitation of proteins by polyethylene glycol. CSH Protoc., 2006 pii: pdb.prot4311.
Smidsrød, O., Skjåk-Bræk, G., 1990. Alginate as immobilization matrix for cells. Trends Biotechnol. 8, 71–78.
Spears, R., 1993. Overview of downstream processing. In: Rehm, H-J, Reed, G (Eds.), Overview of downstream processing. Biotechnology, 40–55.
Sternberg, M., Hershberger, D., 1974. Separation of proteins with polyacrylic acids. Biochim Biophys Acta Protein Struct 342, 195–206.
Straathof, A.J.J., 2011. The proportion of downstream costs in fermentative production processesComprehensive Biotechnology, 2nd Edn. Elsevier, pp. 811–814 Moo-Young, M.
Sutherland, I.W., 1994. Structure-function relationships in microbial exopolysaccharides. Biotechnol. Adv. 12, 393–448.
Teng, S.F., Sproule, K., Hussain, A., Lowe, C.R., 1999. A strategy for the generation of biomimetic ligands for affinity chromatography. Combinatorial synthesis and biological evaluation of an IgG binding ligand. J. Mol. Recognit. 12, 67–75.
Teotia, S., Gupta, M.N., 2001. Reversibly soluble macroaffinity ligand in aqueous two-phase separation of enzymes. J. Chromatogr. A 923, 275–280.
Teotia, S., Gupta, M.N., 2002. Magnetite–alginate beads for purification of some starch degrading enzymes. Mol. Biotechnol. 20, 231–237.
Teotia, S., Gupta, M.N., 2004. Purification of phospholipase D by two-phase affinity extraction. J. Chromatogr. A 1025, 297–301.
Teotia, S., Lata, R., Gupta, M.N., 2001. Free polymeric bioligands in aqueous two-phase affinity extractions of microbial xylanases and pullulanase. Protein Expr. Purif. 22, 484–488.
Teotia, S., Lata, R., Gupta, M.N., 2004. Chitosan as a macroaffinity ligand: purification of chitinases by affinity precipitation and aqueous two-phase extractions. J. Chromatogr. A 1052, 85–91.

Teotia, S., Mondal, K., Gupta, M.N., 2006. Integration of affinity precipitation with partitioning methods for bioseparation of chitin binding lectins. Food Bioprod. Process. 84, 37–43.

Thayer, A.N., 2006. Biocatalysis helps reach a resolution. Enzymatic reactions combined with racemization can generate enantiopure materials in high yields. Chem. Eng. News. 84, 29–31.

Tyagi, R., Kumar, A., Sardar, M., Kumar, S., Gupta, M.N., 1996. Chitosan as an affinity macroligand for precipitation of N-acetyl glucosamine binding proteins/enzymes. Isol Purif 2, 217–226.

van de Velde, F., De Ruiter, G.A., 2002. Carrageenan. In: Steinbüchel, A, de Baets, S, VanDamme, EJ (Eds.), Carrageenan. Biopolymers, 245–274.

van de Velde, F., Lourenco, N.D., Pinheiro, H.M., Bakkered, M., 2002. Carrageenan: a food-grade and bio-compatible support for immobilisation techniques. Adv. Synth. Catal. 344, 815–835.

van den Broek LAM, Boeriu CG. (Eds.). Chitin and Chitosan: Properties and Applications. John Wiley & Sons Ltd., 2019.

van Leemputten, E., Horisberger, M., 1976. Soluble-insoluble complex of trypsin immobilized on acrolein-acrylic acid copolymer. Biotechnol. Bioeng. 18, 587–590.

Vårum, K.M., Anthonsen, M.W., Grasdalen, H., Smidsrød, O., 1991. ^{13}C-N.M.R. studies of the acetylation sequences in partially N-deacetylated chitins (chitosans). Carbohydr. Res. 217, 19–27.

Walter, H., Johansson, G., Brooks, D.E., 1991. Partitioning in aqueous two-phase systems: recent results. Anal. Biochem. 197, 1–18.

Wang, L., Shelton, R.M., Cooper, P.R., Lawson, M., Triffitt, J.T., Barralet, J.E., 2003. Evaluation of sodium alginate for bone marrow cell tissue engineering. Biomaterials 24, 3475–3481.

Wong, K.K., Tan, L.U., Saddler, J.N., 1988. Multiplicity of beta-1,4-xylanase in microorganisms: functions and applications. Microbiol. Rev. 52, 305–317.

Wu, Z., Hu, G., Wang, K., Zaslavsky, B.Y., Kurgan, L., Uversky, V.N., 2017. What are the structural features that drive partitioning of proteins in aqueous two-phase systems? Biochim. Biophys. Acta. Proteins Proteom. 1865, 113–120.

Wuenschell, G.E., Naranjo, E., Arnold, F.H., 1990. Aqueous two phase metal affinity extraction of heme proteins. Bioprocess Eng. 5, 199–202.

Yang, J., Goto, M., Ise, H., Cho, C.S., Akaike, T., 2002. Galactosylated alginate as a scaffold for hepatocytes entrapment. Biomaterials 23, 471–479.

CHAPTER 10

Applications of three phase partitioning and macro-(affinity ligand) facilitated three phase partitioning in protein refolding

Munishwar Nath Gupta[a], Ipsita Roy[b]
[a]Former Emeritus Professor, Department of Biochemical Engineering and Biotechnology, Indian Institute of Technology, New Delhi, India
[b]Department of Biotechnology, National Institute of Pharmaceutical Education and Research, S.A.S. Nagar, Punjab, India

Chapter outline

10.1 Introduction

Protein refolding refers to a process by which an inactivated protein preparation is converted into an active form. Those not familiar with protein sciences (especially basic enzymology) can sometime confuse inactivation with inhibition. A biologically active protein may also appear to be inactive when either a ligand binds to it or when it is chemically modified. In the former case, it may be a reversible process wherein removal of the ligand would restore the activity of the protein. This kind of inhibition is generally discussed in the context of enzyme kinetics as its various classes, e.g. competitive inhibition, uncompetitive inhibition or noncompetitive inhibition, can be distinguished kinetically. A text book example is inhibition of succinate dehydrogenase by the substrate analog malonate. In chemical modification, a chemical reagent can be made to react with amino acid residues of a protein, rendering it inactive. Generally, it is irreversible. Post-translational modifications (PTMs) which occur in vivo have been referred to as reversible covalent modifications. Examples of these are phosphorylation/dephosphorylation, glycosylation/deglycosylation, etc. These are catalyzed mostly by

Three Phase Partitioning
DOI: https://doi.org/10.1016/B978-0-12-824418-0.00002-3

enzymes as a part of signal transduction processes. Inactivation/reactivation occurring in the above cases is *not* a part of protein refolding. In case of protein refolding, inactivation is caused by misfolding or/and aggregation; protein molecules are unfolded (and dissolved) and these unfolded molecules are pushed to the correct folding pathway. While it is the last step which actually constitutes refolding, the whole process of recovery of the biological activity is generally covered under the term protein folding.

This chapter is broadly divided into three sections. To start with, a short introduction to protein structure is provided. This is to ensure that the main theme of protein refolding (by TPP/MLFTPP) is understood by readers who otherwise may not be familiar with the current picture which we have of the protein structure. Given that some significant changes have occurred in the last decade or so (which are yet to make it to general textbooks of biochemistry) and even many practicing enzymologists are not really *au courant* with some of these developments, this section seems necessary. The second section, again briefly, describes many important methods which are used in the academia or the industry for refolding proteins. The last section, of course, discusses the limited but impressive successes published wherein these twin techniques have been used for refolding. It will be helpful for readers if this chapter is read after Chapters 2 and 9 which provide some interesting background for this chapter.

10.1.1 Protein structure

Proteins are said to have four levels of structure: primary, secondary, tertiary and quaternary. This view of looking at proteins was proposed by Linderstrøm-Lang in a conference in USA (Linderstrøm-Lang 1952). In the current language, it went viral! As with most of the things which go 'viral', this has eventually created certain difficulties. To start with, it created a binary of structure or lack of it. As most of our classical picture of proteins is based upon X-ray diffraction, this binary has been reinforced so much that many biologists still interpret their data solely in terms of old structure-function paradigm. This, in turn, considers specificity as being derived from lock and key hypothesis of Emil Fischer (Lichtenthaler 1995).

While the primary structure defines the amino acid sequence, secondary structure refers to local folds of the chain. The tertiary structure covers the global structure of the polypeptide chain in three dimensions, which includes the way these secondary structures further interact. The secondary structures are formed by weak interactions. Tertiary structures exploit these further but formation of disulphide bonds between some cysteine side chains is also an important component of this level of organization. Proteins are frequently characterized by measuring their alpha-helical and beta-sheet structures by CD or FT-IR spectroscopy. Porter and Looger (2018) have recently discovered that a large number of proteins (with structures deposited in PDB) can undergo a switch between folds as a result of mutations. So, secondary structure contents cannot be tied down to "native" conformation. Another insight of recent origin has been that apart from hydrophobic interactions, normal H-bonds, ionic bonds, and van der Waals interactions, other non-covalent interactions play important roles in protein folding/protein-protein interactions. These include CH•••O H-bonding, $n \rightarrow \pi^*$ interactions, C_5 H-bonding, chalcogen bonding, cation–π, π–π, anion–π, X-H•••π and S-arene interactions. Some of these are moderately strong in strength and are more likely to be involved in protein-protein interactions; others, which are weak, play fairly important roles in protein folding (Newberry and Raines, 2019).

Later on, the importance of post-translational modifications was also recognized. If some portions of the polypeptides did not reveal any definite secondary structures when probed by X-ray diffraction, not much attention was paid to them. Structure was tightly correlated with function; hence absence of any structure at the local level was unlikely to have any biological meaning. Such was the thinking; it was also a case of "confirmation bias" from which even scientists suffer frequently.

The static picture of protein structure changed gradually. Koshland's induced fit hypothesis proposed that 'active site' of the enzymes undergoes conformational changes so that it can fit the substrate better. For our discussion later, it is necessary to note that the article 'the' conveys that this is a somewhat unique relationship between an enzyme and the molecule(s) which it helps in getting transformed to products. Over the years, the importance of protein flexibility in determining their functions came to be recognized. All these developments could be accommodated in the pictures of protein structures (those large number of PDB files!), so there did not appear to be any loose ends, nothing to be tidied up.

Two discoveries can be truly said to be 'black swan' events. The first one was the growing evidence of protein multi-specificity (Atkins 2020; Hult and Berglund 2007; Kapoor and Gupta 2012; Kapoor et al. 2015; Khersonsky and Tawfik 2010). It was initially ignored in the context of isoenzymes, multi-specific affinity ligands in affinity chromatography, antibodies and vaccines. Cow pox infected serum acted as a vaccine for small pox because the antibodies produced against cowpox in the blood of cow maids neutralized smallpox virus as a result of non-specificity involved in antigen-antibody interaction. In fact, applications of lipases (one of the most important industrial enzymes) in biotechnology are based upon their non-specificity (Kapoor and Gupta 2012; Ribeiro et al. 2011). The well known area of drug repurposing relies upon the breakdown of the specificity of interactions between a drug and its receptors – a concept outlined by Eherlich and called the magic bullet concept (Strebhardt and Ullrich 2008). Two major phenomena, protein moonlighting and protein promiscuity, are important outcomes of this lack of specificity in biological interactions. These phenomena have become relevant to protein evolution, drug design, signal transduction, use of enzymes for organic synthesis and detoxification.

What is the relevance of this to protein refolding? Upon refolding, the evaluation of the refolded protein naturally includes estimation of the recovery of its biological activity. This estimation is based upon a standard assay which in turn is limited to the classical activity; in case of an enzyme, it is tied to its EC number. So, if one checks a preparation of a refolded lipase, one is mostly checking its esterase activity or its fat/oil hydrolysis capability (Brockerhoff 1969; Desnuelle 1972). Obviously, one does not check, say, its catalytic activity for an aldol condensation. Yet C–C bond formation has been described for many lipases in fairly large number of cases! We have practically no data on evaluation of promiscuous/moonlighting activities of refolded proteins. Our limited information (described in Chapters 2 and 11) indicates that such evaluations can throw some surprises. The second black swan event has been the discovery of the role of intrinsic disorder in biological functions of proteins. Uversky has documented how poor peer reviews fuelled by confirmation bias kept postponing our appreciation of this role (Uversky 2013).

To back up a bit, protein flexibility includes fluctuations like atomic vibrations, relatively faster motions like flipping of aromatic rings of Tyr/Phe, slower movements of domains or hinge bending (all related to kT parameter) but also conformational changes during binding of the ligands to the proteins. Among ligands, proteins present unusual situations. The quaternary structure of multi-subunit proteins has been looked at as the protein-protein interaction between two entities with well defined structures. It turns out that intrinsic disorder may have a role in their function as well. An interesting case is that of hemoglobin; all its subunits have significant levels of disorder. The change of Glu-to Val (which leads to sickle cell anemia) is associated with a possible order to disorder transition locally (Fitzsimmons et al. 2016). This strongly indicates that disorder is as important as order for biological function. Intrinsic disorder mediates many protein-protein interactions involved in functioning of many hub proteins. Many signal transduction processes exploit intrinsic disorder during transient protein-protein interactions. Even many enzymes are now known to have significant level of disorder. In fact, intrinsic disorder occurs in proteins in organisms from all kingdoms.

Again, when we characterize refolded proteins structurally, we are mostly checking the secondary/tertiary structures; we seldom bother to see what happens to the intrinsic disorder! Considering that many neurodegenerative diseases are in fact caused by amyloid formation triggered by intrinsic disorder, we should look at disorder after refolding very closely. We will see why this may be especially important when we use TPP/MLFTPP for protein purification/refolding.

10.1.2 General strategies for protein refolding

Christian Anfinsen received the Nobel Prize in chemistry in 1972 for his work on ribonuclease. His famous experiment, which is described in most textbooks of biochemistry, showed that when the enzyme was unfolded with urea and all its four disulfide bonds were reduced, it regained its activity upon being left as such with just air (for oxidation of -SH groups) after dialyzing out urea. The interpretation was that the primary sequence has enough information for correct refolding of the protein chain (including the correct pairing of the -SH to form the right sets of four disulfide bonds). Actually, it was another case of "luck favoring the prepared mind". Ribonuclease A is a small protein (M_r ~ 14,000 Da) with relatively high stability. More importantly, a very dilute solution of the protein was used in that experiment.

An unfolded protein has mainly two paths to follow. One is aggregation (which is obviously concentration dependent as it is not a unimolecular event); another is to assume its 'native' conformation. We do know that cells have chaperones to assist folding of protein molecules. In the aggregates, the proteins mostly occur in misfolded forms. Aggregation can be prevented by minimizing intermolecular contact between protein molecules till they fold back (correctly). That is what chaperones do (to understate their functions!).

When overexpression in hosts like *E. coli* was attempted, biologists were disappointed that instead of correctly folded soluble proteins, insoluble particles of "inclusion bodies" (these were intracellular, hence the nomenclature) were found (Vallejo and Rinas 2004). The phrase "protein refolding" was actually coined in the context of regaining biologically active proteins from these inclusion bodies. Actually, a lot of synergy between so called applied science and basic science happened. Biologists suddenly realized that there was a need to renew their understanding of proteins in general, but especially phenomena like protein folding and protein aggregation. Not so much later, it was found that protein aggregation is also responsible for many diseases including several key neurodegenerative diseases.

The refolding strategies used more frequently are briefly discussed below.

The first step, of course, is dissolving the inclusion bodies, which also involves unfolding the misfolded protein molecules present in these particles. It may be mentioned that in many cases, inclusion bodies are reported to have significant levels of biological activities (de Marco et al. 2019). It is not clear whether this is due to the presence of some correctly folded molecules or arises because of the molecules having partial activities as a result of some non-native conformations. After all, although denaturation of many proteins is dealt with by a two-step process kinetically, even the so called native conformation is actually a representation of many conformations in equilibria (Englander and Mayne 2014). Denaturation is often a gradual process dependent upon time and concentrations of protein and denaturant (not to mention other denaturing parameters like higher temperature which may co-exist).

Urea (6–8 M), guanidinium hydrochloride (5–6 M), sodium dodecyl sulfate (SDS), alkaline pH (> 9.0), and acetonitrile/propanol cosolvent mixtures are some of the classical reagents/conditions which have been used to solubilize inclusion bodies (Burgess 2009; Marston and Hartley 1990; Vallejo and Rinas 2004). The parameters which one may need to optimize for efficient solubilization are pH, temperature, time, ions and their concentration, concentration reagent and protein, ratio of reagent to protein, redox reagent and blocking of any free -SH groups (Middelberg 2002). Use of surfactants like cetyltrimethylammonium bromide (CTAB) or even SDS is worth mentioning. Among surfactants, using cetyltrimethylammonium chloride (CTAC) is reported to lead to better refolding yields in subsequent step (Puri et al. 1992). A two-minute microwave irradiation at 200 W is reported to solubilize inclusion bodies of seven different kinds of proteins (Datta et al. 2013). Dynamic light scattering (DLS) confirmed complete solubilization and subsequent yields at the refolding steps were not affected due to use of this new solubilization method. As other methods generally require hours for solubilization for most of the cases, it is curious that this method has not been exploited even by the industry. While the industry is often quick to certify their "efficiency", their adoption (or rather lack of it) of efficient strategies is generally not commensurate with that in practice.

When it comes to the actual refolding step, one is again spoilt for choice. Broadly, we can classify these approaches into non-chromatographic and chromatographic methods. The latter are the same as used for protein purification and include ion-exchange, gel filtration, hydrophobic interaction chromatography and even reverse phase columns (which are generally used for peptide separation/purification). In both chromatographic and non-chromatographic methods, it is critical to retain the presence of a reagent to inhibit aggregation. Again, this needs to be optimized as it differs from case to case. Non-chromatographic methods are generally easy to scale-up and are less expensive.

Addition of small molecular weight additives to push the unfolded protein molecules down the correct folding pathway has been fairly popular. Tsumoto et al. (2003) have classified such additives into three main classes. First are the folding enhancers, e.g. sucrose and ammonium sulfate, which work because the folded molecules have higher thermodynamic stability. They however promote protein-protein interactions and hence aggregation competes strongly with folding. There is no data yet on how the presence of large amount of intrinsic disorder would affect their "folding" efficiency; this may be important as intrinsic disorder by 'fly-casting' is known to facilitate protein-protein interactions (Huang and Liu 2009). The second class, called aggregation suppressors, e.g. arginine and mild detergents, do exactly that by reducing protein-protein interactions. While Tsumoto et al. (2003) believe that these do not affect protein stability, arginine especially has such complicated and diverse effects on protein structure (Wen et al. 2015) that this rationale seems a case of oversimplification. The third class consists of strong denaturants like urea, etc. which reduce protein stability and protein-protein interactions. It may be thought that denaturants would have little effect on long regions of intrinsic disorder but more recent work has shown the presence of pre-formed structural motifs which participate in functioning of such proteins (Cukier 2018; Dyson and Wright 2005; Fang et al. 2019). So, if a strong denaturant destroys the conformation of these motifs, the effect would be similar to that on structured regions. In short, at present, we are fairly clueless about how these classes of refolding additives work with proteins having significant levels of intrinsic disorder. That may be at least partly responsible for the observation that "no one shoe fits all" as far as refolding strategies are concerned. Some additives which have been tried during refolding with varying successes are: proline, lysine, polyamines like putrescene, spermidine, and spermine, sugars, polyalcohols and ammonium sulfate (Lange and Rudolph 2005; Tsumoto et al. 2003; Wingfield 2015).

Co-solvents as additives during refolding have attracted considerable attention ever since Cleland and Wang (1990) reported the effect of polyethylene glycol (PEG) derivatives on the refolding of bovine carbonic anhydrase. It is interesting to note that the derivatives of PEG which worked best had neither low nor high hydrophobicity. The polymers with high hydrophobicity apparently bind very tightly to the folding intermediates and hamper refolding. This observation is valuable in the context of choice of polymers to be used in MLFTPP. Jaspard (2000) examined the effect of adding various cosolvents during the refolding of porcine pancreatic elastase at different pH values. DMSO, glycerol and methanol (but not acetonitrile) facilitated the refolding of folding intermediates by increasing their stability. The author believed that the cosolvents increased the packing in the protein molecules.

Lange et al. (2005) have investigated the renaturation of hen egg white lysozyme and a single chain antibody fragment using ionic liquids as additives. Again, the ionic liquids with more hydrophobic imidazolium cations worked less efficiently indicating that an optimum range of hydrophobicity is required. While ionic liquids were originally described as green solvents, this view has been questioned by some of the later workers (Petkovic et al. 2011). In the past decade, synthesis of ionic liquids from renewable resources has been described (Liu et al. 2012). It will be interesting to see whether sustainable processes for protein refolding emerge in the coming years.

It is generally recommended that the proteins present in the inclusion bodies are completely unfolded. However, Jungbauer and Kaar (2007) point out that at least in some cases, retention of some "native-like secondary structures" in the protein molecules of inclusion bodies may present better refolding results. It should be added that the earlier notion of inclusion bodies as being totally inactive has proven wrong in a very large number of cases (Flores et al. 2019; Hrabárová et al. 2015). Many inclusion bodies have, in fact, shown to have a significant level of activity and it has even been proposed that these can be used as such like support free immobilized preparations are used in biocatalysis (de Marco et al. 2019). Crosslinked enzyme aggregates (CLEAs) and crosslinked enzyme crystals (CLECs) are two important examples of support free immobilized preparations which are quoted in this context (Sheldon and Brady 2018). For retaining some correct secondary structures, mild detergents, high pH and low arginine concentrations may be employed for unfolding. Chaperones can be used but they are expensive additives.

Dilution is another method to refold proteins. It is essentially a replay of Anfinsen's famous experiment wherein RNAse molecules refolded on their own in solutions dilute enough to cut down protein-protein interaction (and hence, aggregation). In general, one ends up with protein concentrations in the range of 10–100 μg/ml. That is a problem, as for practical reasons, industry likes to deal with concentrated protein solutions. An additional concentration step not only increases the production cost but is also vulnerable to aggregation. Dilution can be done in a single step or stepwise or even in continuous mode, many different set-ups have been described in the literature (Middelberg 2002; Tsumoto et al. 2003; Yamaguchi et al. 2010).

There have been enough studies made on the effect of high pressure on protein structure and activity; that part of enzymology is called baroenzymology (Alny et al. 1994; Luong et al. 2016). High pressure can also be used to refold proteins. This approach is reported to be especially useful for aggregation-prone proteins; pressure, in the range of 150–200 MPa, not only facilitates refolding while preventing aggregation, but it actually reverses aggregation if it has occurred due to processing conditions such as agitation of the solution. An exceptionally interesting result has been reported by Schoner et al. (2005) about the ligand binding domains of the three different nuclear receptors that could be refolded directly without solubilization/unfolding steps from their inclusion bodies. The domains were checked for their functional activity and high recoveries were reported. Application of high hydrostatic pressure, together with small molecular weight additives, has also been discussed by Qoronfleh et al. (2007). These authors point out that when it comes to applying pressure, there is a "refolding window" – the range of optimal pressure under which the protein molecules favor first order refolding over higher order aggregation. The review lists proteins which have been successfully refolded by applying the pressure within their "refolding window" (Qoronfleh et al. 2007). The list includes bikunin (a Kunitz type protease inhibitor), lysozyme, interleukin receptor antagonist, ligand binding domains of estrogen nuclear receptor, farnesoid X nuclear receptor, prion precursor protein, gamma-interferon, beta-lactamase, human growth hormone and human phosphatases. An important and useful feature of this method is that protein concentration is not a factor as aggregation is not favored. It is also interesting to note that while room temperature works well, varying temperature in either direction can improve the refolding yield. This temperature effect correlates well with melting temperature and surface hydrophobicity of the protein (Qoronfleh et al. 2007). Among other methods, refolding in reverse micelles and hollow-fiber membrane dialysis can be mentioned (Katoh and Katoh 2000).

In case of proteins with disulfide bridges, it is necessary to use redox pairs such as reduced and oxidized forms of glutathione. The thiol form should generally be in 5-fold molar excess so that the overall condition is reducing. Under these conditions, similar to those prevailing in the endoplasmic reticulum, disulfide exchange leads to correct pairing of the sulfhydryl groups. In nature, this is catalyzed by the enzyme disulfide isomerase; addition of this enzyme during refolding also helps to increase the refolding yield. A less expensive approach is to use oxidation conditions in which air/oxygen is passed through the solution. Iodosobenzoic acid or hydrogen peroxide can accelerate disulphide bridge formation. Some oxidation of cysteine and methionine to cysteic acid and methionine sulfone is likely during this approach (Thatcher 1996). However, cyclic bis-(cysteinyl) hexapeptides have been described for oxidative protein refolding of ribonuclease (Cabrele et al. 2002). Woycechowsky et al. (1999) have shown that N-methylmercaptoacetamide or its dithiol analog, trans-1,2-bis-(2-mercaptoacetamido) cyclohexane (which works far better), could substitute for protein disulfide isomerase. They reported that adding the dithiol in the growth medium of *Saccharomyces cerevisiae* led to 3-fold increase in heterologous secretion of the cloned *S. pombe* acid phosphatase, an enzyme with eight disulfide bridges. The design of the dithiol was based upon the low pK_a value and high reduction potential of the active site of protein disulfide isomerase.

All such strategies can be possibly incorporated while planning refolding of proteins by TPP or MLFTPP but all those have not yet been explored.

Nanoparticles of titanium dioxide of about 10 nm size could refold thermally denatured chymotrypsin, ribonuclease A and papain (Raghava et al. 2009). These particles were negatively charged and the proteins were positively charged at the incubation pH. Elution of the refolded proteins with salt confirmed that the binding was electrostatic in nature. Circular dichroism spectra indicated that none of the proteins underwent any changes in their secondary structures as a result of this unfolding/refolding cycle. DLS showed completion of interaction between proteins and nanoparticles within about an hour and elution after this time period led to nearly complete recovery of the enzymatic activities (Raghava et al. 2009). Nanoparticles have been extensively used for protein immobilization. Shah and Gupta (2008) have described the use of multi-walled carbon nanotubes for all three purposes simultaneously; refolding, purification and immobilization. Peptide dendrimers have been reported to inhibit aggregation on the refolding pathway, as seen by DLS (Dubey et al. 2013). The refolding yield correlated positively with the complexity of the dendrons/dendrimers.

The effect of macromolecular crowding during refolding deserves more attention than it has garnered so far (Du et al. 2006; Minton 2000). Li et al. (2001) have explored this factor in the cases of glucose-6-phosphate dehydrogenase and protein disulfide isomerase and found that while overcrowding (with a variety of agents) did not significantly affect the refolding yields, refolding rates were diminished. It is very curious that workers in the area of protein overcrowding use polymers/proteins like dextran, PEG and bovine serum albumin (BSA) but do not refer to the rich data on the effect of such "additives" on protein stability (Gupta 1993). This is another instance of scientists belonging to similar disciplines creating 'silos' between older work and their work by the simple expediency of coining a new term! Chemists and biologists seem to be especially good at it (Gupta 2012).

10.1.3 Smart polymers and protein refolding

Lu et al. (2005b) have discussed the mechanism by which smart polymers help in protein refolding. For this, insights came from the use of surfactants like CTAB in protein refolding (Rozema and Gellman 1996). Monte Carlo simulation showed that the surfactant formed a complex with the denatured protein, allowing it to access pathways and emerge out of the "energy trap". Propelled by the global energy minimum of the native conformation, this led to correct refolding of the protein (Lu et al. 2005a). Again, the binding between the surfactant and the protein has to be in an optimum range. A strong surfactant with high hydrophobicity will lead to a stable denatured protein-surfactant complex; the dissociation of this complex during refolding step is critical (Lu et al. 2005a). A weak complex would not be able to push the protein out of the "energy trap". Early extensive work from Hoffman's group has described the synthesis of thermo-sensitive poly(N-isopropylacrylamide) (PNIPAAm) polymers and how to tune their hydrophobicity (Hoffman 1992, 2014). Such stimuli-sensitive polymers change their hydrophobicity (and hence their binding strength to the denatured protein). So, smart polymers have turned out to be very valuable in protein refolding. This application of smart polymers was reviewed by us earlier (Roy and Gupta 2003a).

Lin et al. (2000) had shown that PNIPAAm can successfully refold β-lactamase. Subsequently, the same group repeated that success with refolding of carbonic anhydrase (Chen et al. 2003). In this approach, one needs to optimize the polymer concentration, as beyond a certain value, higher polymer concentration leads to protein precipitation in the inactive form even without the application of stimulus. Kuboi et al. (2000) incorporated another thermosensitive polymer in the PEG/dextran two-phase system to enhance the refolding yield of carbonic anhydrase by about 1.7 times over the control where the polymer was absent. Again, a process temperature of 52 °C was found to be optimum; this was a function of the temperature dependence of hydrophobicities of the protein and the thermo-sensitive polymer. Umakoshi et al. (2000) have described a random polymer of ethylene oxide and propylene oxide called Breox which was a thermosensitive polymer. Two-phase systems formed with it along with water or dextran were used to refold chymotrypsin inhibitor 2. The inactive protein bound to breox, was refolded and could be separated from the polymer by thermal precipitation of breox (Umakoshi et al. 2000). While the use of thermosensitive polymers was initially thought to be constrained by thermal denaturation of proteins, a rapid progress in synthetic methodologies to prepare thermosensitive smart polymers which have lower critical solution temperature (LCST, the temperature at which the polymer undergoes phase transition) within the moderate temperature range are available (Chen et al. 2002). It may be pertinent to add a few relevant facts about smart polymers. Many smart polymers can respond to more than one stimulus. An example which we encountered is that of alginate. It is generally used as a Ca^{2+}-sensitive polymer but it does precipitate completely below pH 2.0. In another well-known case, Eudragit™ polymers have been generally used as pH-sensitive polymers but as described elsewhere (Guoqiang et al. 1995), their precipitation at high temperature along with Ca^{2+}/organic solvent has been employed to achieve different objectives. Chen et al. (2002) have pointed out the necessity for working with GRAS (generally regarded as safe) polymers only. They point out that many polyol copolymers with the trade name of Pluronic are FDA approved and are GRAS category polymers.

In our lab, we explored the use of Eudragit S-100 in refolding of chymotrypsin inactivated by 8 M urea and+ 0.1 M DTT (dithiothreitol) (Roy and Gupta 2003b). In this work, affinity precipitation was not carried out to recover the refolded protein but the activity recovered due to the presence of the smart polymer was tracked. Refolding was also followed by fluorescence spectroscopy. Complete activity was recovered within 10 min of the addition of the polymer. Fluorescence emission was the maximum at 342 nm after this time. SDS-PAGE analysis showed that multimers of chymotrypsin present in the inactivated preparation had disappeared and only a monomer was present (Roy and Gupta 2003b). In the absence of the polymer, only 10 percent recovery of activity was observed after 24 h As the renaturation conditions involved diluting out urea, this should be due to the renaturation by dilution. In the case of denaturation by dioxane (90 percent), complete renaturation was observed within 1 h after addition of the polymer.

Roy et al. (2005) have reported simultaneous purification and refolding of xylanase present in an industrial preparation and denatured by 8 M urea and DTT. The smart polymer used for affinity precipitation in this case was Eudragit S-100. An interesting feature of this work was that the polymer treated with microwave radiation gave significantly better results than the untreated polymer in terms of the final fold purification. Overall recovered xylanase activity was 96 percent and fold purification was 45. The dissociation of the enzyme from the polymer-protein complex was carried out by simple incorporation of 1 M NaCl in the solution obtained after dissolution of the precipitate by raising the pH (Roy et al. 2005). The data showed that while microwave treatment did not affect the binding of xylanase activity, it eliminated the binding of several contaminating proteins; hence the higher fold purification resulted from the higher selectivity of the polymer was the result of irradiation. The microwave reactor was equipped with a non-contact infra-red continuous feedback temperature control system to maintain the temperature at 65 °C. Heating the polymer in a simple constant temperature bath did not alter the behavior of the polymer indicating that the effect of the irradiation was not due to just thermal exposure. Microwave irradiation of the polymers is known to affect their physicochemical properties. In this case, the pH range for phase transition of the polymer was slightly broadened which confirmed that the microwave treatment had indeed changed the physicochemical behavior of this pH-sensitive polymer. All this information is relevant for the use of such polymers while optimizing a MLFTPP process.

The refolding of urea+DTT-denatured lipases from three diverse sources (bacterial, plant and animal) by affinity precipitation with alginate was reported by Mondal et al. (2006). This investigation also showed that the refolding takes place right at the binding stage which required about 45 min to be completed, as seen by DLS. CD and fluorescence spectra confirmed that the refolded bacterial lipase (with 80 percent recovered activity) was identical to the native structure of the protein. Mondal et al. (2007) showed that a thermally and chemically denatured amylase could be refolded by precipitation with calcium sensitive polymer alginate. In the same study, CcdB-17P mutant obtained as inclusion bodies was refolded by affinity precipitation with Eudragit™. In case of wild type CcdB, the polymer bound protein (obtained by affinity precipitation and dissolved) displayed DNA gyrase inhibition activity and the bioconjugate showed promise for recycling.

The use of affinity precipitation in our laboratory for simultaneous purification and refolding was shown with a recombinant lipase (Singh and Gupta 2008). Intein technology by Creative Biolabs has been extensively used in obtaining recombinant proteins. Their commercial kit uses chitin beads (and reduction by a thiol). Chitosan, the deacetylated form of chitin, is a water soluble polymer. One of the smart polymers used extensively in literature (please see Chapter 9 of this book), chitosan was used to precipitate lipase after dissolving IBs of its intein conjugate in CTAB. Treatment with DTT liberated the enzyme. The recovery was quite comparable to the one obtained with use of the chitin based commercial kit. The refolded /active lipase preparation showed a single band on SDS-PAGE upon analysis (Singh and Gupta 2008). Chitosan is fairly inexpensive and this approach was both economical as well as easily scalable.

10.1.4 Refolding by TPP

One of the early experiments employed a chemically denatured xylanase to establish 'proof of the concept' for this approach (Roy et al. 2004a) (Fig. 10.1). The starting preparation was an industrial grade xylanase, Pectinex™ Ultra SP-L, so that simultaneous purification, if any, could be tracked. The denaturing reagents were 8 M urea and 0.1 M DTT. The refolded enzyme was found in the aqueous layer itself indicating that precipitation *per se* was not necessary for refolding. The recovery of xylanase activity was quite impressive at 93 percent with 21-fold purification; SDS-PAGE analysis showed only two bands in the refolded/purified protein lane (Roy et al. 2004a). To rule out the possibility that this was a mere effect of dilution, the diluted enzyme (with concentration of urea at 2 M) was also subjected to TPP but did not result in any significant recovery of activity. The protein concentration at which refolding was carried out was 2 mg/ml which is higher by a few orders than what simple dilution can generally handle. However, later efforts were towards obtaining the refolded protein in the interfacial layer during TPP so that simultaneous concentration could also be achieved. The same industrial preparation also contains cellulase, cellobiase and beta-glucosidase activities. Roy et al. (2005) carried out refolding by TPP after subjecting the preparation to

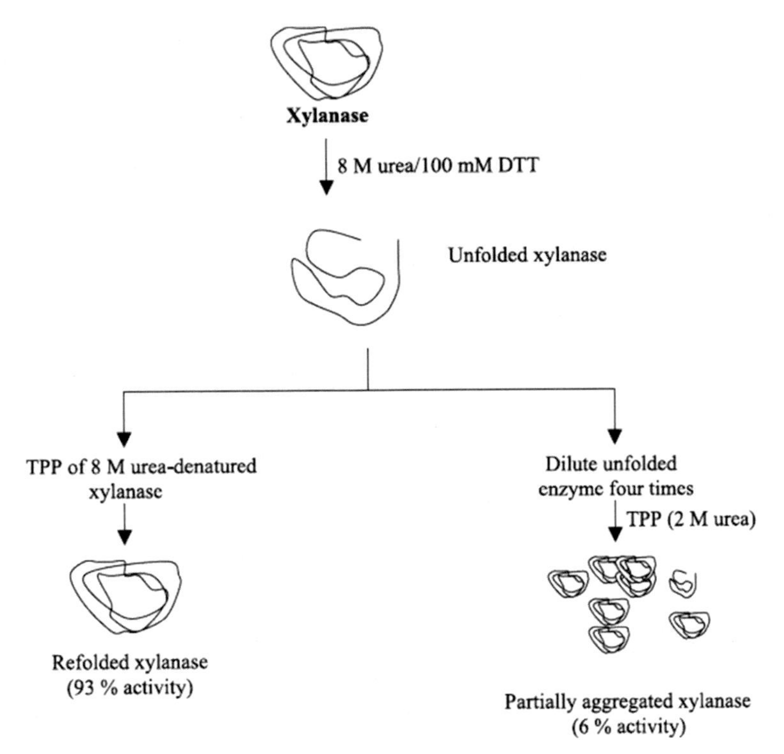

Fig. 10.1 ***Schematic diagram of renaturation process of urea-denatured xylanase by TPP.*** Xylanase was denatured with 8 M urea. The urea-denatured enzyme was saturated with 30 percent (w v^{-1}) ammonium sulfate, followed by the addition of t-butanol [1:1 (v^{-1})]. Following incubation for 1 h, the mixture was centrifuged to separate phases. In another case, the urea-denatured enzyme was diluted four times (final concentration of urea 2 M) and subjected to TPP as described above. Xylanase activity and amount of protein were estimated in the lower aqueous layer and the interfacial precipitate. *Reprinted from Roy et al. 2004a, with permission from Elsevier.*

inactivation by urea+DTT treatment. In this case, the refolded and purified enzymes were obtained in the interfacial layer. For each enzyme, the optimum conditions for carrying out TPP were different. The best results in each case under optimum conditions (for TPP) were cellulase (94 percent activity recovery, 73-fold purification), cellobiase (98 percent activity recovery, 65-fold purification) and β-glucosidase (90 percent activity recovery, 101-fold purification) (Roy et al. 2005). In this instance, the process worked well up to initial protein concentration of 10 mg/ml.

The smart bioconjugates prepared by intermolecular crosslinking of smart polymers with enzymes have found numerous applications in biotechnology (Gupta et al., 2020, Roy et al., 2003). One such bioconjugate of chymotrypsin has been described (Sharma et al. 2003). One key consideration in applications of such enzyme derivatives has been the operational stability and reusability of the biocatalyst. One way to enhance their utility would be to reactivate such biocatalysts once these undergo loss in activity after some cycles. Sardar et al. (2007) showed that a smart bioconjugate of α-chymotrypsin with Eudragit S-100 could be reactivated by TPP. When the thermally inactivated bioconjugate (with no activity) was refolded with TPP, it recovered only 6 percent activity. However, when the thermally inactivated bioconjugate was further unfolded in the presence of 8 M urea and 0.1 M DTT, a second refolding step by TPP led to 90 percent recovery of the biocatalyst activity. Interesting enough, when subjected to TPP, both the native and the completely unfolded free enzymes showed enhancement of catalytic activity over the untreated native enzyme. However, the refolded bioconjugate did not show any such enhancement. This indicated that linkages to the polymer constrained the enzyme and prevented any conformational change during TPP. We will see little later that refolding by MLFTPP (wherein the enzyme is just non-covalently bound to the polymers) differs from refolding by TPP in this respect.

Another major success of TPP was in successful refolding of recombinant proteins overexpressed in *E. coli*. These proteins are of diverse nature and included mutants of CcdB (Controller of Cell Death protein B), MBP (maltose-binding protein) and Trx (thioredoxin) (Raghava et al. 2008b). Almost all these proteins could not be refolded by a few conventional methods which were tried. In each case, the biological activities of the refolded proteins were quite good. Correct refolding was confirmed by fluorescence emission and CD spectra, and melting temperature measurements. SDS-PAGE analysis revealed absence of any aggregated protein molecules. The refolded proteins were obtained as a precipitate in the interfacial layer and hence no concentration step was required (Raghava et al. 2008b).

10.1.5 Refolding by MLFTPP

As discussed in the previous chapter (Chapter 9), there is a lot of conceptual similarity between affinity precipitation and MLFTPP. To start with, both use a macro-sized affinity ligand. The smartness of this macro-(affinity ligand) is exploited in triggering the precipitation of the complex between the target protein and this affinity ligand in case of affinity precipitation. In case of MLFTPP, precipitation occurs by carrying out TPP. It has been found that affinity precipitation could refold chemically denatured proteins or even the ones present in solubilized inclusion bodies (Sharma et al. 2004; Mondal et al. 2006, Gautam et al. 2012). The discovery of MLFTPP prompted us to explore its usefulness in protein refolding.

One key concern was the identification of a suitable smart polymer for refolding of a new protein. For this, a high throughput screening method in 96-well plate was developed (Gautam et al. 2012a). Inclusion bodies (after solubilization/unfolding) of five diverse proteins which were known to be difficult to refold by conventional methods were taken and a panel of some common water soluble polymers was tried as the macro-(affinity ligands). Fluorescence emission maxima were used to identify the success of the refolding due to the presence of the polymer (Table 10.1). In each case, at least

Table 10.1 High-throughput screening of the macro-(affinity ligand) (smart polymer) in 96-well plate for appropriate refolding of the proteins. The promising conditions are shown in bold. Reprinted from Gautam et al. 2012a, with permission from Elsevier.

Proteins	1 Inclusion bodies in 8M urea	2 Dilution control	3 Eudragit I-100	4 Eudragit S-100	5 Protanal LF	6 Alginic Acid	7 Cationic Starch	8 κ-Carrageenan	9 Chitosan
	Fluorescence emission maxima (λ_{max})								
A CcdB-F17P	358 nm	355 nm (1:0.93)	340 nm (1:0.81)	340 nm (1:0.82)	351 nm (1:0.91)	352 nm (1:0.91)	354 nm (1:0.94)	355 nm (1:0.94)	352 nm (1:0.93)
B malETrx	356 nm	352 nm (1:0.80)	342 nm (1:0.50)	342 nm (1:0.50)	350 nm (1:0.78)	352 nm (1:0.80)	352 nm (1:0.80)	354 nm (1:0.82)	351 nm (1:0.77)
C CD4D12	357 nm	354 nm (1:0.84)	340 nm (1:0.62)	342 nm (1:0.63)	352 nm (1:0.82)	353 nm (1:0.82)	355 nm (1:0.85)	353 nm (1:0.82)	355 nm (1:0.85)
D ScFv b12	360 nm	353 nm (1:0.90)	354 nm (1:0.92)	355 nm (1:0.92)	340 nm (1:0.78)	346 nm (1:0.84)	352 nm (1:0.90)	354 nm (1:0.92)	352 nm (1:0.90)
E ScFab b12	360 nm	352 nm (1:0.86)	354 nm (1:0.84)	355 nm (1:0.86)	340 nm (1:0.72)	345 nm (1:0.78)	352 nm (1:0.84)	355 nm (1:0.86)	355 nm (1:0.84)

Columns 1 to 9 indicate inclusion bodies in 8 M urea or dilution control of solubilized inclusion bodies with appropriate buffer instead of the smart polymer or solubilized inclusion bodies with different smart polymers and rows A to E indicate inclusion bodies of different proteins. The full names of the proteins are: A. Controller of cell division or death B (CcdB), B. thioredoxin fused with signal peptide of maltose binding protein (malETrx), C. First two domains of human CD4 105 (CD4D12), D. single chain variable fragment (ScFv) b12, and E. Single chain antigen binding fragment (ScFab) b12. The numbers in the parentheses below the fluorescence emission maxima (λ_{max}) values indicate the ratio of the fluorescence intensity at that λ_{max} (emission) to the fluorescence intensity of the protein solution in 8 M urea at λ_{max} (emission).

one polymer could be identified which facilitated refolding (Gautam et al. 2012a). This incidentally also confirmed one important insight. Folding happens due to the addition of the appropriate polymer, presumably anchoring the protein to facilitate refolding. As aggregation is known to be predominantly caused by hydrophobic interactions, the protein molecules probably bound to the hydrophobic parts of the polymer which acted as hot spots. Hence, whether in case of affinity precipitation or MLFTPP, precipitation merely acts as a separation step for the complex of the target protein and the macro-(affinity ligand). After this screen had identified the suitable polymer, the refolded protein was obtained by affinity precipitation. In all cases, binding was completed within about 50 min and refolded proteins could be dissociated after affinity precipitation (Gautam et al. 2012a). Measurement of biological activity in case of thioredoxin, and change in fluorescence properties upon binding of maltose in case of MBP confirmed the activities of the refolded proteins. Notably, refolding could be scaled up to 58 mg/ml for malETrx (thioredoxin fused with the signal peptide of MBP). This revisiting of the use of affinity precipitation was largely to establish that applications of MLFTPP are unlikely to be constrained by lack of a suitable polymer for a protein. The screen established here can be expanded to the larger set of natural and synthetic polymers to identify a polymer which binds to the chosen protein.

Gautam et al. (2012b) followed up the above results by refolding eight recombinant proteins obtained as inclusion bodies by overexpression in the bacterial host. The proteins were: three mutants of CcdB, malETrx, two mutants of MBP and antibody fragments ScFv and ScFab (against HIV gp120). The water-soluble polymers found to work the best for the proteins were Eudragit L-100, cationic starch and alginate. The parameters to be optimized for MLFTPP were concentration of ammonium sulfate, volumetric ratio of t-butanol to starting aqueous layer and pH of that layer. Refolding of the antibody fragments was tested by SPR using binding against the full length HIV gp120 which had been attached to the chip surface by a standard amine coupling method. The biological activities of refolded CcdB and thioredoxin were evaluated by gyrase inhibition and insulin aggregation assays, respectively (Gautam et al. 2012b). All proteins were found to be refolded correctly. This diverse set consisted of proteins with different sizes, subunit structures, surface charges/hydrophobicities and isoelectric points. To that extent, this indicated that MLFTPP may be worth trying as one among the first set of approaches for refolding of a protein. One of the questions worth addressing is that if binding with the polymer is the common step in affinity precipitation and MLFTPP, why not just use the simpler method of affinity precipitation. The authors have listed the reasons for this after analyzing their data (Gautam et al. 2012b). To start with, in all cases, MLFTPP resulted in higher refolding yields (although to a varying degree). The time involved in MLFTPP is shorter (about 10 min) as compared to 1 h taken for an affinity precipitation. The precipitates obtained in the case of MLFTPP were more compact due to lower hydration. This is not only better for subsequent processes, but lower hydration also results in lower contamination with other proteins in any purification by precipitation. Finally, in MLFTPP, one uses a mixture of t-butanol and ammonium salt as stimulus for precipitation; in affinity precipitation one may have to scout for the appropriate stimulus if a new polymer is being used. A good example is the search for the precipitation stimulus in the work on precipitation of cationic starch for precipitation of fusion proteins with MBP (Raghava et al. 2008a).

10.1.6 Structural changes in proteins due to TPP

Structural (conformational) changes resulting from TPP treatment naturally raise the question whether the refolded molecules obtained after TPP have native structures. This is not a trivial question as the (often) simultaneous enhancements in biological activities were found to be difficult to accept by classical enzymologists. We will discuss this further after reviewing the results related to increase in activities of TPP-treated enzymes in both water as well as nearly anhydrous media. This section covers the small mount of data that is available regarding the accompanying structural changes.

In one of the early and widely cited reviews on TPP (Dennison and Lovrein 1997); increase in both yield and specific activity were reported with several enzymes and protein protease inhibitors. However, this was not highlighted. The authors did mention that overall, TPP is expected to result in more compact conformations (Dennison and Lovrein 1997). The results with Proteinase K showed that "The overall structure of the TPP-treated enzyme is similar to the original structure in an aqueous environment... The most striking observation in respect of the present structure pertains to a relatively higher overall temperature factor (B = 19.7 Å^2 than the value of 9.3 Å^2 for the original enzyme)" (Singh et al. 2001). B-factor generally correlates positively with protein flexibility. The best way to understand how increase in protein flexibility in this case might have resulted in higher activity is to refer to the literature on non-aqueous enzymology (Gupta 1992, 2000). It has been shown that under low water conditions, lyophilized enzymes show poor catalytic activity as lyophilization turns the enzymes into very rigid molecules. This, of course, also renders them highly thermostable under these environments. So, it is logical to believe that higher overall flexibility in Proteinase K resulting from the TPP treatment was responsible for its higher activity.

In case of subtilisin, TPP treatment resulted in four-fold increase in caseinolytic activity of the protease in aqueous buffer and a similar increase in transesterification activity in dry n-octane and t-butanol (Roy et al. 2004b). Similar results were also observed in case of another mammalian protease, α-chymotrypsin (Roy and Gupta 2004). The higher catalytic activities observed for these two proteases in low water containing organic solvents are in line with the above stated argument that increase in conformational flexibility may be the reason for enhanced activities of TPP-treated enzymes. An interesting consequence of TPP treatment was observed in case of α-chymotrypsin (Singh et al. 2005). To start with, the pure enzyme showed 219 percent of the esterase activity of the untreated enzyme, i.e. 119 percent increase in activity, upon being subjected to TPP. The treated enzyme was dissolved in an aqueous buffer and kept for crystallization by vapor diffusion method at 295 K for 4 months.

The above considerations led to a systematic examination of TPP-treated protein structures (Rather and Gupta 2013a). For this purpose, nine proteins (four α-amylases from animal and fungal sources, Proteinase K, β-galactosidase from *A. oryzae*, α-chymotrypsin, subtilisin Carlsberg and bovine serum albumin) were chosen in that study. These were a mix of enzymes with different biological activities and serum albumin. The proteins were of varying purity, the last three of the highest purity commercially available. The idea was to have a representative set of proteins which have been subjected to TPP by different workers. A few general observations could be made. Firstly, the structural changes were subtle rather than large in all aspects. One of the questions we wanted to investigate was whether TPP influenced enzymes from thermophilic and psychrotolerant organisms differently. Enzymes from thermophiles are known to be more rigid and enzymes from psychrotolerant organisms are expected to have more flexibility. It was found that α-amylases from both kinds of organisms did not show any significant difference. Of course, the nature of amino acids in the enzymes from two different kind of organisms is likely to be different and would lead to different responses to TPP in terms of interactions. So, while this needs to be checked more extensively and more rigorously (such as using pure enzymes in both cases and in a more extensive fashion), broadly it appears that TPP cannot be a predictable tool to modulate enzyme flexibility. Only in one case, that of β-galactosidase (*A. oryzae*), the melting point (as determined by CD spectroscopy) decreased from 72 °C to 54 °C. The half-life (determined at the temperature optimum of 45 °C, which did not change) also decreased from 77 min to 39 min. Interesting enough, Proteinase K, with decreased temperature optimum and melting temperature after TPP treatment, had half-life of only 8 min as compared to 95 min for the untreated preparation. These results also have larger implications. It is obvious that results reported in the literature on temperature optimum (or even melting temperature) without accompanying half-life studies at various temperatures should be viewed with caution. There is a general tendency to assume reversibility while measuring melting temperatures by CD spectroscopy; it is necessary that both ramping up and ramping down of the temperature should be carried out to check validity of two step model. The secondary structure contents remained unchanged in most of the cases but showed some changes in either α-helical or β-sheet (or both) content in other cases. No discernible pattern emerged (Rather and Gupta 2013a). TPP treatment resulted in marginal decreases in hydrodynamic radii. On the other hand, changes in surface hydrophobicity (measured by binding of ANS) were significant upon TPP treatment. It decreased by 73 percent in case of α-chymotrypsin but increased by 34 percent in case of subtilisin.

The most peculiar behavior was that of ovalbumin (Rather and Gupta 2013b). Its treatment with TPP resulted in a preparation which behaved somewhat similar to an inclusion body! It was insoluble in aqueous buffers but could be dissolved in 8 M urea. That, in turn, could be refolded by TPP and generated many structural variants depending upon the conditions under which TPP was carried out. One of these showed trypsin inhibitory activity, marginally higher beta-sheet content and higher surface hydrophobicity. More significantly, SEM and AFM showed the presence of amyloid thread like structures.

10.1.7 Some other underexploited applications of TPP treatment of proteins/cells

While Borbás and Kiss (please see Chapter 6 of this book) have looked at the protein precipitate formed at the interface differently from others (Borbás et al. 2001, 2003; Kiss et al. 1998), there is a need to further evaluate the structural and functional aspects of these proteins in situ in precipitate forms. He et al. (2006) have talked of the importance of "Structure-Architecture-Process-Performance (SAPP)" relationship in exploring both top down and bottom-up approaches in using interfacial assemblies in areas such as biosensors, tissue engineering and emulsion processing (Fig. 10.2). The precipitates obtained via TPP are not cases of self-directed assembly considering that these are not formed in the absence of TPP environment. However, these are not assembled in a random way either. Perhaps, we should look at them as a case of triggered self-assembly. He et al. (2006) have listed some methods which have proven useful in looking at films. Ellipsometry (Elwing 1998) is capable of measuring thickness and refractive index of

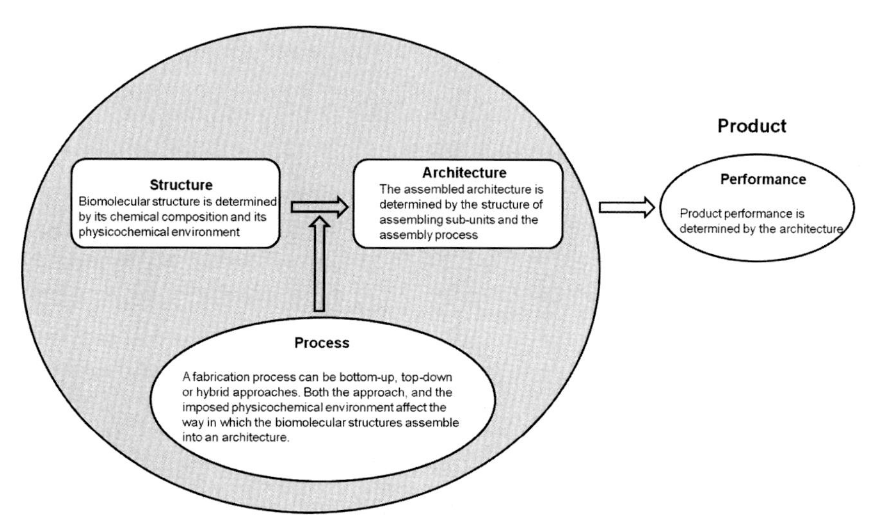

Fig. 10.2 ***The "Structure-Architecture-Process-Performance" (SAPP) concept***. *Reprinted from He et al. 2006, with permission from Elsevier.*

films in real time. Surface plasmon resonance method which can, if combined with fluorescence spectroscopy, be used to track the dynamics of film formation with high sensitivity (Liebermann and Knoll 2000). Quartz crystal microbalance, infrared spectroscopy, atomic force microscopy (AFM), scanning tunneling microscopy (STM) and tunneling electron microscopy (TEM) have also been used (Hashimoto et al. 2010; Hirschmugl 2002; Keller and Kasemo 1998; Müller and Anderson 2002; Weiss et al. 2019). X-ray and neutron reflectometry, while providing low lateral resolution, have been reported to be complementary to AFM (Delcea and Helm 2019). Single molecule FRET is especially useful in observing conformational changes in the protein molecule (König et al. 2015). If the precipitate obtained in TPP is not harvested within a reasonable time, it does stiffen. So, it would be very interesting to learn about the molecular level changes over a period of time after the interfacial layer has been formed. Also, since that review was written, there have been tremendous advances in most of these techniques (and applications of self-assembled systems as well), so these should be able to provide even more valuable information.

The review by He et al. (2006) is also relevant in discussing applications of interfacial nanostructures. While the precipitates obtained with TPP have particle sizes of higher dimensions, many of the applications mentioned should be possible in their cases as well. He et al. (2006) mention that "decreased energy costs in emulsion processing; improved emulsion stability"; this should be of immense interest to technologists in food processing sectors (Owusu-Apenten 2004). The protein gels formed by caseins, whey proteins, egg proteins, gelatin, myosin gels, soybean protein gels and surimi (hydrated protein concentrates of minced fish) gels have been discussed by Aguilera and Rademacher (2004). Of special interest is their comment in the future trends section, "In the development of low-calorie analogs where the singular ability of gels to entrap large quantities of water while imparting desirable textural or functional properties are unsurpassed by other food materials. Phase separated binary gels trapping a liquid phase inside a continuous gel matrix increase the possibilities of creating tailor-made structures for specific food applications." (Aguilera and Rademacher 2004). The gels obtained after TPP are fairly well hydrated. This book also has a chapter "on edible films and coatings from proteins" (Gennadios 2004). Hence, the proteins precipitated by TPP may be underexploited as far as their industrial applications are concerned. It is in this context that insoluble precipitate obtained in case of ovalbumin (Rather and Gupta 2013b) becomes important.

After that diversion into futuristic possibilities about even further applications of TPP; let us revert to the theme of refolding and its structural consequences.

10.2 Conclusion

Much has changed in our understanding of protein structures since the days of that Harry Potter remark by the unknown referee of our early paper on TPP treated enzymes (please see Chapter 2 of this book). In spite of induced fit theory and considerable data on flexibility of proteins (Mukherjee and Gupta 2015; Petsko and Ringe 2004; Purich 2010), the 'lock and key' concept have dominated the overall thinking while looking at enzyme functions, notably their specificities. The last 15 years or so have seen some paradigm shifts in our understanding in this area. It turns out that enzymes are not really so specific. Enzymes can be promiscuous (Atkins 2020; Hult and Berglund 2007; Kapoor and Gupta 2012; Kapoor et al. 2015; Khersonsky and Tawfik 2010). In fact, moonlighting by proteins has shown that enzymes can function as non-catalytic proteins with different biological activities (Gupta et al. 2019; Jeffery 2018). At the structural level, the idea of a protein having a unique conformation is no longer valid. The protein molecule is associated with many kinds of mobility and exists in the form of different conformations which are at equilibrium (Englander and Mayne 2014). Furthermore, most of the regulatory proteins do not have a pre-formed binding site. Most are either intrinsically disordered (IDPs) or have intrinsically disordered regions (IDPRs) (Mittag et al. 2009; Uversky 2013). So, the idea of an enzyme becoming active by being subjected to TPP may not look as far-fetched as it looked decades back. In that context, it is interesting to recall that results of X-ray diffraction of TPP-treated Proteinase K included, "As a result of higher B-factor, a number of residues, particularly their side chains, were found to adopt more than one conformation" (Singh et al. 2001). This was also one of the early observations as a property of intrinsic disorder. So, while we still do not have a complete understanding of the structural consequences of TPP treatment, the experience so far indicates that much depends upon the structure of the protein and to a limited extent, upon the conditions used during TPP treatment. Hence, while TPP seems to work well for refolding of a variety of proteins, MLFTPP, if possible, is more likely to result in a conformation closer to the native structure.

References

Aguilera, J.M., Rademacher, B., 2004. Protein gels. In: Yada, RY (Ed.), Proteins in Food Processing. Woodhead Publishing Ltd., Cambridge, pp. 468–482.

Alny, C.B., Heiber-Langer, I., Inserm, R.L., 1994. New trends in baro-enzymology. High Press Res 12, 187–191.

Atkins, W.M., 2020. Mechanisms of promiscuity among drug metabolizing enzymes and drug transporters. FEBS J. 287, 1306–1322.

Borbás, R., Murray, B.S., Kiss, E., 2003. Interfacial shear rheological behaviour of proteins in three-phase partitioning systems. Colloids Surf. A 213, 93–103.

Borbás, R., Turza, S., Szamos, J., Kiss, E., 2001. Analysis of protein gels formed by interfacial partitioning. Colloid Polym. Sci. 279, 705–713.

Brockerhoff, H., 1969. Esters of phenols as substrate for pancreatic lipase. Biochim. Biophys. Acta 191, 181.

Burgess, R.R., 2009. Refolding solubilized inclusion body proteins. Methods Enzymol. 463, 259–282.

Cabrele, C., Fiori, S., Pegoraro, S., Moroder, L., 2002. Redox-active cyclic bis(cysteinyl)peptides as catalysts for in vitro oxidative protein folding. Chem. Biol. 9, 731–740.

Chen, G., Hoffman, A.S., Kabra, B., 2002. Temperature- and pH-sensitive graft copolymers. In: Galaev, IY, Mattiasson, B (Eds.), Smart Polymers for Bioseparation and Bioprocessing. Taylor and Francis, New York, pp. 1–26.

Chen, Y.-J., Huang, L.-W., Chiu, H.-C., Lin, S.-C., 2003. Temperature-responsive polymer-assisted protein refolding. Enzym Microb Technol 32, 120–130.

Cleland, J.L., Wang, D.I., 1990. Cosolvent assisted protein refolding. Biotechnology (N.Y.) 8, 1274–1278.
Cukier, R.I., 2018. Conformational ensembles exhibit extensive molecular recognition features. ACS Omega 3, 9907–9920.
Datta, I., Gautam, S., Gupta, M.N., 2013. Microwave assisted solubilization of inclusion bodies. Sustain Chem Process 1 (2).
de Marco, A., Ferrer-Miralles, N., Garcia-Fruitós, E., Mitraki, A., Peternel, S., Rinas, U., Trujillo-Roldán, M.A., Valdez-Cruz, N.A., Vázquez, E., Villaverde, A., 2019. Bacterial inclusion bodies are industrially exploitable amyloids. FEMS Microbiol. Rev. 43, 53–72.
Delcea, M., Helm, C.A., 2019. X-ray and neutron reflectometry of thin films at liquid interfaces. Langmuir 35, 8519–8530.
Dennison, C., Lovrien, R., 1997. Three phase partitioning: concentration and purification of proteins. Protein Expr. Purif. 11, 149–161.
Desnuelle, P., 1972. The lipases. In: Boyer, P (Ed.), The Enzymes3rd Ed. Academic Press, New York, pp. 575–616 Vol. 7.
Du, F., Zhou, Z., Mo, Z.Y., Shi, J.Z., Chen, J., Liang, Y., 2006. Mixed macromolecular crowding accelerates the refolding of rabbit muscle creatine kinase: implications for protein folding in physiological environments. J. Mol. Biol. 364, 469–482.
Dubey, P., Gautam, S., Kumar, P.P.P., Sadanandan, S., Haridas, V., Gupta, M.N., 2013. Dendrons and dendrimers as pseudochaperonins for refolding of proteins. RSC Adv. 3, 8016–8020.
Dyson, H.J., Wright, P.E., 2005. Intrinsically unstructured proteins and their functions. Nat. Rev. Mol. Cell Biol. 6, 197–208.
Elwing, H., 1998. Protein absorption and ellipsometry in biomaterial research. Biomaterials 19, 397–406.
Englander, S.W., Mayne, L., 2014. The nature of protein folding pathways. Proc. Natl. Acad. Sci. U. S. A. 111, 15873–15880.
Fang, C., Moriwaki, Y., Tian, A., Li, C., Shimizu, K., 2019. Identifying short disorder-to-order binding regions in disordered proteins with a deep convolutional neural network method. J. Bioinform. Comput. Biol. 17, 1950004.
Fitzsimmons, R., Amin, N., Uversky, V.N., 2016. Understanding the roles of intrinsic disorder in subunits of hemoglobin and the disease process of sickle cell anemia. Intrinsically Disord Proteins 4, e1248273.
Flores, S.S., Nolan, V., Perillo, M.A., Sánchez, J.M., 2019. Superactive β-galactosidase inclusion bodies. Colloids Surf. B Biointerfaces 173, 769–775.
Gautam, S., Dubey, P., Singh, P., Kesavardhana, S., Varadarajan, R., Gupta, M.N., 2012a. Smart polymer mediated purification and recovery of active proteins from inclusion bodies. J. Chromatogr. A 1235, 10–25.
Gautam, S., Dubey, P., Singh, P., Varadarajan, R., Gupta, M.N., 2012b. Simultaneous refolding and purification of recombinant proteins by macro-(affinity ligand) facilitated three-phase partitioning. Anal. Biochem. 430, 56–64.
Gennadios, A., 2004. Edible films and coatings from proteins. In: Yada, RY (Ed.), Proteins in Food Processing. Woodhead Publishing Ltd., Cambridge, pp. 442–467.
Guoqiang, D., Batra, R., Kaul, R., Gupta, M.N., Mattiasson, B., 1995. Alternative modes of precipitation of Eudragit S-100: a potential ligand carrier for affinity based separations. Bioseparation 5, 339–350.
Gupta, M.N., 1992. Enzyme function in organic solvents. Eur. J. Biochem. 203, 25–32.
Gupta, MN (Ed.), 2000. Methods in Non-Aqueous Enzymology. Birkhauser, Basel.
Gupta, MN (Ed.), 1993. Thermostability of Enzymes. Narosa Publishing House/Springer Verlag, New Delhi/Berlin.
Gupta, M.N., 2012. Some deleterious consequences of birth of new disciplines in science: the case of biology. Curr. Sci. 103, 126–127.
Gupta, M.N., Pandey, S., Ehtesham, N.Z., Hasnain, S.E., 2019. Medical implications of protein moonlighting. Indian J. Med. Res. 149, 322–325.
Gupta, M.N., Perwez, M., Sardar, M., 2020. Protein crosslinking: uses in chemistry, biology and biotechnology. Biocatal Biotransf 38, 178–201.
Hashimoto, Y., Minoura, S., Honda, R., Nishiura, A., Hashimoto, Y., Matsumoto, N., Takeda, S., 2010. Development of titanium quartz crystal microbalance sensor by magnetron sputtering. J Oral Tissue Eng 8, 52–59.
He, L., Dexter, A.F., Middleberg, A.P.J., 2006. Biomolecular engineering at interfaces. Chem Engg Sci 61, 989–1003.

Hirschmugl, C.J., 2002. Frontiers in infrared spectroscopy at surfaces and interfaces. Surf. Sci. 500, 577–604.
Hoffman, A.S., 1992. Molecular bioengineering of biomaterials in the 1990s and beyond: a growing liaison of polymers with molecular biology. Artif. Organs 16, 43–49.
Hoffman, A.S., 2014. 4. Poly(NIPAAm) revisited - it has been 28 years since it was first proposed for use as a biomaterial: original research article: applications of thermally reversible polymers hydrogels in therapeutics and diagnostics, 1987; thermally reversible hydrogels: II. Delivery and selective removal of substances from aqueous solutions, 1986; a novel approach for preparation of pH-sensitive hydrogels for enteric drug delivery, 1991. J. Control. Release 190, 36–40.
Hrabárová, E., Achbergerová, L., Nahálka, J., 2015. Insoluble protein applications: the use of bacterial inclusion bodies as biocatalysts. Methods Mol. Biol. 1258, 411–422.
Huang, Y., Liu, Z., 2009. Kinetic advantage of intrinsically disordered proteins in coupled folding-binding process: a critical assessment of the "fly-casting" mechanism. J. Mol. Biol. 393, 1143–1159.
Hult, K., Berglund, P., 2007. Enzyme promiscuity: mechanism and applications. Trends Biotechnol. 25, 231–238.
Jaspard, E., 2000. Role of protein-solvent interactions in refolding: effects of cosolvent additives on the renaturation of porcine pancreatic elastase at various pHs. Arch. Biochem. Biophys. 375, 220–228.
Jeffery, C.J., 2018. Protein moonlighting: what is it, and why is it important? Philos. Trans. R. Soc. Lond. B Biol. Sci. 373, 20160523.
Jungbauer, A., Kaar, W., 2007. Current status of technical protein refolding. J. Biotechnol. 128, 587–596.
Katoh, S., Katoh, Y., 2000. Continuous refolding of lysozyme with fed-batch addition of denatured protein solution. Process Biochem. 35, 1119–1124.
Kapoor, M., Gupta, M.N., 2012. Lipase promiscuity and its biochemical applications. Process Biochem. 47, 555–569.
Kapoor, M., Majumder, A.B., Gupta, M.N., 2015. Promiscuous lipase-catalyzed C–C bond formation reactions between 4 nitrobenzaldehyde and 2-cyclohexen-1-one in biphasic medium: aldol and Morita–Baylis–Hillman adduct formations. Catal. Lett. 145, 527–532.
Keller, C.A., Kasemo, B., 1998. Surface specific kinetics of lipid vesicle adsorption measured with a quartz crystal microbalance. Biophys. J. 75, 1397–1402.
Khersonsky, O., Tawfik, D.S., 2010. Enzyme promiscuity: a mechanistic and evolutionary perspective. Annu. Rev. Biochem. 79, 471–505.
Kiss, É., Szamos, J., Tamás, B., Borbás, R., 1998. Interfacial behavior of proteins in three-phase partitioning using salt-containing water/tert-butanol systems. Colloids Surf. A 142, 295–302.
König, I., Zarrine-Afsar, A., Aznauryan, M., Soranno, A., Wunderlich, B., Dingfelder, F., Stüber, J.C., Plückthun, A., Nettels, D., Schuler, B., 2015. Single-molecule spectroscopy of protein conformational dynamics in live eukaryotic cells. Nat. Methods 12, 773–779.
Kuboi, R., Morita, S., Ota, H., Umakoshi, H., 2000. Protein refolding using stimuli-responsive polymer-modified aqueous two-phase systems. J. Chromatogr. B Biomed. Sci. Appl. 743, 215–223.
Lange, C., Patil, G., Rudolph, R., 2005. Ionic liquids as refolding additives: n'-alkyl and N'-(omega-hydroxyalkyl) N-methylimidazolium chlorides. Protein Sci. 14, 2693–2701.
Lange, C., Rudolph, R., 2005. Production of recombinant proteins for therapy, diagnostics and industrial research by in vitro folding. In: Kiefhaber, T, Buchner, J (Eds.), Protein Folding Handbook. Wiley, Weinheim, Germany, pp. 1245–1280.
Li, J., Zhang, S., Wang, C., 2001. Effects of macromolecular crowding on the refolding of glucose-6-phosphate dehydrogenase and protein disulfide isomerase. J. Biol. Chem. 276, 34396–34401.
Lichtenthaler, F.W., 1995. 100 Years "Schlüssel-Schloss-Prinzip": what made Emil Fischer use this analogy? Angew. Chem. Int. Ed. Engl. 33, 2364–2374.
Liebermann, T., Knoll, W., 2000. Surface-plasmon field-enhanced fluorescence spectroscopy. Colloids Surf. A 171, 115–130.
Lin, S.C., Lin, K.L., Chiu, H.C., Lin, S., 2000. Enhanced protein renaturation by temperature-responsive polymers. Biotechnol. Bioeng. 67, 505–512.
Linderstrøm-Lang, K.U, 1952. Proteins and Enzymes. Lane Medical Lectures 6.
Liu, Q.-P., Hou, X.-D., Li, N., Zong, M.-H., 2012. Ionic liquids from renewable biomaterials: synthesis, characterization and application in the pretreatment of biomass. Green Chem. 14, 304–307.
Lu, D., Liu, Z., Liu, Z., Zhang, M., 2005a. Molecular simulation of surfactant-assisted protein refolding. J. Chem. Phys. 122, 134902.

Lu, D., Liu, Z., Zhang, M., Liu, Z., Zhou, H., 2005b. The mechanism of PNIPAAm-assisted refolding of lysozyme denatured by urea. Biochem. Eng. J. 24, 55–64.
Luong, T.Q., Erwin, N., Neumann, M., Schmidt, A., Loos, C., Schmidt, V., Fändrich, M., Winter, R., 2016. Hydrostatic pressure increases the catalytic activity of amyloid fibril enzymes. Angew. Chem. Int. Ed. Engl. 55, 12412–12416.
Marston, F.A.O., Hartley, D.L., 1990. Solubilization of protein aggregates. Methods Enzymol. 182, 264–276.
Middelberg, A.P.J., 2002. Preparative protein refolding. Trends Biotechnol. 20, 437–443.
Newberry, R.W., Raines, R.T., 2019. Secondary forces in protein folding. ACS Chem. Biol. 14, 1677–1686.
Minton, A.P., 2000. Implications of macromolecular crowding for protein assembly. Curr. Opin. Struct. Biol. 10, 34–39.
Mittag, T., Kay, L.E., Forman-Kay, J.D., 2009. Protein dynamics and conformational disorder in molecular recognition. J. Mol. Recognit. 23, 105–116.
Mondal, K., Bohidar, H.B., Roy, R.P., Gupta, M.N., 2006. Alginate-chaperoned facile refolding of *Chromobacterium viscosum* lipase. Biochim. Biophys. Acta 1764, 877–886.
Mondal, K., Raghava, S., Barua, B., Varadarajan, R., Gupta, M.N., 2007. Role of stimuli-sensitive polymers in protein refolding: α-Amylase and CcdB (controller of cell division or death B) as model proteins. Langmuir 23, 70–75.
Mukherjee, J., Gupta, M.N., 2015. Increasing importance of protein flexibility in designing biocatalytic processes. Biotechnol Rep (Amst) 6, 119–123.
Müller, D.J., Anderson, K., 2002. Biomolecular imaging using atomic force microscopy. Trends Biotechnol. 20, S45–S49.
Owusu-Apenten, R.K., 2004. Testing protein functionality. In: Yada, RY (Ed.), Proteins in Food Processing. Woodhead Publishing Ltd., Cambridge, pp. 217–244.
Petsko, G.A., Ringe, D., 2004. Protein Structure and Function. New Science Press Ltd., London.
Petkovic, M., Seddon, K.R., Rebelo, L.P.N., Silva Pereira, C, 2011. Ionic liquids: a pathway to environmental acceptability. Chem. Soc. Rev. 40, 1383–1403.
Porter, L.L., Looger, L.L., 2018. Extant fold-switching proteins are widespread. Proc. Natl Acad. Sci. USA 115, 5968–5973.
Puri, N.K., Crivelli, E., Cardamone, M., Fiddes, R., Bertolini, J., Ninham, B., Brandon, MR., 1992. Solubilization of growth hormone and other recombinant proteins from *Escherichia coli* inclusion bodies by using a cationic surfactant. Biochem. J. 285, 871–879.
Purich, D.L., 2010. Enzyme Kinetics: Catalysis and Control. Elsevier/Academic Press, San Diego.
Qoronfleh, M.W., Hesterberg, L.K., Seefeldt, M.B., 2007. Confronting high-throughput protein refolding using high pressure and solution screens. Protein Expr. Purif. 55, 209–224.
Raghava, S., Aquil, S., Bhattacharyya, S., Varadarajan, R., Gupta, M.N., 2008a. Strategy for purifying maltose binding protein fusion proteins by affinity precipitation. J. Chromatogr. A 1194, 90–95.
Raghava, S., Barua, B., Singh, P.K., Das, M., Madan, L., Bhattacharyya, S., Bajaj, K., Gopal, B., Varadarajan, R., Gupta, MN., 2008b. Refolding and simultaneous purification by three-phase partitioning of recombinant proteins from inclusion bodies. Protein Sci. 17, 1987–1997.
Raghava, S., Singh, P.K., Rao, A.R., Dutta, V., Gupta, M.N., 2009. Nanoparticles of unmodified titanium dioxide facilitate protein refolding. Mater. Res. 19, 2830–2834.
Rather, G.M., Gupta, M.N., 2013a. Three phase partitioning leads to subtle structural changes in proteins. Int. J. Biol. Macromol. 60, 134–140.
Rather, G.M., Gupta, M.N., 2013b. Refolding of urea denatured ovalbumin with three phase partitioning generates many conformational variants. Int. J. Biol. Macromol. 60, 301–308.
Ribeiro, B.D., de Castro, A.M., Coelho, M.A., Freire, D.M., 2011. Production and use of lipases in bioenergy: a review from the feedstocks to biodiesel production. Enzyme Res 2011, 615803.
Roy, I., Gupta, M.N., 2003a. Smart polymeric materials: emerging biochemical applications. Chem. Biol. 10, 1161–1171.
Roy, I., Gupta, M.N., 2003b. pH-responsive polymer-assisted refolding of urea-and organic solvent-denatured α-chymotrypsin. Protein. Eng. 16, 1153–1157.
Roy, I., Gupta, M.N., 2004. α-Chymotrypsin shows higher activity in water as well as organic solvents after three phase partitioning. Biocatal Biotransf 22, 261–268.

Roy, I., Mondal, K., Sharma, A., Gupta, M.N., 2005a. Simultaneous refolding/purification of xylanase with a microwave treated smart polymer. Biochim. Biophys. Acta 1747, 179–187.
Roy, I., Sharma, A., Gupta, M.N., 2004a. Three-phase partitioning for simultaneous renaturation and partial purification of *Aspergillus niger* xylanase. Biochim. Biophys. Acta 1698, 107–110.
Roy, I., Sharma, A., Gupta, M.N., 2004b. Obtaining higher transesterification rates with subtilisin Carlsberg in nonaqueous media. Bioorg. Med. Chem. Lett. 14, 887–889.
Roy, I., Sharma, A., Gupta, M.N., 2005b. Recovery of biological activity in reversibly inactivated proteins by three phase partitioning. Enzym Microb Technol 37, 113–120.
Roy, I., Sharma, S., Gupta, M.N., 2003. Smart biocatalysts: design and application. Adv. Biochem. Eng. Biotechnol. 86, 159–189.
Rozema, D., Gellman, S.H., 1996. Artificial chaperone-assisted refolding of carbonic anhydrase B. J. Biol. Chem. 271, 3478–3487.
Sardar, M., Sharma, A., Gupta, M.N., 2007. Refolding of a denatured α-chymotrypsin and its smart bioconjugate by three-phase partitioning. Biocatal Biotransf 25, 92–97.
Schoner, B.E., Bramlett, K.S., Guo, H., Burris, T.P., 2005. Reconstitution of functional nuclear receptor proteins using high pressure refolding. Mol. Genet. Metab. 85, 318–322.
Shah, S., Gupta, M.N., 2008. Simultaneous refolding, purification and immobilization of xylanase with multi-walled carbon nanotubes. Biochim. Biophys. Acta 1784, 313–367.
Sharma, S., Kaur, P., Jain, A., Rajeswari, M.R., Gupta, M.N., 2003. A smart bioconjugate of chymotrypsin. Biomacromolecules 4, 330–336.
Sharma, A., Roy, I., Gupta, M.N., 2004. Affinity precipitation and macroaffinity ligand facilitated three-phase partitioning for refolding and simultaneous purification of urea-denatured pectinase. Biotechnol. Prog. 20, 1255–1258.
Sheldon, R.A., Brady, D., 2018. The limits to biocatalysis: pushing the envelope. Chem. Commun. (Camb.) 54, 6088–6104.
Singh, R.K., Gourinath, S., Sharma, S., Roy, I., Gupta, M.N., Betzel, C., Srinivasan, A., Singh, T.P., 2001. Enhancement of enzyme activity through three-phase partitioning: crystal structure of a modified serine proteinase at 1.5 Å resolution. Protein. Eng. 14, 307–313.
Singh, P.K., Gupta, M.N., 2008. Simultaneous refolding and purification of a recombinant lipase with an intein tag by affinity precipitation with chitosan. Biochim. Biophys. Acta 1784, 1825–1829.
Singh, N., Jabeen, T., Sharma, S., Roy, I., Gupta, M.N., Bilgrami, S., Somvanshi, R.K., Dey, S., Perbandt, M., Betzel, C., Srinivasan, A., Singh, T.P., 2005. Detection of native peptides as potent inhibitors of enzymes: crystal structure of the complex formed between treated bovine α-chymotrypsin and an autocatalytically produced fragment, Ile-Val-Asn-Gly-Glu-Glu-Ala-Val-Pro-Gly-Ser-Trp-Pro-Trp, at 2.2 angstroms resolution. FEBS J. 272, 562–572.
Strebhardt, K., Ullrich, A.P., 2008. Ehrlich's magic bullet concept: 100 years of progress. Nat. Rev. Cancer 8, 473–480.
Thatcher, D.R., 1996. Industrial scale purification of proteins. In: Price, NC (Ed.), Industrial scale purification of proteins. Proteins LabFax, 131–137.
Tsumoto, K., Ejima, D., Kumagai, I., Arakawa, T., 2003. Practical considerations in refolding proteins from inclusion bodies. Protein Expr. Purif. 28, 1–8.
Umakoshi, H., Persson, J., Kroon, M., HO, J., Otzen, D.E., Kuboi, R., Tjerneld, F., 2000. Model process for separation based on unfolding and refolding of chymotrypsin inhibitor 2 in thermoseparating polymer two-phase systems. J. Chromatogr. B Biomed. Sci. Appl. 743, 13–19.
Uversky, V.N., 2013. A decade and a half of protein intrinsic disorder: biology still waits for physics. Protein Sci. 22, 693–724.
Vallejo, L.F., Rinas, U., 2004. Strategies for the recovery of active proteins through refolding of bacterial inclusion body proteins. Microb. Cell Fact. 3, 11.
Weiss, A.C.G., Krüger, K., Besford, Q.A., Schlenk, M., Kempe, K., Förster, S., Caruso, F., 2019. In situ characterization of protein corona formation on silica microparticles using confocal laser scanning microscopy combined with microfluidics. ACS Appl. Mater. Interfaces 11, 2459–2469.
Wen, L., Chen, Y., Liao, J., Zheng, X., Yin, Z., 2015. Preferential interactions between protein and arginine: effects of arginine on tertiary conformational and colloidal stability of protein solution. Int. J. Pharm. 478, 753–761.

Wingfield, P.T., 2015. Overview of the purification of recombinant proteins. Curr Protoc Protein Sci 80 6.1.1-6.1.35.
Woycechowsky, K.J., Wittrup, K.D., Raines, R.T., 1999. A small-molecule catalyst of protein folding in vitro and in vivo. Chem. Biol. 6, 871–879.
Yamaguchi, H., Miyazaki, M., Briones-Nagata, M.P., Maeda, H., 2010. Refolding of difficult-to-fold proteins by a gradual decrease of denaturant using microfluidic chips. J. Biochem. 147, 895–903.

CHAPTER 11

Three phase partitioning-based strategies for highly efficient separation of bioactive polysaccharides from natural resources

Jing-Kun Yan
School of Food & Biological Engineering, Institute of Food Physical Processing, Jiangsu University, Zhenjiang, China

Chapter outline

11.1 Introduction

Polysaccharides (PSs) are a class of biological macromolecules which generally consist of at least ten monosaccharides or their derivatives linked by glycosidic bonds in linear or branched chains (Xie et al., 2016;Yu et al., 2018). PSs account for more than 90 percent of natural carbohydrate polymers in nature and are extensively found in animal cell membranes, plant and microbial walls, which are closely related to human life and play a vital role in maintaining life activities (Tolstoguzov, 2004). In the last three decades, PSs have attracted more and more attentions all over the world due to a broad spectrum of biological activities and health benefits, such as immunomodulation, anti-tumor, anti-oxidation, anti-viral, anti-inflammation, anti-radiation and hyperglycemic effects (Hou et al., 2020; Wang et al., 2020a; Yan et al., 2014, 2017a; Yi et al., 2020; Yu et al., 2018). Moreover, PSs, which are non-toxic with high hydrophilicity and easily biodegradable,

Three Phase Partitioning
DOI: https://doi.org/10.1016/B978-0-12-824418-0.00012-6

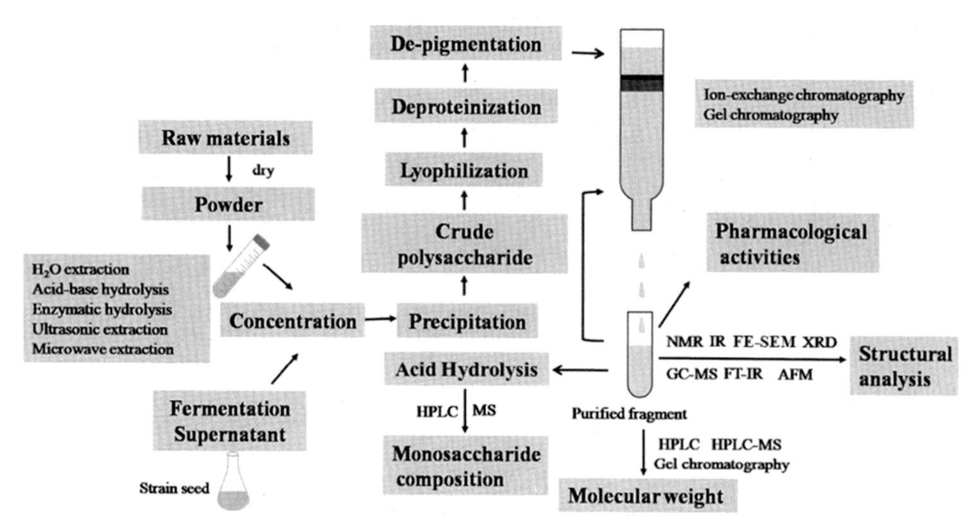

Fig. 11.1 ***Schematic diagram of extraction, separation, purification, structural analysis and pharmacological activities of PSs from natural resources.*** AFM: atomic force microscope, FE-SEM: field emission scanning electron microscope, FT-IR: Fourier-transform infrared spectroscopy, GC–MS: gas chromatography-mass spectrometer, HPLC: high performance liquid chromatography, MS: mass spectrum, NMR: nuclear magnetic resonance, XRD: X-ray diffraction.

have been widely used for the enhancement of functionality of foods in the food industry in recent years (Lu et al., 2019). The extraction and separation of PSs from natural resources is believed to be the necessary prerequisite for carrying out of the research on structural characterization, biological activities and practical applications of PSs. At present, solvent extraction (e.g. with hot water, acid, alkaline, and salt solutions) and assisted extraction (e.g. with enzyme, ultrasound, microwave, ultrahigh pressure) methods have been frequently employed to extract bioactive PSs from animals, plants and microorganisms (Garcia-Vaquero et al., 2017; Leong et al., 2021; Yi et al., 2020; Zhang et al., 2007). Afterwards, the resultant crude PSs must be subjected to a series of treatments, such as ethanol precipitation, deproteinization, decoloration and dialysis prior to purification. Moreover, a number of purification technologies including ultrafiltration, ion exchange chromatography, gel filtration chromatography and affinity chromatography have been used for further fractionation and purification (Bucar et al., 2013). Fig. 11.1 summarizes schematic diagram of extraction, separation, purification, structural analysis and pharmacological activities of PSs from natural resources. However, these existing methods are often expensive and time consuming, requiring enormous amounts of organic solvents and equipment with high investment and maintenance costs. Therefore, it is necessary to develop a simple, fast, green, economical and effective method for the extraction and separation of bioactive PSs, which can be potentially applied in the fields of food, medicine and cosmetics.

Compared with traditional extraction and separation approaches, three phase partitioning (TPP) as a promising alternative strategy, which typically consists of t-butanol, $(NH_4)_2SO_4$ and crude water extracts (or suspensions), is a method for extracting, purifying and concentrating biomolecules under the collective operation of salting out, isoionic precipitation, co-solvent precipitation, osmolytic and kosmotropic precipitation (Dennison and Lovrien, 1997; Roy and Gupta, 2002; Lovrein et al., 1987). Up to now, there are more than 200 papers that have been published on the utilization of TPP for extraction and separation of a variety of bioactive compounds from animals, plants and microorganisms, such as proteins, enzymes and their inhibitors, edible oils or lipids, carbohydrates and small-molecular organic compounds (Chew et al., 2018; Panadare and Rathod, 2017; Yan et al., 2018a). Sharma and Gupta (2002) reported for the first time the successful application of TPP technique in the precipitation and purification of three different commercial preparations of alginates. Subsequently, a variety of PSs, including native and hydrolyzed levan (Coimbra et al., 2010), aloe PS (APS) (Tan et al., 2015), rice bran PS (RBPS) (Wang et al., 2020b), *Corbicula fluminea* PS (Wang et al., 2017; Yan et al., 2017b), *Schizochytrium limacinum* PS (Chen et al., 2020), *Inonotus obliquus* PS (Liu et al., 2019), and *Phellinus baumii exopolysaccharides* (Wang et al., 2019a), have been extracted and separated from natural sources by using TPP approach under different TPP process conditions. Table 11.1 summarizes the TPP and TPP-based methods for extraction and separation of PSs from plants, animals, fungi and fermented broth. Studies have shown that the huge potential capability and efficacy of the TPP technique for extracting and separating bioactive PSs. Fig. 11.2 depicts a typical schematic diagram of extraction and separation of PSs from crude water extracts by using TPP technique. In the TPP process system, the crude water extracts containing PSs and proteins were mixed with $(NH_4)_2SO_4$ and allowed to dissolve completely, followed by the addition of *t*-butanol. After agitation (less than 1 h) and centrifugation, the mixture formed three clear phases, namely, a lower aqueous phase, a protein-rich intermediate phase, and an upper organic phase. The lower $(NH_4)_2SO_4$ phase is mainly used for the enrichment of PSs, and the obtained PSs can be further fractionized and purified by membrane separation or column chromatography; The intermediate phase mainly serves the purpose of deproteinization, which can replace the traditional Sevag method to remove free proteins from PSs; The upper t-butanol phase can be used to dissolve pigments instead of conventional decolorization method, and the *t*-butanol can be reused by rotary evaporation. Collectively, the traditional extraction and separation of PSs including ethanol precipitation, deproteinization and decolorization can be carried out in the TPP system by a single-step procedure.

This chapter is a discussion on TPP as an emerging alternative to extract and separate bioactive PSs from biological materials reported in recent years, including plants, animals and fungi. The main contents of this chapter include: (1) the major factors influencing TPP process for the extraction of PSs, (2) process intensification of TPP system,

Table 11.1 TPP and TPP-based methods for the extraction and separation of PSs from natural resources.

Source	Name	Adopted method	Process parameters	Extraction yield (percent)	Molecular weight (kDa)	Sugar composition	Bioactivity	Reference
Aloe	APS	TPP	$(NH_4)_2SO_4$ 26.35 percent (w/v), *t*-butanol 20.82 percent, 30 °C, pH 6.5		1100			Tan et al., 2015
Rice bran	RBPS	TPP	$(NH_4)_2SO_4$ 28 percent (w/v), slurry to *t*-butanol ratio 1:1.1 (v/v), 40 °C, 1 h, pH 5.10	2.09				Wang et al., 2020b
Peony Stamen		TPP	$(NH_4)_2SO_4$ 10 percent (w/v), *t*-butanol to crude extract 2:1, 60 min, pH 7.0			Rha:GalA:Glc: Gal:Ara = 1.24:1.59:2.0: 9.11:1.68		Luo et al., 2018
Momordica charantia L.	BPS-J	TPP	$(NH_4)_2SO_4$ 20 percent (w/v), *t*-butanol to bitter gourd juice 1.5:1.0 (v/v), 25 °C, 30 min	14.36	550.2(10.18 percent), 360.6 (26.67 percent), 243.1 (63.15 percent)	Ara:Xyl:Gal: Glc: Man:GalA = 9.6: 17.8:1 9.1: 17.4:8.2:1.0	Antioxidant, Hypoglycemic, Cholesterol-lowering	Yan et al., 2021a
Abelmoschus esculentus (L.) Moench	OPS	TPP	$(NH_4)_2SO_4$ 20 percent (w/v), *t*-butanol to bitter gourd juice 1.5:1.0 (v/v), 25 °C, 30 min		5270	Rha:Ara:Gal: Glc:GalA = 1.7:1.0: 16.4:1.9:2.3	Antioxidant, Hypoglycemic	Wang et al., 2020c

Corbicula fluminea		US-assisted TPP	$(NH_4)_2SO_4$ 20 percent (w/v), *t*-butanol 9.8 mL, 35.3 °C, 30 min, pH 6.0	9.32	1311.1, 41.5–92.8	Glc:GlcN: Man = 57.1:5.6 1.0	Antioxidant	Yan et al., 2017b
Corbicula fluminea	PSP	Enzyme-assisted TPP	2 percent papain 60 °C for 2 h; $(NH_4)_2SO_4$ 20 percent (w/v), *t*-butanol: crude extract = 1.5:1.0 (v/v), 35 °C, 30 min	9.0	61.5 (72.4 percent), 2113.4 (27.6 percent)	Glc:GlcN:Man = 10.8:4.4:1.0	Antioxidant	Wang et al., 2017
Corbicula fluminea	PS	US-synergized TPP	$(NH_4)_2SO_4$ 20 percent (w/v), *t*-butanol to crude extract ratio 1:1 (v/v), ultrasonic power 180 W, frequency 40 kHz, duty cycle 100 percent, 35 °C, 10 min	11.22				Yan et al., 2018b
Phellinus baumii	EPS	TPP	$(NH_4)_2SO_4$ 20 percent (w/v), cultured broth to *t*-butanol 1:1.5 (v/v), 35 °C, 30 min, pH 4.6	52.09	41.93 (64.6 percent), 231.3 (35.4 percent)	Ara:Man:Glc: Gal = 2.4: 29.3:3.9:1.0	Antioxidant, Hypoglycemic, Immunostimulatory	Wang et al., 2019a
Phellinus baumii	EPS	TPP	Sodium citrate 19 percent (w/v), dimethyl carbonate to cultured broth ratio 0.5:1.0 (v/v), 30 °C, pH 4.0	71.02	161 (96.1 percent), 1589 (3.9 percent)	Ara:Man: Glc:Gal = 1.4:16.5:5.3:1.0	Antioxidant	Wang et al., 2019b

(continued)

Table 11.1 (Cont'd)

Source	Name	Adopted method	Process parameters	Extraction yield (percent)	Molecular weight (kDa)	Sugar composition	Bioactivity	Reference
Schizochytrium limacinum		Enzyme-assisted TPP	2 percent protamex 45 °C for 2 h at pH 3.0; $(NH_4)_2SO_4$ 34 percent (w/v), slurry to *t*-butanol ratio 1:2.4, 40 °C, 41 min	5.16			Antioxidant	Chen et al., 2020
Inonotus obliquus		TPP	Solid-liquid ratio 1 g to 12 mL, $(NH_4)_2SO_4$ 20 percent (w/v), *t*-butanol 11 mL, 30 °C, 30 min, pH 8.0	2.2	40	Gal:Glc:Xyl: Man = 2.0:3.5: 1.0:1.5	Antioxidant, Immuno-modulatory	Liu et al., 2019

Note: APS: aloe polysaccharide, BPS-J: bitter gourd juice polysaccharide, Ara: arabinose, EPS: exopolysaccharides, Gal: galatose, GalA: galacturonic acid, Glc: glucose, GlcN: glucosamine, Man: mannose, OPS: okra polysaccharide, PSP: polysaccharide-protein complexes, RBPS: rice bran polysaccharides, Rha: rhamnose, Xyl: xylose, US: ultrasound.

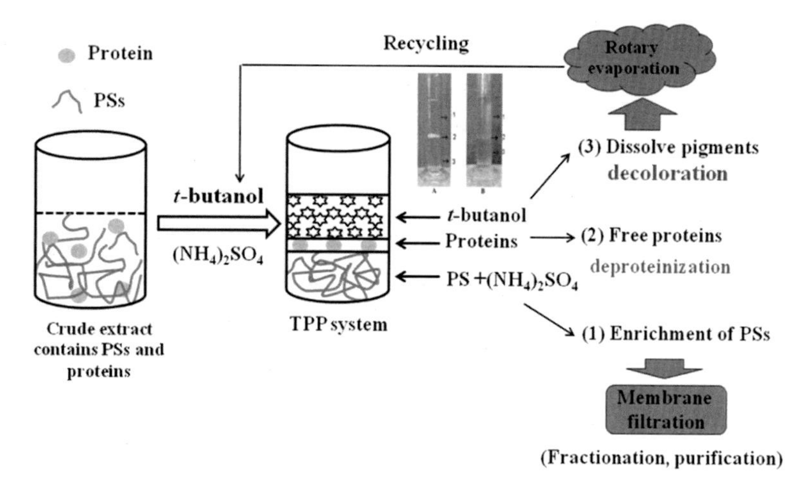

Fig. 11.2 ***Schematic diagram of extraction and separation of PSs by using TPP.***

(3) TPP combined with downstream techniques, and (4) the effects of TPP separation on physicochemical, structural and biological properties of PSs. This chapter intends to provide a meaningful reference for the future research directions of extraction and separation of bioactive PSs from different natural resources by using the TPP technique.

11.2 Factors affecting the TPP process for extraction of PSs

In order to obtain the best separation performance, the effects of process parameters, such as $(NH_4)_2SO_4$ concentration, amount of t-butanol, pH, temperature and time, on the extraction yield (or efficiency) of PSs in the lower aqueous phase should be investigated during the TPP process (Fig. 11.3). Response surface methodology (RSM) has been used to optimize these TPP process parameters (Chen et al., 2020; Wang et al., 2020b;Yan et al., 2017b). In this part, these main factors influencing the TPP process for the extraction of PSs will be discussed.

11.2.1 Ammonium sulfate

It has been reported that $(NH_4)_2SO_4$ as the most popular kosmotropic salt used for "salting out" plays an important role in macromolecule partitioning in the TPP process (Dennison and Lovrein, 1997). Within the measured $(NH_4)_2SO_4$ concentration, the extraction yield of PSs initially increased and then decreased. For example, Yan et al. (2017b) reported that the extraction yield of *C. fluminea* PS (CFPS) firstly increased and then reduced as mass fraction of $(NH_4)_2SO_4$ increased from 10 percent to 60 percent (w/v), and the maximum CFPS extraction yield (9.06 percent) was attained at 20 percent (w/v) $(NH_4)_2SO_4$. At a lower concentration of $(NH_4)_2SO_4$, the effect of "salting in" may be considered as a main drive forcing for the efficient separation of PSs

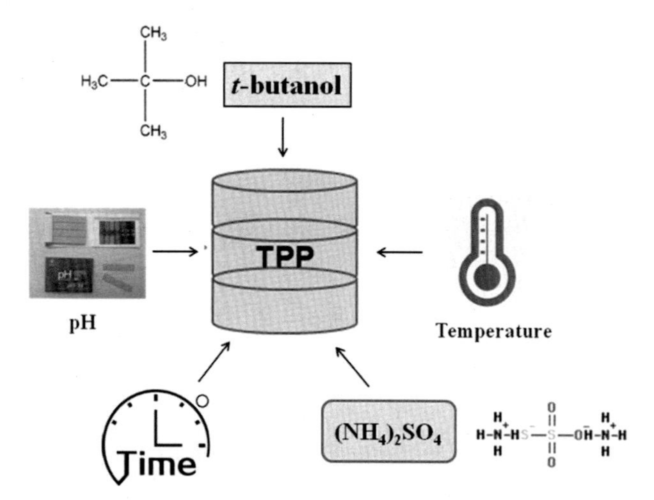

Fig. 11.3 ***Factors affecting the TPP process for extraction of PSs.***

in the lower aqueous phase (Yan et al., 2017b). However, at high salt concentration, the water molecules are attracted by salts through the stronger "salting out" effect. This leads to larger breakage of water-water hydrogen bond networks and weakens the availability of water molecules for solvation of carbohydrates, resulting in an obvious decrease in the extraction yield of PSs (Dutta et al., 2015; Narayan et al., 2008). Also, excessive salt will cause the denaturation of proteins and lowers the extraction yield of PSs (Tan et al., 2015). Thus, $(NH_4)_2SO_4$ must be optimized to obtain the maximum extraction yield of PSs in the TPP system.

11.2.2 *t*-butanol

As described in published literatures (Chew et al., 2018; Panadare and Rathod, 2017; Yan et al., 2018a), *t*-butanol is the most popular organic solvent for partitioning proteins and PSs in the TPP process, due to its unique features, such as dissolving non-polar substances, flotation agent, stabilizing protein structure, and preventing proteolysis and protein-protein interactions (Dennison and Loverian, 1997; Dennison et al., 2000). In the TPP separation system, *t*-butanol as a kosmotrope agent and usually exerts an important synergistic role with $(NH_4)_2SO_4$. Specifically, if the amount of t-butanol is lower, it is not enough to cause separation through synergy with $(NH_4)_2SO_4$. However, if the amount of t-butanol is higher, it is more likely to result in protein denaturation and prevent protein precipitation in the interfacial layer (Sharma and Gupta, 2001). Noticeably, the increased amounts of *t*-butanol do not increase the extraction yield of PSs in the lower aqueous phase during the TPP process since the solution viscosity increases as the amount of t-butanol increases. The increased solution viscosity may decrease molecular mobility, thus reducing the interactions of molecules. Additionally, excess amounts of *t*-butanol lead to reduced selective partitioning of PSs in the lower aqueous phase, hence, lowering the extraction yield of PSs, which may be related with the lack of sufficient water for adequate hydration of SO_4^{2-} (Dennison and Lovrien, 1997). In view of above discussions, an optimal TPP separation process can be well established by selecting the appropriate t-butanol to crude water extract (or suspension) ratio at a constant concentration of $(NH_4)_2SO_4$. Indeed, many studied have revealed that with the increase in the amount of *t*-butanol, the extraction yield of PSs experienced a first increase and then decrease when keeping other parameters, including $(NH_4)_2SO_4$ concentration, pH, temperature and time constant (Chen et al., 2020; Wang et al., 2017, 2019a, 2020b; Yan et al., 2017b).

11.2.3 pH

pH as one of the critical factors changes the distribution and partitioning of macromolecules because of electrostatic interactions between phases and charged macromolecules (Gagaoua and Hafid, 2016). In general, regulating the pH of the TPP system affects the surface charges of ionized proteins, thereby affecting the extraction and separation of PSs (Dennison and Lovrien, 1997). As illustrated in Table 11.1, the TPP system at the weak acidic pH condition is more efficient for extraction of PSs, which may be attributed to the acidic PSs or glycoproteins mainly extracted from natural resources. For example, Wang et al. (2020b) found that the extraction yield of rice bran PS (RBPS) from rice bran increased in the range of pH 3.0–5.0, but there was no significant change in the RBPS extraction yield when the pH above 5.0 was used. Similar results have been also reported in previous studies (Tan et al., 2015; Wang et al., 2019a; Yan et al., 2017b).

11.2.4 Temperature

Temperature as another physical parameter has an important impact on the rate of mass transfer, and thus influencing the separation performance of TPP experiments. Coimbra et al. (2010) demonstrated that the temperature of the TPP system controlled at 15 °C−35 °C is more conducive to the enrichment of PSs in the lower aqueous phase. This may be because this temperature region is beneficial for the expansion of PSs chains, resulting in the exposure of a large number of hydroxyl groups, promoting the formation of more hydrogen bonds, and thereby making PSs more hydrophilic and more concentrated in the lower aqueous phase. Similarly, Tan et al. (2015) and Yan et al. (2017b) also found that the TPP separation experiment operated below 35 °C, resulting in the maximum aloe PS (APS) extraction efficiency and *C. fluminea* PS (CFPS) extraction yield. In addition, the temperature at 40 °C for efficient bioseparation of starch and chitosan in the TPP system was selected by Mondal et al. (2004) and Sharma et al. (2003), respectively. Moreover, a higher temperature will lead to excessive energy consumption. Therefore, in consideration of both extraction efficiency and energy consumption, the TPP system consisting of

$(NH_4)_2SO_4$ and *t*-butanol for extraction and separation of bioactive PSs is generally advised to operate at low temperatures (<40 °C) (Table 11.1).

11.2.5 Time

Extraction time is one of important physical parameters affecting the TPP separation process, which has appreciable influences on the yield, quality and production cost of the extracted PSs. Most of studies have shown that extraction time had no apparent impacts on the extraction yield of PSs in the lower aqueous phase, and 30 min of extraction time was deemed to be enough for the TPP to achieve equilibrium in extracting bioactive PSs (Wang et al., 2019a, 2020b; Yan et al., 2017b).

11.3 Process intensification of TPP system for PSs extraction

To improve partitioning performance, several assisted or synergized methods as process intensification tools, such as enzyme-assisted TPP, ultrasound (US)-assisted TPP and US-synergized TPP, have been developed as variations on the standard TPP separation process, and have been applied for the efficient extraction and separation of PSs from various biomaterials (Table 11.1). In the following, these assisted or synergized TPP systems for highly efficient extraction and separation of PSs are briefly introduced by taking examples of their uses and underlying mechanisms.

11.3.1 Enzyme-assisted TPP

Enzyme-assisted extraction of plant biomolecules, as an efficient, benign, sustainable and environmentally friendly extraction technology, has become a potential alternative to traditional solvent extraction methods and has attracted increasing attention (Nadar et al., 2018). At present, enzyme-assisted TPP has been widely adopted by many researchers for the efficient recovery of polyphenols, oils, PSs, flavors and colorants from various natural sources (Marathe et al., 2017). Wang et al. (2017) developed an enzyme-assisted TPP for the extraction and separation of polysaccharide-protein complexes (PSP) from *C. fluminea* (Fig. 11.4, Table 11.1). The highest extraction yield of PSP was 9.0 percent under the optimized conditions: 2 percent (w/v) papain, 20 percent (w/v) $(NH_4)_2SO_4$ concentration, 1.5: 1.0 (v/v) *t*-butanol to crude extract ratio, 30 min and 35 °C, which was higher than the yield (7.6 percent) of E-PSP extracted with traditional enzyme-assisted extraction. Moreover, PSP showed higher antioxidant activities in vitro than that of E-PSP. Very recently, Chen et al. (2020) also used enzyme-assisted TPP to simultaneously extract oil and PSs from *S. limacinum*. Results showed that *S. limacinum* was pretreated with 2.0 percent protamex at pH 3.0 and 45 °C for 2 h, and the maximum PS yield (5.16 percent) was obtained under the following optimal TPP conditions through Box-Behnken design: 34 percent $(NH_4)_2SO_4$ mass fraction, 1:2.4 slurry to *t*-butanol ratio, 41 min and 40 °C. Furthermore, the algal PSs that was extracted by enzyme-assisted TPP exhibited potential antioxidant activity in vitro.

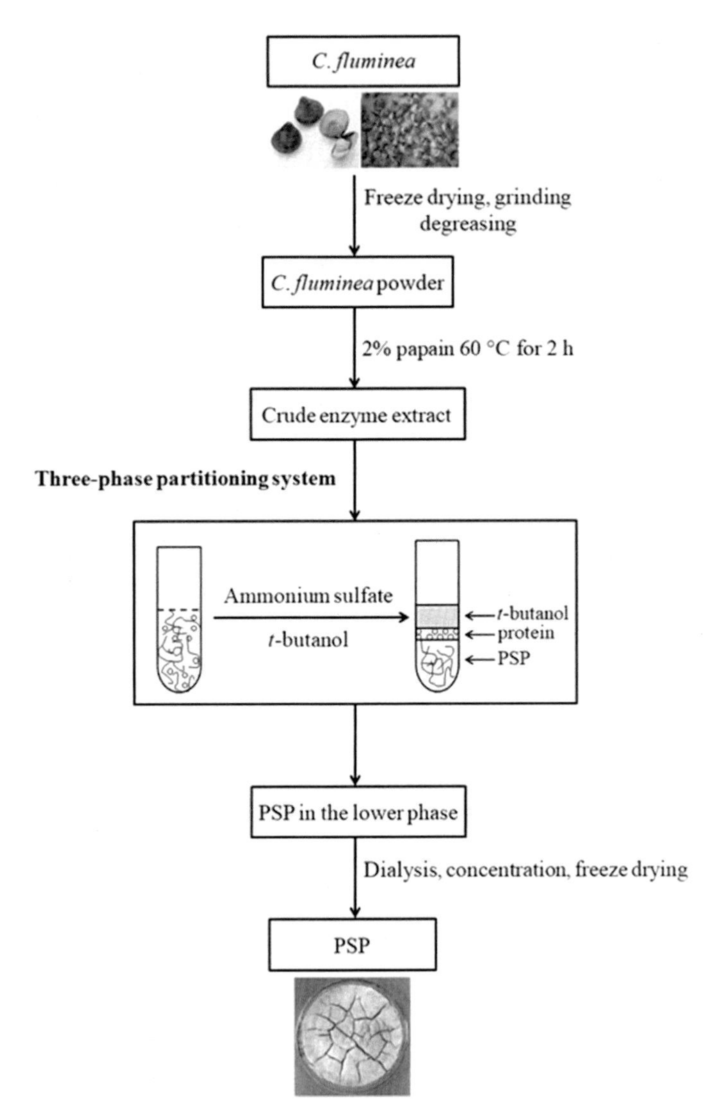

Fig. 11.4 ***Experimental procedures for the extraction and separation of polysaccharide-protein complexes (PSP) from C. fluminea by using enzyme treatment and TPP process*** (Wang et al., 2017).

11.3.2 US-assisted TPP

Compared with conventional solvent extraction, US-assisted extraction as a non-thermal extraction technique has been widely utilized for the extraction of biologically active molecules from plant, animal and marine sources in much shorter times, at low temperature, with less energy and solvent consumption (Ojha et al., 2020). In general, ultrasonic waves used for improving extraction efficiency are mainly in the frequency range of 20–100 kHz, where large cavitation bubbles are produced and the collapse of these bubbles causes extreme mechanical shear forces. Prior to TPP separation experiments, US treatment can effectively destroy the cell wall of biological material and promote the dissolution of bioactive molecules including PSs through cavitation and mechanical effects. For instance, Yan et al. (2017b) has used US pretreatment (20 kHz, 300 W, 60 min, and 25 °C) and TPP technique in sequence for the extraction and separation of CFPS from *C. fluminea.* The highest extraction yield of CFPS (9.32 percent) was obtained under optimized process parameters of 20.0 percent (w/v) mass fraction of $(NH_4)_2SO_4$, 9.8 mL t-butanol, 35.3 °C, 30 min and pH 6.0, which was slightly higher than that of PSP (9.0 percent) extracted with enzyme-assisted TPP (Wang et al., 2017). The obtained CFPS were proteoglycans with high molecular weight (MW) of 1311.1 kDa and low MW of 41.5–92.8 kDa and mainly contained glucose (Glc), glucosamine (GlcN) and mannose (Man) in a molar ratio of 57.1: 5.6: 1.0, possessing strong free-radical scavenging abilities and antioxidant property in vitro.

11.3.3 US-synergized TPP

As we know, TPP is a mass transfer based phenomenon and process intensification in mass transfer results into not only improved partitioning efficiency of the product but also enhanced yield of the target product (Yan et al., 2018a). Meanwhile, US as a simple and feasible process intensification tool has been applied in a great number of separation protocols to enhance mass transfer and yield during extraction, fermentation, adsorption, bioremediation and biocatalysis processes (Gogate and Kabadi, 2009; Santos et al., 2009). For these reasons, TPP coupled with US has been successfully employed to extract and recovery various biomolecules from natural sources, such as custard apple seed oil (Panadare et al., 2020), curcuminoids (Patil and Rathod, 2020), mangiferin (Kulkarni and Rathod, 2014), ursolic acid and oleanolic acid (Vetal et al., 2014), astaxanthin (Chougle et al., 2014), fibrinolytic enzyme (Avhad et al., 2014), and serratiopeptidase (Pakhale and Bhagwat, 2016). As for the extraction of PSs, Yan et al. (2018b) pointed out that ultrasound performed in ultrasonic water bath (180 W, 40 kHz, 10 min, 35 °C) during extraction and separation of *C. fluminea* PSs via TPP resulted in about 20.39 percent increase in extraction yield and about 66.67 percent reduction in extraction time compared with conventional TPP protocol, and further showed that the enhanced or synergized effects of US on TPP separation is mainly attributed to the ultrasonic cavitation, as well as the accompanied physical and chemical effects (Fig. 11.5). Moreover, Wang et al. (2019a) demonstrated that a dual-frequency power ultrasound in a sequential mode coupled with TPP resulted in about 9.12 percent increase in extraction yield of *P. baumii* exopolysaccharides (EPSs) and about 80 percent reduction in extraction time as compared to the traditional TPP protocol. Thus, US-synergized TPP can be explored as a promising and time-saving option for the high-efficient extraction of bioactive PSs from different natural sources.

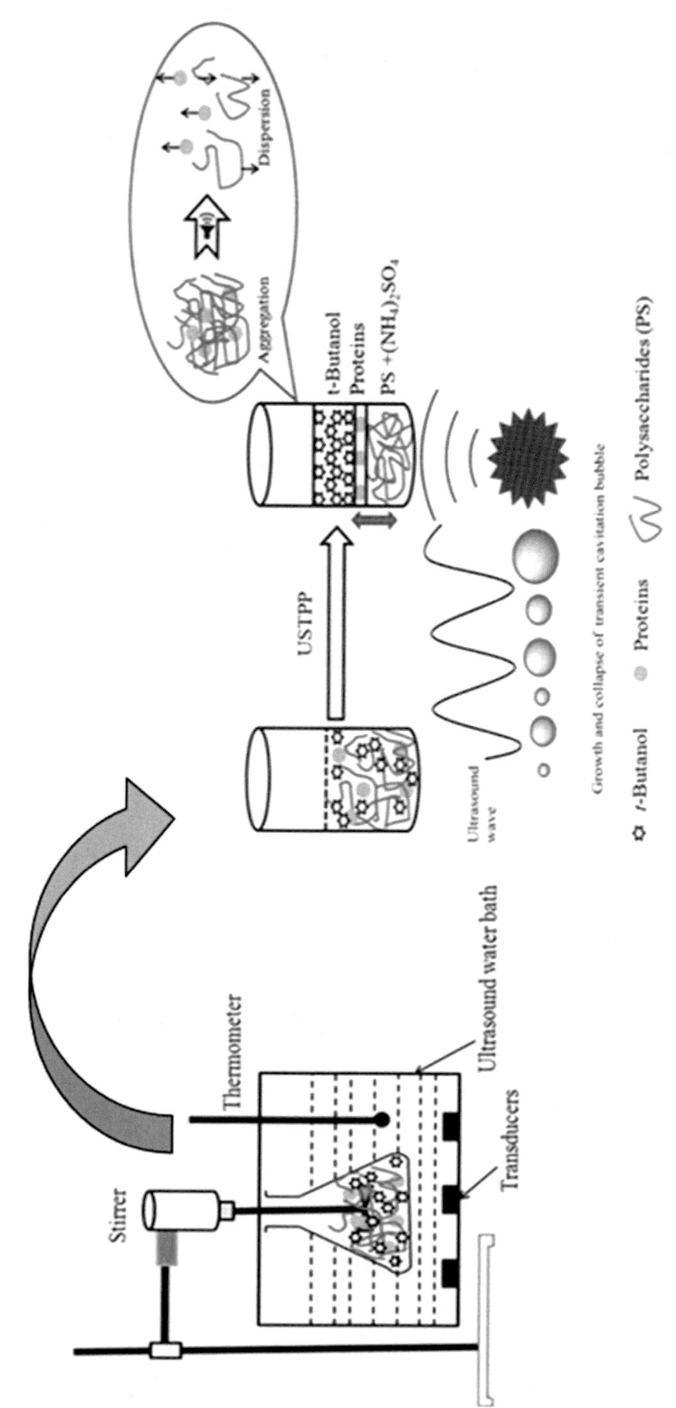

Fig. 11.5 Schematic diagram of the extraction and separation of PSs from C. fluminea and its corresponding mechanism by using TPP coupled with ultrasound (Yan et al., 2018b).

11.4 TPP combined with downstream techniques

TPP separation process can not only improve the extraction yield (or efficiency) of PSs by combining with upstream technologies such as enzyme and US pretreatments, but also further fractionate and purify PS by combining with downstream technologies such as column chromatography, membrane separation and non-solvent gradient precipitation. For example, Yan et al. (2017c) reported that the CFPS, which was extracted from *C. fluminea* via TPP approach, was further fractionated and purified through column chromatographies to obtain a novel proteoglycan (CFPS-11). The CFPS-11 was composed of Glc and GlcN in a molar ratio of 12.2:1.0 with a MW of 807.7 kDa. The CFPS-11 structure comprised of a (1→4)-α-D-glucopyranosyl (Glcp) backbone, and the (1→3)-α-D-Glcp residues formed branches at the O-6 position with 1-linked-α-D-Glcp and 1-linked-α-D-*N*-acetylglusamine terminal residues (Fig. 11.6). Moreover, the protein moiety (~5 percent) was covalently bonded to the polysaccharide chain of CFPS-11 in *O*-linkage via both serine and thereonine residues. Very recently, Wang et al. (2020c) for the first time used TPP coupled with gradient $(NH_4)_2SO_4$ precipitation (GNP) method to rapidly separate and purify bioactive polysaccharides (OPSs) from fresh okra (*Abelmoschus esculentus* (L.) Moench) pods (Fig. 11.7). Results demonstrated that OPS-50 obtained via GNP at a saturation of 50 percent $(NH_4)_2SO_4$ after TPP had higher precipitation yield, carbohydrate (93.47 percent) and uronic acid (22.34 percent) contents with similar monosaccharide composition and different molar ratios than the other OPSs, but showed lower MW (3650 kDa), more flexible chain and looser microstructures. Furthermore, OPS-50 exhibited stronger antioxidant activity and α-amylase and α-glucosidase inhibitory activities in vitro. Meanwhile, Yan et al. (2021b) developed a method combining TPP with gradient ethanol precipitation (GEP) to extract and fractionate garlic PSs (GPSs: GPS35, GPS50, GPS65, GPS80) from raw garlic (*Allium sativum* L.) bulbs (Fig. 11.8). GPS80 as a fructose polymer showed higher sugar (86.68 percent) and uronic acid (12.89 percent) contents, lower MW (8.93 kDa) and looser surface morphology compared with the other three GPSs. Among the four GPSs, GPS80 possessed the best antioxidant potential, α-amylase and α-glucosidase inhibitory activities, and NO stimulatory activity on RAW264.7 macrophages in vitro. Consequently, TPP combined with non-solvent gradient precipitation (e.g. GNP and GEP) has been developed as a promising and feasible bioseparation strategy for high-efficient separation

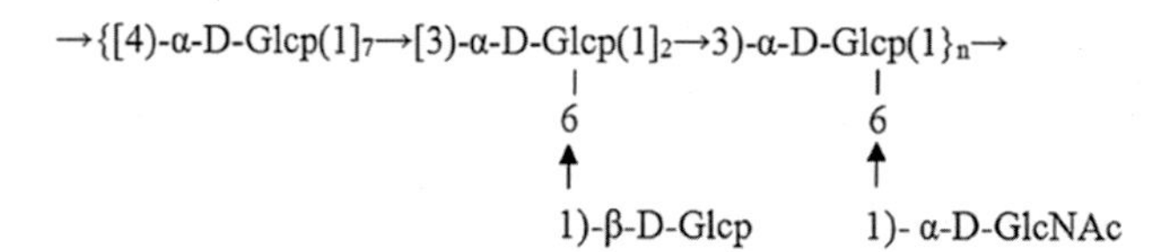

Fig. 11.6 ***Possible chemical structure of repeating unit of CFPS-11 purified from Corbicula fluminea*** (Yan et al., 2017c).

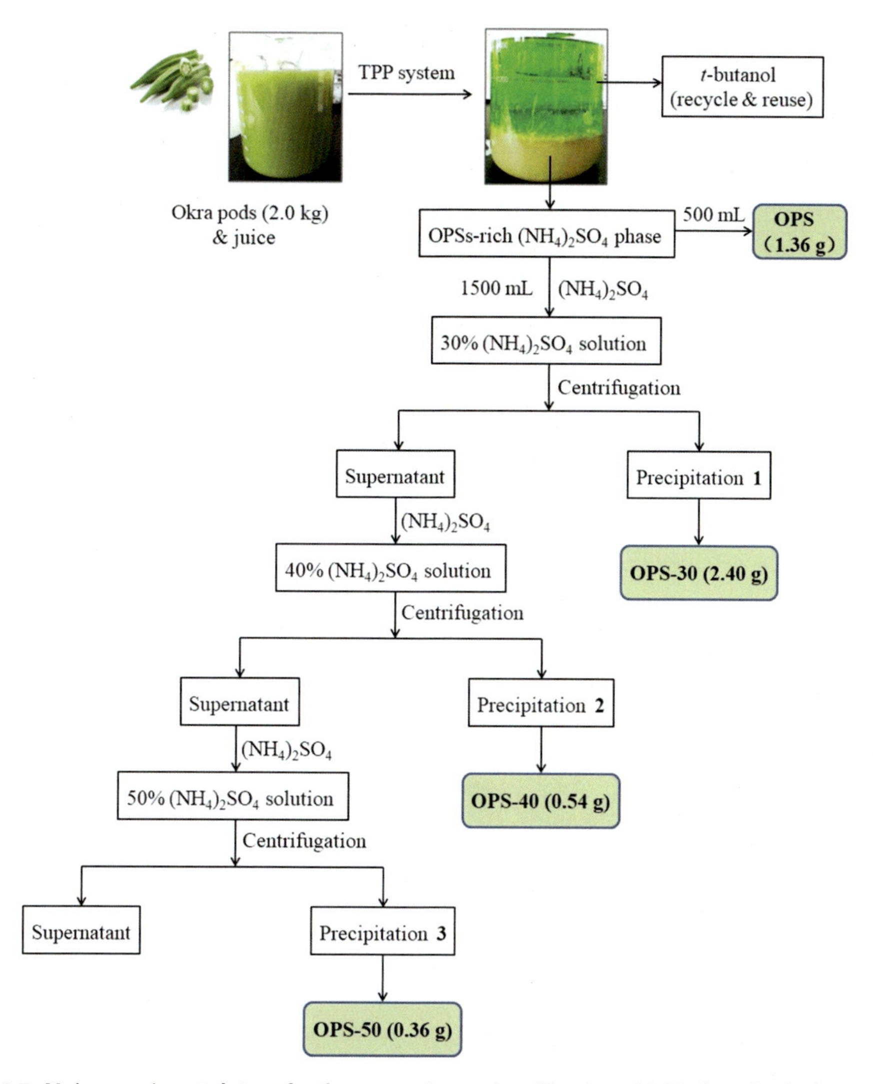

Fig. 11.7 ***Main experimental steps for the separation and purification of OPSs from fresh okra pods by using TPP and TPP coupled with gradient $(NH_4)_2SO_4$ precipitation methods*** (Wang et al., 2020c).

and fractionation of bioactive PSs with high purity from fresh biomaterials for different food, cosmetic, and pharmaceutical applications.

11.5 TPP separation influencing the properties of PSs

It has been well documented that the tertiary structure of Proteinase K underwent evident changes after subjecting it to TPP separation system (Singh et al., 2001). Correspondingly, it was necessary to explore the physicochemical and biofunctional

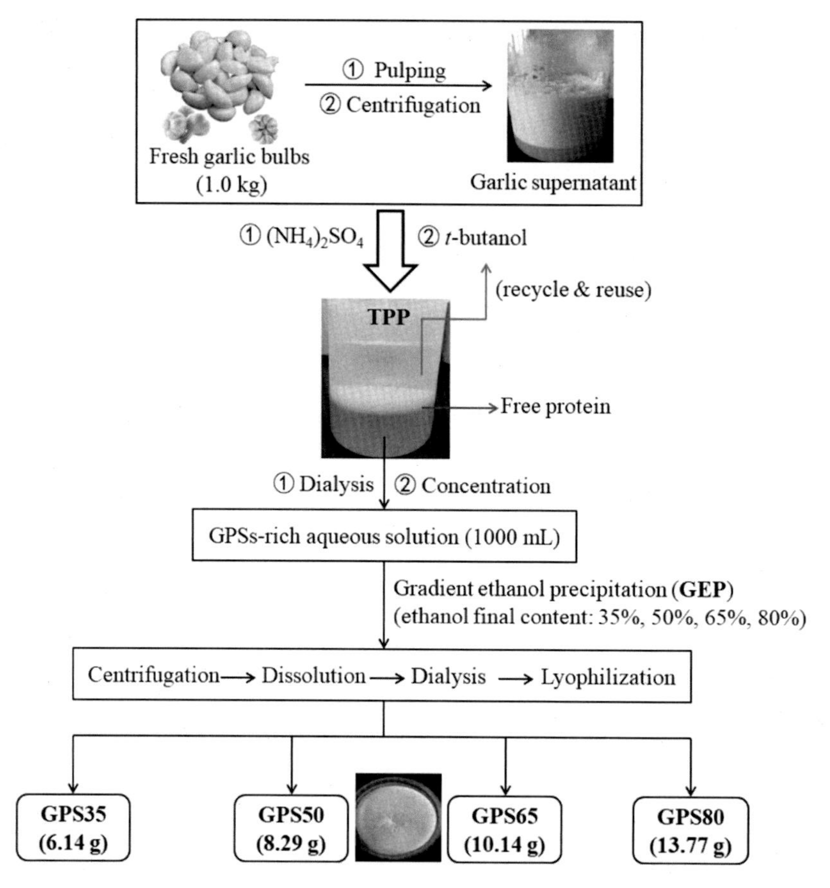

Fig. 11.8 ***Extraction and isolation procedures for GPSs from raw garlic bulbs by using TPP combined with GEP approach*** (Yan et al., 2021b).

consequences of conducting TPP with PSs. Sharma et al. (2003) investigated the changes in physicochemical properties and biodegradability of chitosan before and after TPP treatment. They found that TPP-treated chitosan was completely soluble at pH 6.0, but the solubility behavior of TPP-treated chitosan showed distinct changes. They also mentioned that both lyophilization and TPP treatment led to structural changes of chitosan, which was not conducive to the interaction with magnesium ions. Additionally, TPP treatment can also result in decrease in the biodegradability of chitosan. Similarly, Mondal et al. (2004) demonstrated that TPP treatment resulted in decrease in susceptibility of potato starch towards amylolytic hydrolysis, which was mainly attributed to the decreased intra-molecular hydrogen bonding. Therefore, TPP can be developed as a feasible and powerful tool to achieve desirable changes in the properties of PSs for potential applications in functional foods, medicine and cosmetics.

To highlight the advantages of TPP separation technique in the green and efficient preparation of bioactive PSs from natural sources, some conventional extraction methods (e.g. enzyme-assisted extraction, ultrasonic-assisted extraction) and separation techniques (e.g. ethanol precipitation, deproteinization, decolorization, and dialysis) have also been used to extract and separate PSs from the same materials (Fig. 11.1), and the physicochemical and biological properties of these PSs are evaluated and compared. For instance, Wang et al. (2017) have extracted and separated PSP from *C. fluminea* via enzyme-assisted TPP approach, and comparatively investigated physicochemical characteristics and in vitro antioxidant activities of PSP with that of E-PSP separated by using traditional enzyme extraction and a series of separation protocols. Compared with E-PSP, PSP had higher extraction yield (9.0 percent) and carbohydrate content (81.7 percent), but possessed lower MW (61.5 kDa, 72.4 percent). Both PSP and E-PSP are proteoglycans with different monosaccharide and amino acid compositions, whereas their primary structures were not significantly destroyed after subjecting to TPP separation. PSP exhibited better free radical scavenging ability and antioxidant capacity in vitro than that of E-PSP. Likewise, Yan et al. (2018b) reported that PS from *C. fluminea* extracted with US-synergized TPP had higher yield (11.22 percent) than those of PSs separately extracted with TPP (9.32 percent) and US extraction (USE, 6.05 percent). They revealed that the synergistic effect of US in TPP separation process might be attributed to the accelerated mass transfer rate by US, which was beneficial for the separation of PSs and protein, thereby resulting in an increase in extraction yield of PSs in the lower phase, as well as a reduction in extraction time (Fig. 11.5). Recently, Wang et al. (2019a) directly extracted and separated EPSs from the fermentation broth of *Phellinus baumii* by using TPP technique, and its physicochemical characteristics and biological activities was compared with those of EPS-C obtained by conventional ethanol precipitation and separation protocols. EPS and EPS-C both showed similar structural features and different monosaccharide compositions and MWs. EPS with higher carbohydrate (88.21 percent) and uronic acid (3.37 percent) contents exhibited better antioxidant, hypoglycemic and immunostimulatroy activities in vitro than those of EPS-C. Furthermore, Wang et al. (2019b) found that EPS-D obtained from the cultured broth of *P. baumii* via the TPP system with dimethyl carbonate (DMC) and sodium citrate (SC) had higher extraction yield (66.90 percent), carbohydrate (89.67 percent) and uronic acid (3.56 percent) contents as compared to the EPS obtained via the TPP system with t-butanol and $(NH_4)_2SO_4$, as well as different monosaccharide compositions and MWs. EPS-D possessed stronger antioxidant potential in vitro than that of EPS-D. These outcomes demonstrated that the newly developed TPP system with DMC as a greener organic solvent has great potential to replace t-butanol for the highly -efficient separation of bioactive PSs from various natural sources.

11.6 Conclusions

TPP, as a promising alternative bioseparation technology, has attracted considerable attention and has been extensively used for the separation and recovery of various biomolecules such as proteins, enzymes and their inhibitors, lipids and oils, carbohydrates, and small-molecular organic compounds from their natural sources (Yan et al., 2018a). In this chapter, TPP has been discussed as an efficient, economical and simple method to extract bioactive PSs from natural resources. Process parameters such as $(NH_4)_2SO_4$ concentration, t-butanol amount, pH, temperature and time have important influences on the extraction yields of PSs during the TPP experiments. Moreover, the extraction efficiency of TPP technique for PSs can be improved by adopting process intensification tools including use of enzymes and ultrasonics. Several emerging physical processing technologies, such as microwave, ultrahigh pressure, pulsed electric field, and cold plasma, can also be used as potential intensification tools in the TPP process for the extraction of PSs. Additionally, the TPP separation can be combined with downstream technologies, like non-solvent gradient precipitation and membrane separation, to obtain homogenous PSs. Compared with traditional extraction and separation methods, TPP can not only enhance the yield and purity of PSs, but also improve their biological activities. Overall, it would certainly be beneficial to develop rapid and direct bioseparation platforms such as TPP for the extraction and separation of bioactive PSs from different natural sources in the context of reducing operating costs and simplifying the techniques required to obtain the targeted PSs.

Acknowledgements

This work was supported financially by the National Natural Science Foundation of China (31,671,812) and Jiangsu Overseas Research & Training Program for University Prominent Young & Middle-aged Teachers and Presidents.

References

Avhad, D.N., Niphadkar, S.S., Rathod, V.K., 2014. Ultrasound assisted three phase partitioning of a fibrinolytic enzyme. Ultrason. Sonochem. 21, 628–633.

Bucar, F., Wube, A., Schmid, M., 2013. Natural product isolation-how to get from biological material to pure compounds. Nat. Prod. Rep. 30, 525–545.

Chen, W., Jia, Z., Huang, G., Hong, Y., 2020. Global optimization for simultaneous extraction of oil and polysaccharides from *Schizochytrium limacinum* by enzyme-assisted three-phase partitioning. J. Food Process Preserv. 44, e14824.

Chew, K.W., Ling, T.C., Show, P.L., 2018. Recent developments and applications of three-phase partitioning for the recovery of proteins. Sep. Purif. Rev. 48 (1), 52–64.

Chougle, J.A., Singhal, R.S., Baik, O.D., 2014. Recovery of astaxanthin from Paracoccus NBRC 101723 using ultrasound-assisted three phase partitioning (UA-TPP). Sep. Sci. Technol. 49, 811–818.

Coimbra, C.G.O., Lopes, C.E., Calazans, G.M.T, 2010. Three-phase partitioning of hydrolyzed levan. Bioresource Technol. 101, 4725–4728.

Dennison, C., Lovrien, R., 1997. Three phase partitioning: concentration and purification of proteins. Protein Expres. Purif. 11, 149–161.

Dennison, C., Moolman, L., Pillay, C.S., Meinesz, R.E., 2000. *t*-Butanol: nature gift for protein isolation. S. Afr. J. Sci. 96, 159–160.

Dutta, R., Sarkar, U., Mukherjee, A., 2015. Process optimization for the extraction of oil from *Crotalaria juncea* using three phase partitioning. Ind. Crop. Prod. 71, 89–96.

Gagaoua, M., Hafid, K., 2016. Three phase partitioning system, an emerging non-chromatographic tool for proteolytic enzymes recovery and purification. Biosensors J. 5 (1), 1–4.

Garcia-Vaquero, M., Rajauria, G., O'Doherty, J.V., Sweeney, T., 2017. Polysaccharides from macroalgae: recent advances, innovative technologies and challenges in extraction and purification. Food Res. Int. 99, 1011–1020.

Gogate, P., Kabadi, A., 2009. A review of applications of cavitation in biochemical enginnering/biotechnology. Biochem. Eng. J. 44, 60–72.

Hou, C., Chen, L., Yang, Z., Ji, X., 2020. An insight into anti-inflammatory effects of natural polysaccharides. Int. J. Biol. Macromol. 153, 248–255.

Kulkarni, V.M., Rathod, V.K., 2014. Extraction of mangiferin from Manifera indica leaves using three phase partitioning coupled with ultrasound. Ind. Crop. Prod. 52, 292–297.

Leong, Y.K., Yang, F.C., Chang, J.S., 2021. Extraction of polysaccharides from edible mushrooms: emerging technologies and recent advances. Carbohyd. Polym. 251, 117006.

Liu, Z., Yu, D., Li, L., Liu, X., Zhang, H., Sun, W., Lin, C.C., Chen, J., Chen, Z., Wang, W., Jia, W., 2019. Three-phase partitioning for the extraction and purification of polysaccharides from the immunomodulatory medicinal mushroom *Inonotus obliquus*. Molecules 24, 403.

Lovrein, R., Goldensoph, C., Anderson, P.C., Odegaard, B., 1987. Three phase partitioning (TPP) via t-butanol: enzyme separation from crudes. In: Burgess, R (Ed.), Protein Purification: Micro to Macro. A.R. Liss, Inc., New York, pp. 131–148.

Lu, X., Chen, J., Guo, Z., Zheng, Y., Rea, M.C., Su, H., Zheng, X., Zheng, B., Miao, S., 2019. Using polysaccharides for the enhancement of functionality of foods: a review. Trends Food Sci. Tech. 86, 311–327.

Luo, L., Xue, Y., Yang, Y., Zhu, W., 2018. Study on the purification of polysaccharide from *Peony Stamen* by three phase partitioning and physicochemical property. Food Machinery 34 (8), 123–134 (in Chinese).

Marathe, S.J., Jadhav, S.B., Bankar, S.B., Singhal, R.S., 2017. Enzyme-assisted extraction of bioactives. In: Puri, M (Ed.), Food Bioactives: Extraction and Biotechnology Applications. Springer International Publishing, Cham, pp. 171–201.

Mondal, K., Sharma, A., Gupta, M.N., 2004. Three phase partitioning of starch and its structural consequences. Carbohyd. Polym. 56, 355–359.

Nadar, S.S., Rao, P., Rathod, V.K., 2018. Enzyme assisted extraction of biomolecules as an approach to novel extraction technology: a review. Food Res. Int. 108, 309–330.

Narayan, A.V., Madhusudhan, M.C., Raghavarao, K.S.M.S, 2008. Extraction and purification of *Ipomoea* peroxidase employing three-phase partitioning. Appl. Biochem. Biotech. 151, 263–272.

Ojha, K.S., Aznar, R., O'Donnell, C., Tiwari, B.K., 2020. Ultrasound technology for the extraction of biologically active molecules from plant, animal and marine sources. TrAC-Trend Anal. Chem. 122, 115663.

Pakhale, S.V., Bhagwat, S.S., 2016. Purification of serratiopeptidase from Serratia marcescens NRRL B 23112 using ultrasound assisted three phase partitioning. Ultrason. Sonochem. 31, 532–538.

Panadare, D.C., Gondaliya, A., Rathod, V.K., 2020. Comparative study of ultrasonic pretreatment and ultrasound assisted three phase partitioning for extraction of custard apple seed oil. Ultrason. Sonochem. 61, 104821.

Panadare, D.C., Rathod, V.K., 2017. Three phase partitioning for extraction of oil: a review. Trends Food Sci. Tech. 68, 145–151.

Patil, S.S., Rathod, V.K., 2020. Synergistic effect of ultrasound and three phase partitioning for the extraction of curcuminoids from *Curcuma longa* and its bioactivity profile. Process Biochem. 93, 85–93.

Roy, I., Gupta, M.N., 2002. Three-phase partitioning of proteins. Anal. Chem. 300, 11–14.

Santos, H.M., Lodeiro, C., Capelo-Martienz, J.L., 2009. The power of ultrasound. In: Capelo-Martienz, JL (Ed.), Ultrasound in Chemistry: Analytical Applications. Wiley-VCH Verlag GmbH cCo. KgaA, Weinheim, Germany, pp. 1–10.

Sharma, A., Gupta, M.N., 2001. Three phase partitioning as a large-scale separation method for purification of a wheat germ bifunctional protease/amylase inhibitor. Process Biochem. 37, 193–196.

Sharma, A., Gupta, M.N., 2002. Three phase partitioning of carbohydrate polymers: separation and purification of alginates. Carbohyd. Polym. 48, 391–395.

Sharma, A., Mondal, K., Gupta, M.N., 2003. Some studies on characterization of three phase partitioned chitosan. Carbohyd. Polym. 52, 433–438.

Singh, R.K., Gourinath, S., Sharma, S., Roy, I., Gupta, M.N., Betzel, C., Srinivasan, A., Singh, T.P., 2001. Enhancement of enzyme activity through three phase partitioning: crystal structure of a modified serine protease at 1.5 A resolution. Protein. Eng. 14, 307–313.

Tan, Z.J., Wang, C.Y., Yi, Y.J., Wang, H.Y., Zhou, W.L., Tan, S.Y., Li, F.F., 2015. Three phase partitioning for simultaneous purification of aloe polysaccharide and protein using a single-step extraction. Process Biochem. 50, 482–486.

Tolstoguzov, V., 2004. Why are polysaccharides necessary? Food Hydrocolloid 18, 873–877.
Vetal, M.D., Shirpurkar, N.D., Rathod, V.K., 2014. Three phase partitioning coupled with ultrasound for the extraction of ursolic acid and oleanolic acid from Ocimum sanctum. Food Bioprod. Process. 92, 402–408.
Wang, Y.Y., Qiu, W.Y., Wang, Z.B., Ma, H.L., Yan, J.K., 2017. Extraction and characterization of anti-oxidative polysaccharide-protein complexes from *Corbicula fluminea* through three-phase partitioning. RSC Adv. 7, 11067–11075.
Wang, Y.Y., Ma, H., Ding, Z.C., Yang, Y., Wang, W.H., Zhang, H.N., Yan, J.K., 2019a. Three-phase partitioning for the direct extraction and separation of bioactive exopolysaccharides from the cultured broth of *Phellinus baumii*. Int. J. Biol. Macromol. 123, 201–209.
Wang, Y.Y., Ma, H., Yan, J.K., Wang, K.D., Yang, Y., Wang, W.H., Zhang, H.N., 2019b. Three-phase partitioning system with dimethyl carbonate as organic phase for partitioning of exopolysaccharides from *Phellinus baumii*. Int. J. Biol. Macromol. 131, 941–948.
Wang, W., Xue, C., Mao, X., 2020a. Radioprotective effects and mechanisms of animal, plant and microbial polysaccharides. Int. J. Biol. Macromol. 153, 373–384.
Wang, H., Geng, H., Chen, J., Wang, X., Li, D., Wang, T., Yu, D., Wang, L., 2020b. Three phase partitioning for simultaneous extraction of oil, protein and polysaccharide from rice bran. Innov. Food Sci. Emerg. 65, 102447.
Wang, C., Yu, Y.B., Chen, T.T., Wang, Z.W., Yan, J.K., 2020c. Innovative preparation, physicochemical characteristics and functional properties of bioactive polysaccharides from fresh okra (*Abelmoschus esculentus* (L.) Moench). Food Chem. 320, 126647.
Xie, J.H., Jin, M.L., Morris, G.A., Zha, X.Q., Chen, H.Q., Yi, Y., Li, J.E., Wang, Z.J., Gao, J., Nie, S.P., Shang, P., Xie, M.Y., 2016. Advances on bioactive polysaccharides from medicinal plants. Crit. Rev. Food Sci. 56, S60–S84.
Yan, J.K., Wang, W.Q., Wu, J.Y., 2014. Recent advances in *Cordyceps sinensis* polysaccharides: mycelial fermentation, isolation, structure, and bioactivities: a review. J. Funct. Foods 6, 33–47.
Yan, J.K., Pei, J.J., Ma, H.L., Wang, Z.B., Liu, Y.S., 2017a. Advances in antitumor polysaccharides from *Phellinus sensu lato*: production, isolation, structure, antitumor activity, and mechanisms. Crit. Rev. Food Sci. 57 (6), 1256–1269.
Yan, J.K., Wang, Y.Y., Qiu, W.Y., Shao, N., 2017b. Three-phase partitioning for efficient extraction and separation of polysaccharides from *Corbicula fluminea*. Carbohyd. Polym. 163, 10–19.
Yan, J.K., Wang, Y.Y., Qiu, W.Y., Wu, L.X., Ding, Z.C., Cai, W.D., 2017c. Purification, structural characterization and bioactivity evaluation of a novel proteoglycan produced by *Corbicula fluminea*. Carbohyd. Polym. 176, 11–18.
Yan, J.K., Wang, Y.Y., Qiu, W.Y., Ma, H., Wang, Z.B., Wu, J.Y., 2018a. Three-phase partitioning as an elegant and versatile platform applied to nonchromatographic bioseparation process. Crit. Rev. Food Sci. 58 (14), 2416–2431.
Yan, J.K., Wang, Y.Y., Qiu, W.Y., Wang, Z.B., Ma, H., 2018b. Ultrasound synergized with three-phase partitioning for extraction and separation of *Corbicula fluminea* polysaccharides and possible relevant mechanisms. Ultrason. Sonochem. 40, 128–134.
Yan, J.K., Yu, Y.B., Wang, C., Cai, W.D., Wu, L.X., Yang, Y., Zhang, H.N., 2021a. Production, physicochemical characteristics, and in vitro biological activities of polysaccharides obtained from fresh bitter gourd (*Momordica charantia* L.) via room temperature extraction techniques. Food Chem. 337, 127798.
Yan, J.K., Wang, C., Yu, Y.B., Wu, L.X., Chen, T.T., Wang, Z.W., 2021b. Physicochemical characteristics and in vitro biological activities of polysaccharides derived from raw garlic (*Allium sativum* L.) bulbs via three-phase partitioning combined with gradient ethanol precipitation method. Food Chem. 339, 128081.
Yi, Y., Xu, W., Wang, H.X., Huang, F., Wang, L.M., 2020. Natrual polysaccharides experience physiological and functional changes during preparation: a review. Carbohyd. Polym. 234, 115896.
Yu, Y., Shen, M., Song, Q., Xie, J.H., 2018. Biological activities and pharmaceutical applications of polysaccharide from natural resources: a review. Carbohyd. Polym. 183, 91–101.
Zhang, M., Cui, S.W., Cheung, P.C.K., Wang, Q, 2007. Antitumor polysaccharides from mushrooms: a review on their isolation process, structural characteristics and antitumor activity. Trends Food Sci. Tech. 18, 4–19.

CHAPTER 12

Technologies for oil extraction from oilseeds and oleaginous microbes

S.P.Jeevan Kumar[a,b], Vijay Kumar Garlapati[c], Lohit Kumar Srinivas Gujjala[d], Rintu Banerjee[e]

[a]Seed Biotechnology Laboratory, ICAR-Indian Institute of Seed Science, Mau, Uttar Pradesh, India
[b]ICAR-Directorate of Floricultural Research, Pune, Maharashtra, India
[c]Department of Biotechnology and Bioinformatics, Jaypee University of Information Technology, Waknaghat, Himachal Pradesh, India
[d]Advanced Technology Development Centre, Indian Institute of Technology, Kharagpur, West Bengal, India
[e]Microbial Biotechnology and Downstream Processing Laboratory, Agricultural and Food Engineering Department, Indian Institute of Technology, Kharagpur, West Bengal, India

Chapter outline

12.1 Introduction

12.1.1 Oilseeds and nutritional security

Global goals, also known as sustainable development goals have been enshrined by adoption of United Nations Member States (193 countries) in 2015 to end poverty, stop climate change and fight inequality by 2030 (Bill & Melinda Gates Foundation 2020). In order to fulfill these goals, particularly on ending the poverty, oilseeds play an important role in ensuring the food and nutritional security of increasing global population. Oilseeds are primarily rich in oil content and used as a source of edible oil. In addition,

Three Phase Partitioning
DOI: https://doi.org/10.1016/B978-0-12-824418-0.00011-4

these oilseeds are also used to produce pharmaceuticals, emulsifiers, surfactants, plasticizers, detergents, adhesives, lubricants, oleochemicals, fuels and cosmetics etc., (Qiu et al., 2011; Rahman and Jimenez, 2016). Oilseed crops such as sunflower (*Helianthus annus* L.), peanut/groundnut (*Arachis hypogaea* L.), safflower (*Carthamus tinctorius* L.), soybean (*Glycine* max L.), rapeseed/canola (*Brassica napus* L.), cotton (*Gossypium hirsutum* L.), flax (*Linum usitatissimum*), castor (*Ricinus communis* L.) and sesame (*Sesamum indicum* L.) are predominantly cultivated for oilseeds production.

The United States, China, Brazil, Argentina and India are major oilseed producing countries in the world. Moreover, soybean, rapeseed, palm oil and sunflower have been identified as potential vegetable oils for biodiesel production (Qiu et al., 2011). Owing to their uses in nutrition and fuel production, oilseed crops cultivation areas have been significantly increased; as a result, increased production was observed due to higher yields per unit area (Lu et al., 2011). In addition to oilseeds, oleaginous microorganisms have gained significant attention owing to the capacity to accumulate higher lipid content. Oleaginous microorganisms (OM) are defined as microbes that synthesize lipids greater than 20 percent of lipid content on dry weight basis (Kumar and Banerjee, 2013). These OM have gained greater attention for biodiesel production due to similarity of lipid composition with vegetable oils, easy to scale up production processes and having capacity to synthesize lipids using agro-residues and industrial byproducts (Kumar and Banerjee, 2018). Among OM, microalgae is one of the vital microbes that play role in carbon sequestration and concomitantly synthesize lipids (Kumar et al., 2017a, 2020a; Garlapati et al., 2017). Oleaginous fungi and yeast are also widely explored due to their lipid synthesizing capability under stress conditions in a few days (Banerjee et al., 2019).

Oilseeds produced in most of the countries are predominantly used for oil extraction (Table-12.1). Similarly, oil extraction technologies pertinent to oleaginous microbes are a key determinant for the efficiency with which the products can be obtained. In addition to predominant oilseed crops, recent studies on new plants have been investigated that might act as new sources of edible vegetable oils (Araujo et al., 2018; Bourgou et al., 2020; Santos et al., 2020). Effective usage of oils is facilitated with the efficient oil

Table 12.1 Oil content and specific energy density of varied oilseeds.

Oilseed Crop	Oil Content (percent)	Specific Energy Density (MJ kg^{-1})
Groundnut (*Arachis hypogaea*)	39–48	–
Rapeseed (*Brassica napus*)	37–46	37.8
Soybean (*Glycine* max)	17–20	37.8
Sunflower (*Helianthus annus*)	39–48	40.0
Sesame (*Sesamum indicum*)	38–40	–
Cotton (*Gossypium hirsutum*)	16–18	–
Castor (*Ricinus communis*)	42–45	39.5
Oil Palm (*Elaeis guineensis*)	18–26	–

extraction method in oilseeds as well as OM. To optimize the efficient oil extraction method and solvent pertinent to different crop oilseeds and OM, a basic understanding on structure of oilseeds is required.

Oilseeds comprises of three basic units such as seed coat, embryo and one or more food structures. Seed coat in oilseeds is marked either with a seed scar or hilum, whose main function is to protect the embryo from bacterial and fungal infection. Oilseeds cotyledon accumulates lipids and proteins. Predominantly, proteins occupy a major proportion of 60–70 percent with a size ranging from 2 to 20 μm in various oilseeds. In contrary, lipids are available as lipid reserves in oilseeds in sizes ranging from 1 to 2 μm, which primarily differ from one species to another. The structural parts of the groundnut seed are shown as an illustration (Fig. 12.1).

12.2 Importance of oil and lipid extraction

Conventionally, oils are extracted from oilseeds through either mechanical process or solvent extraction. Mechanical extraction of oil involves exertion of sufficient force on confined seed to rupture the cells and release the oil from the seed. These extractions are performed in a container with tiny perforations which are slotted or round enough to allow the liquid content to escape from the seeds (Ben-Youssef et al., 2017). Mechanical oil extraction is conventionally applied for oil extraction from varied oilseeds due to ease in operation; however, this method results in low yields and requires solvent extractions to recover the residual oil from the oilseeds. To overcome the disadvantages associated with the mechanical press, chemical extraction methods have been developed by Folch

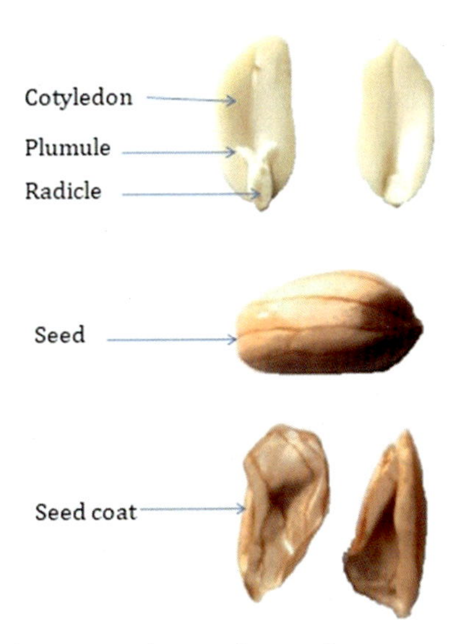

Fig. 12.1 ***Diagram depicting the structural parts of groundnut.***

and Bligh and Dyer using chloroform and methanol solvents. Folch method used chloroform and methanol organic solvents at 2:1 ratio for selective extraction of oils/lipids from biological samples (Folch et al., 1957; Kumar et al., 2017b). Bligh and Dyer (1959) have introduced an efficient and mild procedure using chloroform and methanol at 1:2 ratio for oil extraction from frozen fish (Bligh and Dyer, 1959). These solvent mixtures have shown great potential in extracting oils/fats from biological samples but use of hazardous organic solvents impedes the quality of the oil extraction (Kumar et al., 2017a). In solvent extractions, hexane is one of the most frequently used solvents owing to higher oil yield recovery (Fine et al., 2013; Kumar et al., 2017c). As per the estimate of Environmental Protection Agency (EPA) Toxic release inventory, more than 20,000 t of hexane are released to the atmosphere each year derived from vegetable oils (DeSimone, 2002). Although, hexane is a good solvent in extraction of oil from oilseeds but it has been listed as hazardous air pollutant by the US Environmental Protection Agency (Kumar et al., 2017b). Besides, hexane is neurotoxic. Owing to the environmental and toxicological concerns, finding alternative and safe techniques is essential. Recent studies emphasize that the green extraction techniques and solvents are gaining increasing importance for their higher efficiency, eco-friendly nature and lower wastes.

Production of biodiesel from oleaginous microbes is a five-step process consisting of cultivation of microbes, drying of the microbial biomass, extraction and transesterification of oils and purification of biodiesel. According to the report of Lardon et al. (2009), 90 percent of the total energy expended for the production of biodiesel is incurred due to extraction of lipids and hence it directly influences the economic viability of the process. Majorly fossil fuel derived solvents *viz.,* hexane, combination of hydrocarbons etc., are used for lipids extraction (Santos Reis et al., 2020). Thus, it is a challenge to develop less energy intensive methods along with selection of green solvents (bio-based) which would be environment friendly. Green solvents have been classified in to five major types *viz.,* aqueous solvent systems, ionic liquids, deep eutectic solvents, bio-based solvents and switchable solvents (Schuur et al., 2019). In order to select appropriate solvents for extraction of lipids from oleaginous microbes, theoretical tools *viz.,* Hansen solubility parameters (HSP) and Conductor like screening model for real solvents (COSMO-RS) have to be used (Klamt, 2011; Paduszyński, 2018).

12.3 Green solvents and techniques for oil and lipid extraction

The theme of green chemistry concept is mainly aimed to utilize the energy resources with concomitant reduction in wastes. Oil/lipid extraction from oilseeds/oleaginous microbes is mainly governed by some principles, as listed below:

- Usage of green solvents derived from renewable sources
- Deployment of technologies that aim for reduction in energy consumption
- Green solvents particularly agro-solvents or water
- Development of biorefinery concept *i.e.,* formation of co-products instead of waste

- Higher efficiency and quality product without any contaminants
- To minimize the number of unit operations and development of robust and controlled processes

12.3.1 Selection of extraction solvent

Selection of solvents depends mainly on the solute and the processing conditions of extraction. An extraction solvent should be such that the partition coefficient (as shown in Eq. (12.1)) between solute and solvent should be high in magnitude for a specific solvent.

$$K = C_{solute\ in\ organic\ phase} / C_{solute\ in\ aqueous\ phase} \tag{12.1}$$

Based on the "like-dissolves-like" principle, oil content in the oilseeds/neutral lipids present within the oleaginous microbes would be extracted efficiently in non-polar solvents. For polar lipids, non-polar solvent is not an efficient selection considering its linkage with the matrix of the oilseeds/biomass. Hence, co-solvents are used to break the linkage between oilseeds/biomass matrix and polar molecules which are part of the oils/lipids. A mixture of solvents *viz.*, chloroform and methanol have been traditionally used for extraction of total lipids as described by Bligh and Dyer (1959) and Folch (1957), whereas hexane is widely used for oil extraction from oilseeds. Although, these methods and solvent have been used popularly, still the use of chloroform, hexane and methanol has safety issues considering their toxic and flammable nature. In this context, green solvents can be useful considering their biological origin, eco-friendly and sustainable nature.

Further, to select the perfect combination of solute and solvent, Hansen solubility parameters (HSP) is quite helpful, wherein, three specific parameters are involved *viz.*, non-polar (dispersive), polar (dipole-dipole, dipole induced dipole) and hydrogen bonding aspects (Hansen, 1969). It has an inherent assumption that the total cohesive energy is equal to the energy corresponding to all the three above-mentioned aspects. These provide theoretical estimates of the possibilities for a combination of solute and solvent by calculating relative energy difference (RED) which is a ratio of Hansen distance between solute and solvent to the radius of the solubility sphere. If RED < 1, then the combination will have high affinity while values > 1 will have low affinity. Other than HSP, screening model for real solvents (COSMO-RS) is also an efficient tool to select appropriate solvents. The main advantages of COSMO-RS process are the simplicity of its algorithm, numerical stability and it is highly accurate in terms of calculating solvation energy (Klamt, 2011).

The basic principle behind COSMO-RS process is the quantum chemical considerations and statistical thermodynamics, which are used to predict the thermodynamic feasibility without the need of experimental data. Initially the solute is embedded in a continuum of dielectric such that polarization charge density can be induced on its surface. Then, from the COSMO calculations, the solute is converged to its optimal state (energetically) based on its geometry and density of electron around the individual atoms/molecules. From the three-dimensional configuration, interaction energy between the interacting molecules is calculated in terms of σ potential. This parameter explains the affinity between the solute and solvent in terms of polarity and hydrogen bonding (Aissou et al., 2017). Ben-Youssef et al. (2017) have reported that 2-methyl tetrahydrofuran can be used as an alternative to hexane for extraction of oil from Tunisian date palm seeds based on HSP study. Cascant et al. (2017) have reported that D-limonene and p-cymene can be used as solvents for extraction of triacylglycerides from Salmon fish based on HSP study.

12.4 Green solvents for oil/ lipid extraction

12.4.1 Bio-derived solvents

Terpenes extracted from citrus peels, pine trees, etc., *viz.,* D-limonene, p-cymene, gum turpentine, pinene, etc., and other bio-based solvents *viz.,* 2-methyltetrahydrofuran, cyclopentyl methyl ether, isopropanol, ethanol, ethyl lactate, ethyl acetate, dimethyl carbonate have high affinity for oils/lipids that can be used for their extraction. Sicaire et al. (2015) have reported that 2-methyltetrahydrofuran can be used as a solvent for extraction of oils from food crops. Briel et al. (2016) have reported that cyclopentyl methyl ester, 2-methyltetrahydrofuran (MeTHF) and ethyl acetate can be used as solvents for extraction of lipids from oleaginous *Yarrowia lipoltica.* Yang et al. (2014) have reported that ethanol can be used as an extraction solvent for lipids from *Picochlorum* sp. using mild extraction conditions while obtaining similar lipid yield when compared with conventional solvents.

The Hansen solubility parameters of bio-derived solvents has been shown in Table 12.2, using these parameters relative energy difference values have been calculated by Sicaire et al. (2015). Based on these values, they have reported that MeTHF, D-limonene and p-cymene are superior solvents for extraction of TAGs than hexane (RED values of

Table 12.2 Hansen solubility parameters of the bio-derived solvents used for extraction of oils/lipids.

Solvents	δ_D	δ_P	δ_H
MeTHF	16.4	4.7	4.6
D-limonene	16.7	1.8	3.1
P-cymene	17.3	2.3	2.4
Methyl acetate	15.5	7.2	7.6
Ethyl acetate	15.8	5.3	7.2
Butanol	16	5.7	15.8
Isopropyl alcohol	15.8	6.1	16.4
Ethanol	15.8	8.8	19.4

Data adopted from Sicaire et al. (2015).

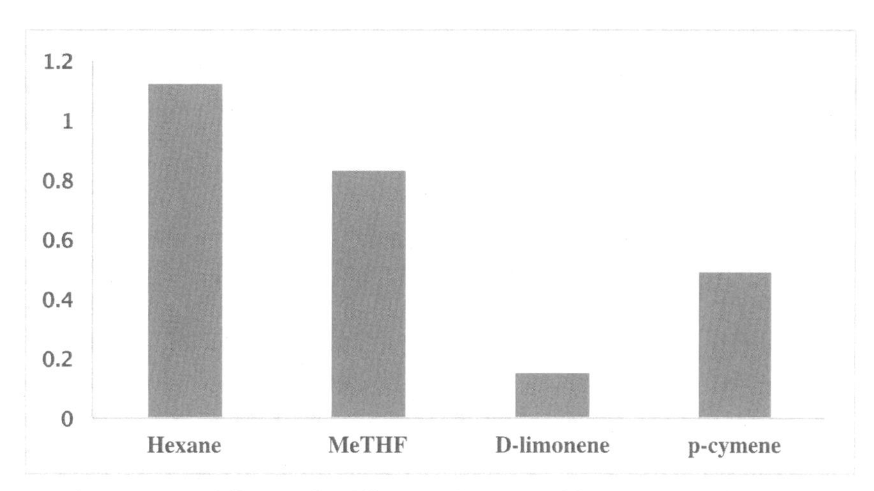

Fig. 12.2 ***Relative energy difference for different solvents used for extraction of lipids.*** *Data adopted from Sicaire et al. (2015).*

bio-derived solvents < hexane as shown in Fig. 12.2.) Advantages of using bio-derived solvents are their low toxicity, biodegradable nature, renewable nature etc. Although these solvents have potential, still validation at large scale and economic evaluation of the process has not been widely reported.

12.4.2 Supercritical fluid technology

Supercritical fluids (SCFs) have been used extensively in food and pharmaceutical industries since their physical properties can be influenced by variation in temperature and pressure. SCFs are highly selective towards the solute with low extraction time and are non-toxic in nature when compared to conventional petro-based solvents. Supercritical fluid extraction (SFE) is a viable option for oil extraction from oilseeds owing to the properties of environment friendly nature, low toxicity, cost and flammability (Herrero et al., 2018). Carbon dioxide (CO_2) is a widely used supercritical fluid not only for oil extraction but also for extraction of other neutral lipids. *Sc*-CO_2 extraction coupled with tetra hydrofuran (THF) extraction resulted in 16 percent of oil yield in carvi seeds at 40 °C and 200 bar. Similar results have been observed in rice bran and rosehip seeds (Soares et al., 2016; Salgın et al., 2016). Some studies have reported that by increasing pressure while keeping temperature constant may enhance the yield of extraction. In addition, supplementation with modifiers *viz.,* ethanol can improve the yield of extraction. Mouahid et al. (2013) have reported that supercritical CO_2 is efficient in extracting lipids from *Nannochloropsis oculata*, *Cylindrotheca closterium*, *Chlorella vulgaris* and *Spirulina platensis*. The following process parameters were maintained: pressure 40 MPa, temperature 333 K, flow rate of carbon dioxide at 0.3–0.5 kg/h which led to an extracted fraction containing > 90 percent triglycerides content. However, large scale application of SCF technology has not been reported widely and hence there is a scope for further investigation.

12.4.3 Ionic liquids and deep eutectic solvents

Ionic liquids (ILs) are organic salts which melt below 100 °C and contain an organic cation. These solvents are eco-friendly in nature, non-flammable, remain in liquid state for a wide range of temperature and hence are categorized as green solvents. Since they are composite in nature, tailor made properties can be attributed to these solvents *viz.,* conductivity, hydrophobicity, polarity, solubility etc. Application of ILs and Deep eutectic solvents (DES) for oil extraction from oilseeds are yet to be explored. However, for lipid extraction from oleaginous microbes, they have showed significant potential in extracting quality lipids. Kim et al. (2012) have reported extraction of lipids from *Chlorella vulgaris* using a mixture of [Bmim][CF_3SO_3] and methanol, which led to extraction of 19 percent (w/w) of the lipids and were observed to yield more lipids than the conventional Bligh and Dyer's method. (Chua et al., 2018) have reported that cholinium arginate is an efficient solvent for extraction of lipids from *Nannochloropsis* sp. where only 1.4 percent (w/w) of lipids was left over in the microalgal biomass. Using ILs, cell walls of microbes gets lysed and thus two immiscible layers are formed and from the organic phase, lipids can be obtained. In spite of these advantages, ILs are not widely used owing to their costly nature.

New generation ionic liquids consisting of choline chloride and a hydrogen bond donor (amides/amines/alcohols/carboxylic acids) are deep eutectic solvents (DES). These solvents have melting point lower than the corresponding components. When compared with conventional ionic liquids, DESs are simple to prepare and economical; have low toxicity, and are biodegradable in nature. Tommasi et al. (2017) have reported that combination of choline chloride and carboxylic acids (oxalic acid, levulinic acid, urea, ethylene glycol, sorbitol) were efficient in extraction of lipids from *Phaeodactylum tricornutum*. Cicci et al. (2018) have reported that DES with a combination of 1,2-propanediol, choline chloride and water in the molar ratio of 1:1:1 led to efficient lipid extraction from *Scenedesmus dimorphus*.

12.4.4 Switchable solvents

Switchable solvents are solvents, whose miscibility with water can be switched according to the need by adding or stripping carbon dioxide from the system. They can be classified into switchable polarity and hydrophobic solvents. Application of switchable solvents in oil industry has great potential and the process development for soy oil extraction from soybean seeds is underway. Soybean oil extraction from soybean seeds consumes large amount of solvent and energy due to distillate removal from the solvent (hexane). Pham et al. (2008) have attempted switchable solvent strategy using amidine and excess water. Results showed efficient oil/solvent separation and quality extraction without any hindrance to adventitious water. However, separation of water from amidine is difficult, identification of switchable hydrophilicity solvent that readily separate water is quintessential.

Pertinent to OM lipid extraction, a mixture of 1,8-diazabicycloundec-7-ene (DBU) and an alcohol (switchable polar solvent, SPS) has been reported to be used for extraction of lipids from microalgae (Jessop et al., 2005; Pham et al., 2008). Inherently, DBU has low polarity in alcohol but upon addition of CO_2, its solubility increases. After the extraction of lipids, solvents can be separated without the need of distillation which makes the process economical. However, a major disadvantage with the use of SPS is that both the solvents and microalgae have to be dried properly to prevent unwanted bicarbonate formation.

Switchable hydrophobic solvents (SHS) consist of tertiary amines which are conventionally immiscible with water and hence they form biphasic mixture. However, after addition of carbon dioxide at 1 atm, the tertiary amines get miscible with water since they form bicarbonate salts. In presence of bicarbonate, SHS can function efficiently and hence solvents and microalgae don't have to be dried prior to use. This advantage of SHS, makes extraction of lipids from wet microbial biomass a possibility. Boyd et al. (2012) have reported that N,N-dimethylcyclohexylamine (SHS) is an efficient solvent for extraction of lipids from *Botryococcus braunii*, where 22 percent w/w of crude lipids have been extracted. Cicci et al. (2018) have developed a strategy of complete extraction of biomolecules *viz.*, proteins, carbohydrates and lipids from *Scenedesmus dimorphus*. In this process, two different modes were used, first mode comprises of hydrophobic solvent treatment followed by switching to hydrophilic and vice-versa in the second mode. In the first mode, 96.1 percent and in second mode, 92.6 percent of lipid yield were obtained.

12.5 Conventional and green extraction techniques for oil/ lipid extraction

12.5.1 Oil/Expeller pressing

Oil/Expeller pressing is one of the simplest methods to extract oil by using mechanical crushing technique. Advantages of this technique includes its hand free nature and low maintenance issues. However, there is a requirement for large amounts of biomass and long incubation period for extraction. Topare et al. (2011) have reported extraction of lipids from algae using expeller press where 75 percent of the lipids have been extracted.

12.5.2 Bead milling

Bead milling is an age-old technique which has been used predominantly in cosmetics industry for reducing the particle size of paint/lacquer. It has also been used for microbial cell disruption for extraction of intracellular products. Advantages of this technique includes a single pass of operation, high efficiency of cell disruption, easy operation for loading of biomass, can be operated at mild operating conditions, etc. Montalescot et al.

(2015) have reported that the use of continuous bead milling led to efficient cell disruption of *Nannochloropsis oculata* and *Porphyridium cruentum*. *The efficiency of disintegration also depends upon the type of beads as it has been reported that* ZrO_2 *beads were superior to glass beads* (Doucha and Lívanský 2008). Muellemiestre et al. (2016) have reported that out of ultrasound, microwave, and bead milling process, bead milling led to highest lipid extraction from *Y. lipolytica*. However, its high energy consumption and poor transfer of energy from the rotating shaft to the microbial cells are its disadvantages.

12.5.3 Enzyme assisted oil/lipid extraction from oilseeds/biomass

Aqueous enzymatic extraction (AEE) is one of the promising approaches for oil extraction from oilseeds. Besides, this approach also aides in efficient extraction of protein from oilseeds. Unlike solvent and mechanical press methods, AEE method is highly preferable owing to superior quality of the oil and its healthy consumption (Yusoff et al., 2015; Chabrand and Glatz, 2009;Yang et al., 2014). Generally, oil extraction carried out in water is known as aqueous extraction, which results in poor oil yield. Addition of enzymes in water in AEE facilitates easy separation of targeted compounds without any damage of their properties particularly on smell and taste (Kumar et al., 2017b; Latif and Anwar, 2011).

AEE is significantly gaining momentum by virtue of environmental and safety regulatory concerns. Besides, it is environment friendly, economical, much safer, facilitates flexible operation with lower energy consumption and operational costs coupled with meagre capital investment than the solvent extraction (Yang et al., 2014; Wu et al., 2009). Owing to milder operating conditions of AEE, co-products such as non-toxic meal, protein and value added fiber are produced with superior quality, particularly useful in soybean crop and other crop residues used as feed for cattle. Another great advantage of the AEE is to separate the phospholipids simultaneously from the oil, which concomitantly abate the degumming step and therefore reduce the cost of downstream processing (Latif and Anwar, 2011). AEE has several advantages in comparison to conventional techniques like mechanical press and solvent extraction. However, on the other hand, AEE has some disadvantages such as longer incubation time, higher cost for drying and procurement/availability of suitable formulations of enzymes, and requirement of post-extraction de-emulsification step for higher oil extraction and efficiency (Nadar et al., 2018)

Efficiency of AEE depends on several factors such as selection of enzyme/enzyme mixture, type of solvent (buffer composition), pre-treatment of oleaginous materials, particle size of oilseeds, pH, enzyme concentration/substrate ratio, temperature, oil:water ratio and shaking regime. In addition to these parameters, process optima parameters depends on the type of oilseeds and its composition. Some of the studies pertinent to extraction and optimization of oil yields have been illustrated in Table-12.3. Effect of various parameters on oil extraction from oilseeds has been discussed in recent literature (Kumar et al., 2017b; Nadar et al., 2018; Nde and Foncha, 2020).

Table 12.3 Green solvents/extraction techniques for oil extraction from oilseeds.

Green solvent/extraction technique	Species	Extraction conditions	Oil content (wt percent)	References
Supercritical carbon dioxide extraction	Favela seed oil (*Cnidoscolus quercifolius*)	Supercritical carbon dioxide at 60 °C and 30 MPa	41.0	Santos et al. (2020)
Enzyme assisted extraction	Yellow mustard flour	pH 4.5, Temp-40 °C, Time- 1 h, Enzyme –Protex 6 L®, substrate loading-3percent (w/w)	92.5	Boulila et al. (2015)
Enzyme assisted extraction	Borage seeds (*Borago officinalis*)	pH 4.5, Temp-45 °C, Time- 9 h, Enzyme –Celluloblast®, substrate loading-1 percent (w/w)	42.50 g/kg	Soto et al. (2008)
Enzyme assisted extraction	Watermelon seeds	pH-7.89, Temp-47.13 °C, Time- 7.8 h, Enzyme –Protex 6 L®: 2.63 percent,	97.92	Sui et al. (2011)
Enzyme assisted extraction	*Irvingiagabonensis* seed kernels	pH-3.5–5.5, Temp-55 °C, Time- 18 h, Enzyme –Viscozyme L®: 2 percent,	68.00	Womeni et al. (2008)
Ultrasound assisted extraction	Radish seed oil	Solvent to seed ratio of 12 mL $g-{}^{1}$, Temp-60 °C, Extraction time of 60 min	25.00	Stevanato and da Silva (2019)
Crude multi-enzymatic extracts (CME) coupled with hydrodistillation	Ginger essential oil (*Zingiberofficinale*Roscoe)	Temp-40 °C, Incubation-130 min, CME/H2O (75/425 mL mL^{-1})	47.95	dos Santos et al. (2020)
Supercritical carbon dioxide and Tetra hydro furan extraction	*Carumcarvi* seeds	Temp-40 °C, Pressure-200 bar	16.00	Bourgou et al., (2020)

(continued)

Table 12.3 (Cont'd)

Green solvent/extraction technique	Species	Extraction conditions	Oil content (wt percent)	References
Supercritical carbon dioxide extraction	Rosehip seeds	Temp-40 °C, Pressure-30 MPa Volumetric flow rate of supercritical solvent- 0.75 mL/min, particle size-355 < Dp < 500 μm	16.50 g/oil/100 g dry solid	Salgin et al. (2016)
Ultrasound assisted extraction	Hempseeds	Exposure time- 25min, Power output- 200 W Solvent-to-solid ratio-7:1 (v/w) Acting on–off ratio: 20:20 (s/s);	86.50	Lin et al. (2012)
Ultrasound assisted extraction	Chia seeds (*Salvia hispanica*)	Extraction time- 90 min Ultrasonic bath at 40 kHz	79.30	Rosas-Mendoza et al. (2017)
Microwave assisted extraction	Castor beans	Treatment time-120 s, Electric power- 280 W	44.34	Mgudu et al. (2012)
Microwave assisted extraction	Palms of Colombian Amazon region	Treatment time-15 min, Sample to solvent ratio-2:3	45.00	Mosquera et al. (2013)
Microwave assisted extraction	Cottonseed oil	Treatment time 3.57 min, Moisture content of cottonseed oil-14 percent Cottonseed to solvent ratio-1:4	32.60	Taghvaei et al. (2014)

The enzymatic treatment for extraction of lipids from microbial biomass depends upon the composition of the cell wall. Conventionally, cellulases, xylanases, amylase, pectinase, papain, etc., are used for cell wall degradation (Kumar et al., 2020b). Advantages of this technique include mild operating conditions, specificity towards the cell wall, low energy requirements, absence of harsh chemicals and simple physical conditions (Gujjalla et al., 2019). Jin et al. (2012) have reported that a recombinant β−1,3-glucomannanase plMAN5C has been used for pretreatment of *Rhodosporidium toruloides* where 96.6 percent of lipids have been extracted using ethyl lactate as the solvent at room temperature and atmospheric pressure. Although this technique is a green and eco-friendly approach, it suffers from long processing times and high capital costs. Role of green solvents and extraction techniques in lipid extraction from OM is given in Table-12.4.

12.5.4 Microwave treatment

Microwave assisted extraction (MAE) uses the energy of microwaves to disrupt the cell membrane and thus the intracellular lipids gets released into the organic solvent. During microwave treatment, polar compounds align themselves in the direction of electric field and rotate at high speed which causes heat leading to cell disruption. Major advantage with MAE is its energy efficiency when compared with the conventional heating process. The determinants for MAE efficiency are partition co-efficient, solubility of desired compounds in the solvent, di-electric properties of solvent and oilseeds/biomass and mass diffusivity (Isopencu et al., 2019). Treatment of castor beans with MAE at 280 W for 120 s resulted in optimum yield of 44.34 percent (Mgudu et al., 2012). In another study, higher oil yield of palm oil with desirable lower unsaturated fatty acid was obtained in comparison to solid-liquid extraction and expeller press methods (Mosquera et al., 2013). Taghvaei et al. (2014) have obtained 32.6 percent of oil yield in 3.5 min; whereas in soxhlet extraction 34.7 percent of oil was extracted in 16 h. This study shows the efficiency of MAE in improved oil yields and quality at very meager times rather than the conventional longer incubation time.

MAE has been proved efficient for lipid extraction from OM. Guerra et al. (2014) used microwave treatment as the pretreatment strategy, which led to enhanced lipid extraction than the conventional Bligh and Dyer's method and this process further aids in savings of chemicals and energy. Although, MAE has its associated advantages, still it is limited to polar solvents and specifically not for volatile solvents. Owing to the thermal nature, the technique remained unsuitable for polyunsaturated fatty acids (PUFA) extraction, as the latter are prone to oxidation and degradation (Yousuf et al., 2018). In addition, MAE formation of free radicals and heat during the processing is an associated disadvantage. Some of these disadvantages can be avoided by using temperature controlled microwave reactors; however, such reactors for large scale processing are expensive.

Table 12.4 Green solvents/extraction techniques for lipid extraction from oleaginous microbes.

Green solvent/extraction technique	Principle	Extraction conditions	Lipid content (wt percent)	References
Bligh and Dyer and Folch extraction method	Phase separation	0.3 ml buffer, 0.35 ml chloroform and methanol Folch method- 0.66 ml chloroform and 0.33 ml methanol Shaking regime-20 min Evaporation time – 30 min	23.5 percent in *R. glutinis*	Vasconcelos et al. (2018)
Soxhlet method using bio-based solvent	Solvent reflux	Biomass added to 50 ml petroleum based and bio based solvent for 8 h	25.22 percent−26.47 percent in *C. vulgaris* and *Nannochloropsis* sp.	Mahmood et al. (2017)
Acid treatment and hexane used in extraction process	Biphasic mixture	Treatment of 1 wt percent H_2SO_4 and 8 wt percent yeast solids at 170 °C for 60min. after this add equal amount of hexane and stirred for 1 h	88.5 percent−93 percent	Kruger et al. (2018)
Super critical carbon dioxide and ethanol as solvent	Supercritical fluids as extractant at higher temperature and pressure above its critical points	Two cyclone separator with 6 kg/h, CO_2 flow rate operating pressure 20 MPa at 40 °C.	69 percent	Hegel et al. (2011)
Solvent extraction	Solute is add to heterogeneous mixture	Hexane and chloroform used as solvent Time −4 h	33.3 percent−44.3 percent	Prommuak et al. (2012)
Supercritical extraction	Supercritical fluids as extractant at higher temperature and pressure above its critical points	Solvent pressure- 210 bar Temp.−55 °C CO_2 flow rate–6–6.5 kg/h	50 percent−60 percent	Priyadarshini et al. (2008)

Ultrasound assisted	Ultrasound treatment with pulse different mode	Chloroform methanol solvent (1:1) Ultrasound power density-0.6 W/ml Temp.- 30 °C for 30 min.	99.3 percent	Selvakumar and Sivashanmugam, (2019)
Solvent extraction	Solute is added to heterogeneous mixture	Hexane as solvent Extraction time–5–25 min. Temp.–40–110°C in central composite design	13.72 g oil/100 g seeds	Kraujalis et al. (2013)
Solvent extraction	Solute is added to heterogeneous mixture	Mixture of solvents hexane and methanol and chloroform and methanol	5.16 percent−5.51 percent, respectively	Shin et al. (2018)
Solvent extraction	Solute is added to heterogeneous mixture	Chloroform, methanol, ethanol and dichloromethane in 1:5:1:1 (v/v) Stirring- 3 h at 400 rpm	19.4 percent lipids g^{-1}DCW	Mustapha and Isa, (2020)
Enzymatic hydrolysis	Soften the cell wall and lipids easily extracted through solvents	Cellulase and β-glucosidase used as enzyme hydrolyser for 72 h pH-4.8 and temp.- 50 °C Solvents used-hexane, methanol and chloroform	FAME productivity-59.4 mg FAME/g cell	Cho et al. (2013)
Enzymatic hydrolysis	Soften the cell wall and lipids easily extracted through solvents	Immobilized cellulase enzyme on polyacrylonitrilenanofibrous membrane pH- 4.6 and temp.- 50 °C	63.4 percent	Fu et al. (2010)
Enzymatic hydrolysis	Soften the cell wall and lipids easily extracted through solvents	Dialyzed solution of recombinant plMAN5C treated with *R.toruloides* at 200 rpm pH- 4.5 Solvents used- chloroform, hexane and ethyl acetate	96.6 percent	Jin et al. (2012)
Enzymatic hydrolysis	Soften the cell wall and lipids easily extracted through solvents	Papain, neutrase, cellulase, pectinase, alcalase and snailase enzymes used for hydrolysis Temp.- 40–60 °C	Oil recoveries-104.6 percent (pectinase and papain)	You et al. (2011)

(continued)

Table 12.4 (Cont'd)

Green solvent/ extraction technique	Principle	Extraction conditions	Lipid content (wt percent)	References
Enzymatic hydrolysis followed by cold pressing	Soften the cell wall and lipids easily extracted through solvents	Cellulase, pectinase, hemicellulase used for hydrolysis Enzymes crushed seed ratio-0.25 percent Temp.- 45 °C Moisture content-20 percent Time – 9 h	95 percent by double pressing	Soto et al. (2007)
Enzymatic hydrolysis	Soften the cell wall and lipids easily extracted through solvents	Enzymatic mixture cellulase, pectinase, hemicellulase Temp.- 35–50 °C pH- 3.5–4.5	86.1 percent	Huo et al. (2015)
Enzymatic hydrolysis	Soften the cell wall and lipids easily extracted through solvents	Cellulase, xylanase, pectinase Stirring speed-250 rpm	13.8 g/100 g	Zhang et al. (2018)
Green solvents	Ecofriendly, extracted from renewable raw materials	Cyclopentyl methyl ether (CPME) and ethanol as solvents Temp.- 80 °C Time- 60 min.	39.4 percent	Santoro et al. (2019)
Green solvents	Eco-friendly, extracted from renewable raw materials	2-methyl-tetra hydrofuran, ethanol, water, isoamyl alcohol Time-15 min.	99.6 ± 7.6 mg g^{-1} of fatty acids	de Jesus et al. (2018)
Green solvents	Eco-friendly, extracted from renewable raw materials	2-methyl-tetra hydrofuran and n-hexane as solvent Time- 30 min.	7.24–5.97 percent	Ben-Youssef et al. (2017)
Ultrasound aided green solvents	Eco-friendly, extracted from renewable raw materials	Ultrasonic frequency- 24kHz, 400 W Solvents-hexane and isopropanol Time- 20 min.	80 percent and 79 percent, respectively	Perrier et al. (2017)
Green solvents	Eco-friendly, extracted from renewable raw materials	Terpenes as green solvents Time- 8 h	0.91 percent−1.52 percent	Dejoye et al. (2012)

Combined bead milling and enzymatic hydrolysis	Disruption of cell wall	Bead miller agitation speed-2039 rpm and temp. 25 °C Enzymatic hydrolysis by lipase enzyme at pH-7.4 Temp.−37 °C for 24 h	8 percent	Alavijeh et al. (2020)
Switchable solvents	Wet biomass disruption	N,N-dimethylcyclohexylamine (DMCHA),dipropylamine,N-ethylbutylamine (EBA) solvents used Time – 30min.	13.6 percent, 7.0 percent, 12.3 percent, respectively	Al-Ameri and Al-Zuhair, (2019)
Enzyme assisted aqueous extraction	Sonication and enzymatic treatment	Enzymes cocktail of more than 30 enzymes Temp.−95 °C Ultrasonic power-600W	49.82 percent lipids recovery	Liang et al. (2012)
Ultasonication followed by solvent and soxhlet extraction	Cell disruption	Ultrasound frequency-40 kHz Amplitude −70 percent for 15 min. n-hexane as solvent stir at 300 rpm at 60 °C for 8 h	6.6 percent−9.4 percent	Fattah et al. (2020)
Ultrasound assisted extraction	Cell disruption	Ultrasonic power-1000 W Time- 30 min.	Oil recovery 0.21 percent	Adam et al. (2012)
Bio-based solvents	Eco friendly, extracted from agriculture resource	Hexane, ethyl acetate, ethanol, IPA, DMC, ethyl lactate, p-cymene, α-pinene, D-limonene and MeTHF as solvent Time – 10 min at 25 °C centrifugation- 5000 rpm	12.63 percent−15.94 percent	Breil et al. (2016)
Ultrasound based extraction	Cell disruption	Ultrasonic power-7.7 W/cm^2 Temp.- 40 °C	60.5 percent−97 percent of total oil extraction yield	Sicaire et al. (2016)

12.5.5 Ultrasound assisted extraction (UAE)

Ultrasonication is an efficient technique for oil/lipids extraction due to cavitation and acoustic streaming which are generated during the process. The working principle involves in UAE is the behavior of cavitation i.e., formation, growth and collapse of bubbles. As a result, due to mechanical and physico-chemical pressure the material matrix disrupts and facilitates in the release of extractable compounds (Gutte et al., 2015). UAE is preferred due to eco-friendly nature, takes lesser extraction time than conventional soxhlet extraction with efficient product recovery. Efficiency of UAE depends on several factors such as ultrasonic power, extraction time, solid loading, ultrasonic intensity and temperature etc.

Lin et al. (2012) have reported an increase in oil yield upto 86.5 percent at an exposure time of 30 min. Further increase of exposure time resulted in decrease of oil content. This might have been due to destruction of cell material to the oscillations generated from the solvent and oil. In another study, enhanced oil yield of 79.3 percent was obtained in chia seeds after incubation of 90 min; whereas the conventional stirring treatment resulted in 69.2 percent of oil yield in 90 min (Rosas-Mendoza et al., 2017). Similarly, optimization of other parameters pertinent to UAE for oil extraction has to be done for maximum efficiency of the technique.

UAE is not only effective in oil extraction but has substantiated its potential for lipid extraction from oleaginous microbes. Prabakaran and Ravindran (2011) have reported that ultrasonication is efficient in extraction of lipids from *Phaeodactylum tricornutum*, *Nannochloropsis gaditana*, and *Chaetoceros calcitrans*. Zhang et al. (2014) have reported extraction of lipids from *Trichosporon oleaginosus* and observed that ultrasonication at 520 kHz and 40 W for 15 min incubation with an extraction solvent of chloroform to methanol ratio of 1:1 v/v led to the maximum lipid extraction. There are two different theories for selecting operational parameters for ultrasound extraction; one is low frequency and high power and the other vice-versa. Wu et al. (2012) have suggested that at low frequency and high power, the efficiency of cell disruption is superior when compared to high frequency and low power during ultrasound treatment and they have attributed the superiority to the direct application of shear force towards the cells while applying low frequency ultrasound treatment.

12.6 Conclusion

Oilseeds used for production of edible oils and in various other industries have necessitated towards alternative green solvents and techniques. Besides, oilseed cake, which is rich in protein, can be supplemented as feed to animals provided that the extract is non-toxic and harmless. These objectives could be met with green solvents and techniques that are in great demand owing to their eco-friendly and non-toxic nature, reduction in energy and solvent consumption, extraction in lesser time with high efficiency.

Although these solvents and techniques are showing promising results, economic viability of these solvents and processes at large/industrial scale are yet to be validated. Application of green extraction technique coupled with green solvents and focus on development of integrated process could be interesting options to be explored further (Chandel et al., 2020; Garlapati et al., 2020). In biodiesel industry, lipid extractions of wet biomass/in situ transesterifications of oleaginous lipids could reduce the cost and make the process economically viable. Thus exploration of green solvents and extraction techniques not only ensure food and nutritional security but also enrich the quality assurance and economically viability of the products.

Author's contributions

SPJK has conceived and drafted the chapter, GVK surveyed review literature and drafted few portions of the chapter, LKSG drafted the chapter and RB designed, drafted and edited the chapter. All authors read and approved the final manuscript.

Competing interest

The authors declare that they have no competing interests.

Acknowledgements

The authors are grateful to Prof. Maryline Abert Vian, GREEN (Groupe de Recherche en Eco-Extraction de Produits Naturels), Université d'Avignon et des Pays de Vaucluse, INRA, UMR 408, 84,000 Avignon, France, for giving permission to cite the table in the chapter.

References

Adam, F., Abert-Vian, M., Peltier, G., Chemat, F., 2012. Solvent-free ultrasound-assisted extraction of lipids from fresh microalgae cells: a green, clean and scalable process. Bioresour. Technol. 114, 457–465.

Aissou, M., Chemat-Djenni, Z., Yara-Varón, E., Fabiano-Tixier, A.S., Chemat, F., 2017. Limonene as an agro-chemical building block for the synthesis and extraction of bioactive compounds. C. R. Chim. 20, 346–358.

Al-Ameri, M., Al-Zuhair, S., 2019. Using switchable solvents for enhanced, simultaneous microalgae oil extraction-reaction for biodiesel production. Biochem. Eng. J. 141, 217–224.

Alavijeh, R.S., Karimi, K., Wijffels, R.H., van den Berg, C., Eppink, M., 2020. Combined bead milling and enzymatic hydrolysis for efficient fractionation of lipids, proteins, and carbohydrates of *Chlorella vulgaris* microalgae. Bioresour. Technol. 12, 3321.

Araújo, A.C.M.A., Oliveira, RdÉ, Menezes, E.G.T., Dias, B.O., Terra, A.W.C., Queiroz, F., 2018. Solvent effect on the extraction of soluble solids from murici and pequi seeds. J. Food Process Eng. 41, e12813.

Banerjee, R., Kumar, S.P.J., Mehendale, N., Sevda, S., Garlapati, V.K., 2019. Intervention of microfluidics in biofuel and bioenergy sectors: technological considerations and future prospects. Renew. Sust. Energ. Rev. 101, 548–558.

Ben-Youssef, S., Fakhfakh, J., Breil, C., Abert-Vian, M., Chemat, F., Allouche, N., 2017. Green extraction procedures of lipids from Tunisian date palm seeds. Ind. Crops. Prod. 108, 520–525.

Bill & Melinda Gates Foundation. The Goalkeepers Report 2020. https://www.gatesfoundation.org/goalkeepers/report. Accessed 10 Sep, 2020.

Bligh, E.G., Dyer, W.J., 1959. A rapid method of total lipid extraction and purification. Can. J. Biochem. 37, 911–917.

Boulila, A., Hassen, I., Haouari, L., Mejri, F., Amor, I.B., Casabianca, H., Hosni, K., 2015. Enzyme-assisted extraction of bioactive compounds from bay leaves (*Laurus nobilis* L.). Ind. Crops. Prod. 74, 485–493.

Bourgou, S., Bettaieb Rebey, I., Dakhlaoui, S., Msaada, K., Saidani Tounsi, M., Ksouri, R., Fauconnier, M.L., Hamrouni-Sellami, I., 2020. Green extraction of oil from *Carum carvi* seeds using bio-based solvent and supercritical fluid: evaluation of its antioxidant and anti-inflammatory activities. Phytochem. Anal. 31, 37–45.

Boyd, A.R., Champagne, P., McGinn, P.J., MacDougall, K.M., Melanson, J.E., Jessop, P.G., 2012. Switchable hydrophilicity solvents for lipid extraction from microalgae for biofuel production. Bioresour. Technol. 118, 628–632.

Breil, C., Meullemiestre, A., Vian, M., Chemat, F., 2016a. Bio-based solvents for green extraction of lipids from oleaginous yeast biomass for sustainable aviation biofuel. Molecules 21, 196.

Breil, C., Meullemiestre, A., Vian, M., Chemat, F., 2016b. Bio-based solvents for green extraction of lipids from oleaginous yeast biomass for sustainable aviation biofuel. Molecules 21, 196.

Cascant, M.M., Breil, C., Garrigues, S., de la Guardia, M., Fabiano-Tixier, A.S., Chemat, F., 2017. A green analytical chemistry approach for lipid extraction: computation methods in the selection of green solvents as alternative to hexane. Anal. Bioanal. Chem. 409, 3527–3539.

Chabrand, R.M., Glatz, C.E., 2009. Destabilization of the emulsion formed during the enzyme-assisted aqueous extraction of oil from soybean flour. Enzyme Microb. Technol. 45, 28–35.

Chandel, A.K., Garlapati, V.K., Jeevan Kumar, S.P., Hans, M., Singh, A.K., Kumar, S., 2020. The role of renewable chemicals and biofuels in building a bioeconomy. Biofuel. Bioprod. Bior. https://doi.org/10.1002/bbb.2104.

Cho, H.S., Oh, Y.K., Park, S.C., Lee, J.W., Park, J.Y., 2013. Effects of enzymatic hydrolysis on lipid extraction from *Chlorella vulgaris*. Renew. Energy. 54, 156–160.

Chua, E.T., Brunner, M., Atkin, R., Eltanahy, E., Thomas-Hall, S.R., Schenk, P.M., 2018. The ionic liquid cholinium arginate is an efficient solvent for extracting high-value *Nannochloropsis* sp. lipids. ACS Sustain. Chem. Eng. 7, 2538–2544.

Cicci, A., Sed, G., Jessop, P.G., Bravi, M., 2018. Circular extraction: an innovative use of switchable solvents for the biomass biorefinery. Green Chem. 20, 3908–3911.

de Jesus, S.S., Ferreira, G.F., Fregolente, L.V., Maciel, F.R., 2018. Laboratory extraction of microalgal lipids using sugarcane bagasse derived green solvents. Algal Res 35, 292–300.

Dejoye, T.C., Abert, V.M., Ginies, C., Elmaataoui, M., Chemat, F., 2012. Terpenes as green solvents for extraction of oil from microalgae. Molecules 17, 8196–8205.

DeSimone, J.M., 2002. Practical approaches to green solvents. Science 297, 799–803.

dos Santos Reis, N., de Santana, N.B., de Carvalho Tavares, I.M., Lessa, O.A., dos Santos, L.R., Pereira, N.E., Soares, G.A., Oliveira, R.A., Oliveira, J.R., Franco, M., 2020. Enzyme extraction by lab-scale hydrodistillation of ginger essential oil (*Zingiber officinale* Roscoe): chromatographic and micromorphological analyses. Ind. Crop. Prod. 146, 112210.

Doucha, J., Lívanský, K., 2008. Influence of processing parameters on disintegration of Chlorella cells in various types of homogenizers. Appl. Microbiol. Biotechnol. 81, 431.

Fattah, I.M., Noraini, M.Y., Mofijur, M., Silitonga, A.S., Badruddin, I.A., Khan, T.M., Ong, H.C., Mahlia, T.M., 2020. Lipid extraction maximization and enzymatic synthesis of biodiesel from microalgae. Appl. Sci. 10, 6103.

Fine, F., Vian, M.A., Tixier, A.S.F., Carre, P., Pages, X., Chemat, F., 2013. Les agro-solvents pour l'extraction des huiles végétales issues de graines oléagineuses. OCL 20, A502.

Folch, J., Lees, M., 1957. Sloane Stanley GH. A simple method for the isolation and purification of total lipids from animal tissues. J. Biol. Chem. 226, 497–509.

Fu, C.C., Hung, T.C., Chen, J.Y., Su, C.H., Wu, W.T., 2010. Hydrolysis of microalgae cell walls for production of reducing sugar and lipid extraction. Bioresour. Technol. 101, 8750–8754.

Garlapati, V.K., Chandel, A.K., Kumar, S.J., Sharma, S., Sevda, S., Ingle, A.P., Pant, D., 2020. Circular economy aspects of lignin: towards a lignocellulose biorefinery. Renew. Sust. Energ. Rev. 130, 109977.

Garlapati, V.K., Gour, R.S., Sharma, V., Roy, L.S., Prashant, J.K.S., Prashant, J.K.S., Kant, A., Banerjee, R., 2017. Current status of biodiesel production from microalgae in India. Advances in Biofeedstocks and Biofuels: Production Technologies for Biofuels 2, 129–154.

Guerra, M.E., Gude, V.G., Mondala, A., Holmes, W., Hernandez, R., 2014. Extractive-transesterification of algal lipids under microwave irradiation with hexane as solvent. Bioresour. Technol. 156, 240–247.

Gujjala, L.K., Kumar, S.J., Talukdar, B., Dash, A., Kumar, S., Sherpa, K.C., Banerjee, R., 2019. Biodiesel from oleaginous microbes: opportunities and challenges. Biofuels 10, 45–59.

Gutte, K.B., Sahoo, A.K., Ranveer, R.C., 2015. Effect of ultrasonic treatment on extraction and fatty acid profile of flaxseed oil. OCL 22 (6), D606.

Hansen, C.M., 1969. The universality of the solubility parameter. Ind. Eng. Chem. 8, 2–11.

Hegel, P.E., Camy, S., Destrac, P., Condoret, J.S., 2011. Influence of pretreatments for extraction of lipids from yeast by using supercritical carbon dioxide and ethanol as co-solvent. J. Supercrit. Fluids. 58 (1), 68–78.

Herrero, M., Ibañez, E., 2018. Green extraction processes, biorefineries and sustainability: recovery of high added-value products from natural sources. J. Supercrit. Fluids. 134, 252–259.

Huo, S., Wang, Z., Cui, F., Zou, B., Zhao, P., Yuan, Z., 2015. Enzyme-assisted extraction of oil from wet microalgae *Scenedesmus sp.* G4. Energies 8, 8165–8174.

Isopencu, G., Stroescu, M., Brosteanu, A., Chira, N., Pârvulescu, O.C., Busuioc, C., Stoica-Guzun, A., 2019. Optimization of ultrasound and microwave assisted oil extraction from sea buckthorn seeds by response surface methodology. J. Food Process Eng. 42, e12947.

Jessop, P.G., Heldebrant, D.J., Li, X., Eckert, C.A., Liotta, C.L., 2005. Reversible nonpolar-to-polar solvent. Nature 436, 1102.

Jin, G., Yang, F., Hu, C., Shen, H., Zhao, Z.K., 2012. Enzyme-assisted extraction of lipids directly from the culture of the oleaginous yeast *Rhodosporidium toruloides*. Bioresour. Technol. 111, 378–382.

Kim, Y.H., Choi, Y.K., Park, J., Lee, S., Yang, Y.H., Kim, H.J., Park, T.J., Kim, Y.H., Lee, S.H., 2012. Ionic liquid-mediated extraction of lipids from algal biomass. Bioresour. Technol. 109, 312–315.

Klamt, A., 2011. The COSMO and COSMO-R solvation models. Wiley Interdiscip. Rev. Comput. 1, 699–709.

Kraujalis, P., Venskutonis, P.R., Pukalskas, A., Kazernavičiūtė, R., 2013. Accelerated solvent extraction of lipids from *Amaranthus spp.* seeds and characterization of their composition. LWT-Food Sci. Technol. 54, 528–534.

Kruger, J.S., Cleveland, N.S., Yeap, R.Y., Dong, T., Ramirez, K.J., Nagle, N.J., Lowell, A.C., Beckham, G.T., McMillan, J.D., Biddy, M.J., 2018. Recovery of fuel-precursor lipids from oleaginous yeast. ACS Sustain. Chem. Eng. 6, 2921–2931.

Kumar, S.P.J., Avanthi, A., Chintagunta, A.D., Gupta, A., Banerjee, R., 2020a. Oleaginous lipid: a drive to synthesize and utilize as biodiesel. Practices and Perspectives in Sustainable Bioenergy. Springer, New Delhi, pp. 105–129.

Kumar, S.P.J., Kumar, N.S.S., Chintagunta, A.D., 2020b. Bioethanol production from cereal crops and lignocelluloses rich agro-residues: prospects and challenges. SN Appl. Sci. 2, 1673. https://doi.org/10.1007/s42452-020-03471-x.

Kumar, S.P.J., Gujjala, L.K.S., Dash, A., Talukdar, B., Banerjee, R., 2017c. Biodiesel production from lignocellulosic biomass using oleaginous microbesLignocellulosic Biomass Production and Industrial Applications. John Wiley & Sons Inc, New York, pp. 65–92.

Kumar, S.P.J., Banerjee, R., 2013. Optimization of lipid enriched biomass production from oleaginous fungus using response surface methodology. Indian J. Exp. Biol. 51, 979–983.

Kumar, S.P.J., Kumar, G.V., Dash, A., Scholz, P., Banerjee, R., 2017a. Sustainable green solvents and techniques for lipid extraction from microalgae: a review. Algal Res. 21, 138–147.

Kumar, S.P.J., Prasad, S.R., Banerjee, R., Agarwal, D.K., Kulkarni, K.S., Ramesh, K.V., 2017b. Green solvents and technologies for oil extraction from oilseeds. Chem. Cent. J. 11, 9–15.

Lardon, L., Helias, A., Sialve, B., Steyer, P., Bernard, O., 2009. Life-cycle assessment of biodieselproduction from microalgae. Environ. Sci. Technol. 43, 6475–6481.

Latif, S., Anwar, F., 2011. Aqueous enzymatic sesame oil and protein extraction. Food Chem. 125, 679–684.
Liang, K., Zhang, Q., Cong, W., 2012. Enzyme-assisted aqueous extraction of lipid from microalgae. J. Agric. Food Chem. 60, 11771–11776.
Lin, J.Y., Zeng, Q.X., An, Q.I., Zeng, Q.Z., Jian, L.X., Zhu, Z.W., 2012. Ultrasonic extraction of hempseed oil. J. Food Process Eng. 35, 76–90.
Lu, C., Napier, J.A., Clemente, T.E., Cahoon, E.B., 2011. New frontiers in oilseed biotechnology: meeting the global demand for vegetable oils for food, feed, biofuel, and industrial applications. Curr. Opin. Biotech. 22, 252–259.
Mahmood, W.M.A.W., Theodoropoulos, C., Gonzalez-Miquel, M., 2017. Enhanced microalgal lipid extraction using bio-based solvents for sustainable biofuel production. Green Chem. 19, 5723–5733.
Meullemiestre, A., Breil, C., Abert-Vian, M., Chemat, F., 2016. Microwave, ultrasound, thermal treatments, and bead milling as intensification techniques for extraction of lipids from oleaginous *Yarrowia lipolytica* yeast for a bio-jetfuel application. Bioresour. Technol. 211, 190–199.
Mgudu, E.M., Kabuba, J., Belaid, M., 2012. Microwave–assisted extraction of castor oil, Proceedings of the International Conference on Nanotechnology and Chemical Engineering ICNCS. Bangkok, Thailand 21–22 December 2012.
Montalescot, V., Rinaldi, T., Touchard, R., Jubeau, S., Frappart, M., Jaouen, P., Bourseau, P., Marchal, L., 2015. Optimization of bead milling parameters for the cell disruption of microalgae: process modeling and application to *Porphyridium cruentum* and *Nannochloropsis oculata*. Bioresour. Technol. 196, 339–346.
Mosquera, D.M., Carrillo, M.P., Gutiérrez, R.H., Diaz, R.O., Hernández, M.S., Fernández-Trujillo, J.P., 2013. Microwave technology applied to natural ingredient extraction from Amazonian fruits. Foods 1, 1–8.
Mouahid, A., Crampon, C., Toudji, S.A.A., Badens, E., 2013. Supercritical CO2 extraction of neutral lipids from microalgae: experiments and modelling. J. Supercrit. Fluids. 77, 7–16.
Mustapha, S.I., Isa, Y.M., 2020. Utilization of quaternary solvent mixtures for extraction of lipids from *Scenedesmus obliquus* microalgae. Cogent Eng 7, 1788877.
Nadar, S.S., Rao, P., Rathod, V.K., 2018. Enzyme assisted extraction of biomolecules as an approach to novel extraction technology: a review. Food Res. Int. 108, 309–330.
Nde, D.B., Foncha, A.C., 2020. Optimization methods for the extraction of vegetable oils: a review. Processes 8, 209.
Paduszyński, K., 2018. Extensive evaluation of the conductor-like screening model for real solvents method in predicting liquid-liquid equilibria in ternary systems of ionic liquids with molecular compounds. J. Phys. Chem. B. 122, 4016–4028. doi:10.1021/acs.jpcb.7b12115.
Perrier, A., Delsart, C., Boussetta, N., Grimi, N., Citeau, M., Vorobiev, E., 2017. Effect of ultrasound and green solvents addition on the oil extraction efficiency from rapeseed flakes. Ultrason. Sonochem. 39, 58–65.
Pham, L., Chiu, D., Heldebrant, D.J., Huttenhower, H., John, E., Li, X., Pollet, P., Wang, R., Eckert, C.A., Liotta, C.L., Jessop, P.G., 2008. Switchable solvents consisting of amidine/alcohol or guanidine/alcohol mixtures. Ind. Eng. Chem. 47, 539–545.
Prabakaran, P., Ravindran, A.D., 2011. A comparative study on effective cell disruption methods for lipid extraction from microalgae. Lett. Appl. Microbiol. 53, 150–154.
Priyadarshini, S.B., Balaji, A., Murugan, V., 2008. Extraction of microbial lipids by using supercritical carbon dioxide as solvent. Asian J. Chem. 20, 51–53.
Prommuak, C., Pavasant, P., Quitain, A.T., Goto, M., Shotipruk, A., 2012. Microalgal lipid extraction and evaluation of single-step biodiesel production. Eng. J. 16, 157–166.
Qiu, F., Li, Y., Yang, D., Lib, X., Sun, P., 2011. Biodiesel production from mixed soybean oil and rapeseed oil. Appl. Energy 88, 2050–2055.
Rahman, M., de Jiménez, M.M., 2016. Designer oil cropsBreeding Oilseed Crops for Sustainable Production. Academic Press, pp. 361–376.
Rosas-Mendoza, M.E., Coria-Hernández, J., Meléndez-Pérez, R., Arjona-Román, J.L., 2017. Characteristics of chia (*Salvia hispanica* L.) seed oil extracted by ultrasound assistance. J. Mex. Chem. 61, 326–335.

Salgın, U., Salgın, S., Ekici, D.D., UludaĽ, G., 2016. Oil recovery in rosehip seeds from food plant waste products using supercritical CO2 extraction. J. Supercrit. Fluids. 118, 194–202.

Santoro, I., Nardi, M., Benincasa, C., Costanzo, P., Giordano, G., Procopio, A., Sindona, G., 2019. Sustainable and selective extraction of lipids and bioactive compounds from microalgae. Molecules 24, 4347.

Santos, K.A., da Silva, E.A., da Silva, C., 2020. Supercritical CO_2 extraction of favela (*Cnidoscolus quercifolius*) seed oil: yield, composition, antioxidant activity, and mathematical modeling. J. Supercrit. Fluids. 165, 104981.

Schuur, B., Brouwer, T., Smink, D., Sprakel, L.M., 2019. Green solvents for sustainable separation processes. Curr. Opin. Green Sustain. Chem. 18, 57–65.

Selvakumar, P., Sivashanmugam, P., 2019. Ultrasound assisted oleaginous yeast lipid extraction and garbage lipase catalyzed transesterification for enhanced biodiesel production. Energy Convers. Manag. 179, 141–151.

Shin, H.Y., Shim, S.H., Ryu, Y.J., Yang, J.H., Lim, S.M., Lee, C.G., 2018. Lipid extraction from *Tetraselmis* sp. microalgae for biodiesel production using hexane-based solvent mixtures. Biotechnol. Bioprocess Eng. 23, 16–22.

Sicaire, A.G., Vian, M., Fine, F., Joffre, F., Carré, P., Tostain, S., Chemat, F., 2015. Alternative bio-based solvents for extraction of fat and oils: solubility prediction, global yield, extraction kinetics, chemical composition and cost of manufacturing. Int. J. Mol. Sci. 16, 8430–8453.

Sicaire, A.G., Vian, M.A., Fine, F., Carré, P., Tostain, S., Chemat, F., 2016. Ultrasound induced green solvent extraction of oil from oleaginous seeds. Ultrason. Sonochem. 31, 319–329.

Soares, J.F., Dal Prá, V., de Souza, M., Lunelli, F.C., Abaide, E., da Silva, J.R., Kuhn, R.C., Martínez, J., Mazutti, M.A., 2016. Extraction of rice bran oil using supercritical CO_2 and compressed liquefied petroleum gas. J. Food Eng. 170, 58–63.

Soto, C., Chamy, R., Zuniga, M.E., 2007. Enzymatic hydrolysis and pressing conditions effect on borage oil extraction by cold pressing. Food Chem. 102, 834–840.

Soto, C., Concha, J., Zuniga, M.E., 2008. Antioxidant content of oil and defatted meal obtained from borage seeds by an enzymatic-aided cold pressing process. Process Biochem. 43, 696–699.

Stevanato, N., da Silva, C., 2019. Radish seed oil: ultrasound-assisted extraction using ethanol as solvent and assessment of its potential for ester production. Ind. Crop. Prod. 132, 283–291.

Sui, X., Jiang, L., Li, Y., Liu, S., 2011. The research on extracting oil from watermelon seeds by aqueous enzymatic extraction method. Procedia Eng 15, 4673–4680.

Taghvaei, M., Jafari, S.M., Assadpoor, E., Nowrouzieh, S., Alishah, O., 2014. Optimization of microwave-assisted extraction of cottonseed oil and evaluation of its oxidative stability and physicochemical properties. Food Chem. 160, 90–97.

Tommasi, E., Cravotto, G., Galletti, P., Grillo, G., Mazzotti, M., Sacchetti, G., Samorì, C., Tabasso, S., Tacchini, M., Tagliavini, E., 2017. Enhanced and selective lipid extraction from the microalga *P. tricornutum* by dimethyl carbonate and supercritical CO_2 using deep eutectic solvents and microwaves as pretreatment. ACS Sustain. Chem. Eng. 5, 8316–8322.

Topare, N.S., Raut, S.J., Renge, V.C., Khedkar, S.V., Chavanand, Y.P., Bhagat, S.L., 2011. Extraction of oil from algae by solvent extraction and oil expeller method. Int. J. Chem. Sci. 9, 1746–1750.

Vasconcelos, B., Teixeira, J.C., Dragone, G., Teixeira, J.A., 2018. Optimization of lipid extraction from the oleaginous yeasts *Rhodotorula glutinis* and *Lipomyces kononenkoae*. AMB Express 8, 126.

Womeni, H.M., Ndjouenkeu, R., Kapseu, C., Mbiapo, F.T., Parmentier, M., Fanni, J., 2008. Aqueous enzymatic oil extraction from *Irvingia gabonensis* seed kernels. Eur. J. Lipid Sci. Technol. 110, 232–238.

Wu, J., Johnson, L.A., Jung, S., 2009. Demulsification of oil-rich emulsion from enzyme-assisted aqueous extraction of extruded soybean flakes. Bioresour. Technol. 100, 527–533.

Wu, X., Joyce, E.M., Mason, T.J., 2012. Evaluation of the mechanisms of the effect of ultrasound on *Microcystis aeruginosa* at different ultrasonic frequencies. Water Res. 46, 2851–2858.

Yang, F., Xiang, W., Sun, X., Wu, H., Li, T., Long, L., 2014. A novel lipid extraction method from wet microalga *Picochlorum* sp. at room temperature. Mar. Drugs 12, 1258–1270.

You, J.Y., Peng, C., Liu, X., Ji, X.J., Lu, J., Tong, Q., Wei, P., Cong, L., Li, Z., Huang, H., 2011. Enzymatic hydrolysis and extraction of arachidonic acid rich lipids from *Mortierella alpina*. Bioresour. Technol. 102, 6088–6094.

Yousuf, O., Gaibimei, P., Singh, A., 2018. Ultrasound assisted extraction of oil from soybean. Int. J. Curr. Microbiol. Appl. Sci. 7, 843–852.

Yusoff, M.M., Gordon, M.H., Niranjan, K., 2015. Aqueous enzyme assisted oil extraction from oilseeds and emulsion de-emulsifying methods: a review. Trends Food Sci. Tech. 41, 60–82.

Zhang, X., Yan, S., Tyagi, R.D., Drogui, P., Surampalli, R.Y., 2014. Ultrasonication assisted lipid extraction from oleaginous microorganisms. Bioresour. Technol. 158, 253–261.

Zhang, Y., Kong, X., Wang, Z., Sun, Y., Zhu, S., Li, L., Lv, P., 2018. Optimization of enzymatic hydrolysis for effective lipid extraction from microalgae *Scenedesmus sp.* Renew. Energy. 125, 1049–1057.

CHAPTER 13

Three phase partitioning (TPP) as an extraction technique for oleaginous materials

Sandesh J. Marathe[a], Nirali N. Shah[b], Rekha S. Singhal[a]
[a]Department of Food Engineering and Technology, Institute of Chemical Technology, Mumbai, India
[b]Centre for Technology Alternatives for Rural Areas, IIT Bombay, Mumbai, India

Chapter outline

13.1 Introduction

Oils and lipids form an important component of various daily usages such as cooking, bakery, and confectionery, as well as the canned food industry. Oils are also an essential element of the non-food industries such as detergents, paints, varnishes, pharmaceuticals, and cosmetics (Mariana et al., 2013). The content of oil in oleaginous materials such as seeds, nuts, kernels, or fruit pulps ranges between 3–70 percent of the total weight of the material (Nde and Foncha, 2020). The industrial uses of oil calls for its optimal extraction. Various techniques have been used traditionally to extract oil from oleaginous materials such as seeds (sunflower, rapeseed, soybean, etc.) and fruits (coconut, olive, and palm) (Mariana et al., 2013). However, each method has its pros and cons, and the choice of the extraction method depends greatly on the nature of the source and the oil content. The method used for the extraction of oil also decides its final quality and safety. This has driven the interest of researchers towards developing various hyphenated and novel techniques for the extraction of oil.

Three Phase Partitioning
DOI: https://doi.org/10.1016/B978-0-12-824418-0.00014-X

The physical methods used for extraction may not always ensure the complete recovery of oil. The extraction of the remainder drives the need for developing additional techniques. Owing to its simplicity and cost-effectiveness, solvent extraction has also been employed for oil extraction. However, despite the high extraction efficiency, the use of solvents raises health and environmental safety concerns (Mwaurah et al., 2020). The solution to the concerns associated with the practice of these common extraction techniques is the use of non-conventional techniques. Non-conventional techniques offer benefits such as improved quality of extracted oil, time efficiency, high yield, environmental safety, and cost-effectiveness (Mwaurah et al., 2020). Various new techniques are being developed for efficient extraction of oil such as enzyme-assisted extraction, microwave-assisted extraction, ultrasound-assisted extraction, and supercritical fluid CO_2-extraction (Mwaurah et al., 2020).

Yet another novel technique for extraction of oil from oleaginous materials that is gaining the interest of researchers is three phase partitioning (TPP). This chapter focuses on the use of TPP as a novel technique for the extraction of oleaginous material, the mechanism of TPP, and the advantages of TPP. Various hyphenated TPP techniques have also been discussed.

13.2 Conventional extraction techniques for oleaginous material

The traditionally used extraction methods for oleaginous material include chemical (solvent) extraction, mechanical extraction (hydraulic and screw press), and distillation (Mariana et al., 2013). Mechanical extraction is one of the oldest known techniques used for the extraction of oil, where oil is essentially forced out of the oleaginous material by applying pressure (Nde and Foncha, 2020). The presses used for the mechanical extraction of oil include hydraulic press or screw press driven by a motor (Bhuiya et al., 2015; Nde and Foncha, 2020). The yield of the oil by mechanical extraction is majorly governed by the force applied during extraction, the duration of the force applied, and the pretreatment of the oleaginous material at elevated temperatures (Nde and Foncha, 2020). Although the mechanical extraction process is efficient and gives a safer product, it often gives lower yields than solvent extraction.

Due to its ease of use and cost-efficiency, the solvent extraction method has been preferred over others. Furthermore, solvent extraction has been reported to give higher yields of oil (Bhuiya et al., 2015). The efficacy of the solvent extraction method is based on the ability of solvents to effectively dissolve oils from the substrates. Solvent extraction is carried out either as a batch or a continuous process and involves three major steps—cleaning and conditioning the oilseed, extraction of oil, and separation of miscella. The pretreatment of the oilseed, the type of solvent used, and the extraction temperature greatly affect the quality of crude oil (Nde and Foncha, 2020). For instance, polyunsaturated oil is highly sensitive to oxidation. Thus, elevated temperatures during extraction can cause alterations in the polyunsaturated oil quality (Guédé et al., 2017).

Although solvent extraction offers higher extraction yields, the final product is not free of solvent residues. For instance, due to its significant solubility, non-corrosive nature, and cost-effectiveness, hexane is a commercially used extraction solvent. However, it is considered a hazardous and flammable air pollutant (Tan et al., 2016). Thus, from the food safety point of view, researchers are preferring safer (green) alternatives of solvents for the extraction of oil such as water, ethanol, and ionic liquids.

There has been a growing interest in the use of modern techniques such as supercritical carbon dioxide extraction, microwave-assisted extraction, ultrasound-assisted extraction, and enzyme-assisted extraction. Another promising technique in recent times is three phase partitioning (TPP). In comparison to other techniques, TPP has been reported to give better quality of oil with higher yields (Sharma et al., 2019). In addition to the extraction of oil, TPP has been reported as an efficient method for fractionating and concentrating various proteins/enzymes. TPP has also been coupled with other methods to give hyphenated techniques such as ultrasound assisted-TPP (UTPP) or microwave-assisted TPP (MTPP) (Dutta et al., 2015).

13.3 Mechanism of extraction using TPP

Briefly, TPP involves mixing the powdered oleaginous source with water to prepare a slurry of raw material, followed by addition of salt. Finally, *t*-butanol is added and the resultant mixture is stirred, and allowed to stand to form three separate layers. The oil seeds may be separated by sieving, followed by separation of the three phases by gravimetry or centrifugation (Kurmudle et al., 2011; Dutta et al., 2015; Mulchandani et al., 2015). The upper layer of *t*-butanol contains dissolved oil, middle layer contains proteins precipitated with salt, and the lower aqueous layer contains hydrophilic constituents such as carbohydrates, fibers, gums, etc. The schematic representation of TPP is shown in Fig. 13.1. After the process of TPP is complete, the oil extracted in the organic phase is recovered by evaporating the solvent.

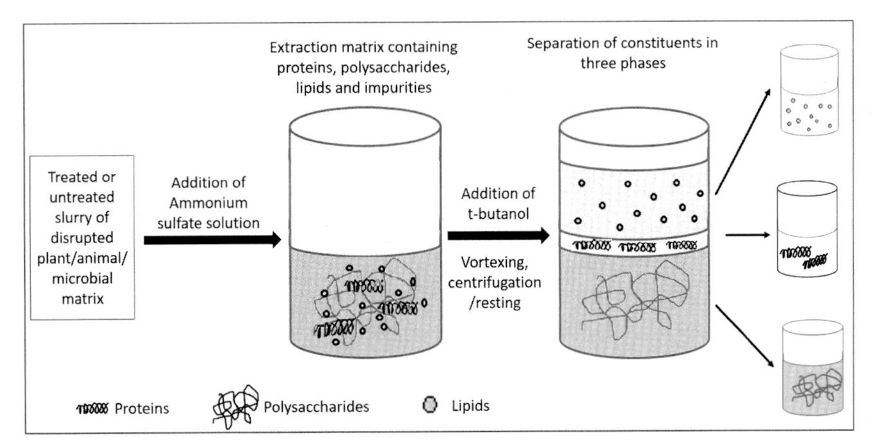

Fig. 13.1 ***Schematic representation of three phase partitioning for the extraction of constituents from plant, animal or microbial cells/matrix.***

The functioning of TPP is essentially based on the 'order' of the water molecules due to the intermolecular interactions. Chaotropes disorder the hydrogen bonding between water molecules, whereas kosmotropes are order-forming agents which contribute to the stability of water-water interactions. Normally *t*-butanol is completely miscible with water and forms a solution. However, upon addition of a kosmotrope such as ammonium sulfate which contributes to the stability of intermolecular interactions in water, the solution gets separated into two layers of water and *t*-butanol. If the water-butanol system is fed with protein component, a third distinct layer of protein forms between the water and butanol layers. Depending on the chemical properties of the various components of the feed added to the water-butanol-ammonium sulfate system, they are segregated into the polar water phase, or the non-polar *t*-butanol phase. The low interfacial tension between the phases results in high mass transfer (Ketnawa et al., 2017). TPP can be considered as a combination of the various phenomena such as salting out, co-solvent precipitation, isoionic precipitation, alcohol precipitation, and osmolytic and kosmotropic precipitation of proteins (Ketnawa et al., 2017; Yan et al., 2018). In case of oil extraction using TPP, lipids are extracted into the *t*-butanol phase whereas polar compounds are pulled into the aqueous phase, and a third layer of protein is formed between the polar and non-polar phases (Dutta et al., 2015). The dehydration property of ammonium sulfate and the ability of *t*-butanol to enhance the protein buoyancy together account for the accumulation of protein in the middle layer (Yan et al., 2018). Thus, TPP gives relatively contaminant-free products (such as oil) making the process feasible and lucrative.

Sulfate anions which are Hofmeister kosmotrope and a water structure promoter, form electrostatic interactions with protein. On the other hand, the co-solvent (*t*-butanol) acts as a neutral molecule. Thus, an optimum concentration of both is required to achieve the desired physiochemical behavior of the phases. For this reason, optimization of the process parameters is necessary to employ the process successfully (Dutta et al., 2015). Although *t*-butanol is the typically used solvent, methanol or ethanol may be employed for the purpose of a co-solvent (Dutta et al., 2015). However, ethanol or methanol may not behave like *t*-butanol, except at temperatures near or below zero (Dennison and Lovrien, 1997). The phenomenon of TPP is not entirely based on the kosmotropic nature of the constituents, but also on the capacity of the salt to enforce appropriate electrostatic forces, protein hydration shifts, and the ability to force protein conformation tightening (Dennison and Lovrien, 1997). Perhaps, the most crucial factor is the binding tendency of the divalent sulfate anion to a few cationic sites of various proteins. This binding consequently leads to a major shrinkage of conformation from an expanded state (Dennison and Lovrien, 1997). This is the main reason of the pH dependency of the salting out process. Thus, although other Hofmeister kosmotropes such as phosphate anions might behave similar to sulfate anions, the pH dependency of the separation phenomenon restricts their use, especially when the process is based on inorganic anions binding to the proteins (Dennison and Lovrien, 1997).

Apart from the various interactions mentioned above, colloidal interactions also play an important role in TPP. Oilseeds show presence of core organelles called lipid bodies or oleosomes which contain triglycerides and free fatty acids and protein bodies which show high protein content. Studies based on soybean and peanuts have revealed that the oleosomes are embedded in a cytoplasmic protein network. The spaces between protein bodies is filled with oleosomes and cytoplasmic protein network (Marathe et al., 2017). This explains the importance of separating proteins from the oilseeds in order to achieve maximum liberation/extraction of oil.

13.4 Advantages of TPP

The safety concerns associated with the conventional extraction techniques have driven the development of newer, efficient and green extraction techniques such as TPP. Considering the use of solvents such as hexane in conventional solvent extraction, there has been an increasing interest in developing a safer technique for extraction of oil which can also be scalable for practical applications. Hexane is commonly used for the extraction of oil from oilseed. However it is flammable. Additionally it also contributes to the volatile organic compounds and can react with air pollutants to form photochemical oxidants making it a potent pollutant (Vidhate and Singhal, 2013). In this regard, TPP is greener and rapid, and scaling up can be carried out even using crude suspensions. In contrast to the use of solvents such as hexane, methanol and ethanol for conventional extraction, TPP uses *t*-butanol which has a higher boiling point, thus reducing the addition of volatile organic compounds to the atmosphere even with the use of open systems (Kurmudle et al., 2011). Furthermore, the separation in TPP is facilitated by chilling rather than heating (Yan et al., 2018) which further reduces the chances of contributing to atmospheric pollutants.

The separation of the three phases in TPP occurs in a relatively short time (1 h). The separation of proteins is easy as it precipitates as a separate layer between the upper organic and lower aqueous layers (Kurmudle et al., 2011). Polar contaminants such as saccharides get concentrated in the lower aqueous layer. The time required for TPP is less and its efficiency is comparable to that of Soxhlet extraction technique (Vidhate and Singhal, 2013). Researchers are now showing interest in hyphenated TPP techniques which further improve its efficiency of extraction.

Thus, TPP is a straightforward technique and the equipment used is rather simple and easy to operate. Moreover, the scaling up of TPP seems promising considering the similarity in results of large-scale and small-scale operation, making it reliable (Ketnawa et al., 2017). These advantages of TPP make it an efficient, lucrative and eco-friendly technique for the extraction of oleaginous material.

13.5 Factors affecting TPP

13.5.1 Salt

The proportion of salt in TPP plays a crucial role in the amount of oil extracted from oilseed. Due to the fact that salt plays a direct role in separation of proteins by precipitation which eventually sets the oleosomes free, the kind and proportion of salt influences extraction yield of TPP. For instance, various salts such as ammonium sulphate, ammonium carbonate, ammonium chloride and ammonium acetate have been studied for the separation of oil from Sunn hemp seeds (Dutta et al., 2015). Amongst these, although ammonium carbonate shows high values of partition coefficient, it shows lower solubility than other salts. Moreover, even though ammonium chloride and acetate showed comparable results, they showed weak kosmotropic effect. In this regard, ammonium sulfate is the most suitable salt for protein precipitation in TPP. This could be due to the favorable kosmotropic effect exerted by sulfate ions, making it preferable amongst the other salts of Hofmeister series (such as Cl^-, Br^-, SO^{-2}, CH COO^-, CO_3^{-2} etc.). In comparison to salts of sodium and potassium, the substantial solubility of ammonium salts in water, and the higher partitioning area in the phase diagram, makes ammonium salts suitable for TPP (Dutta et al., 2015). Ammonium sulfate also shows higher ionic strength in comparison to common salt (NaCl). The sulfate anions (SO_4^{2-}) tend to bind to the cationic sites of these proteins contributing to the 'salting out' process (Yan et al., 2018). The salt concentration in TPP is usually kept at the minimum of 20 percent (w/v), and is increased further as per the requirement. Higher salt concentrations lead to the depletion of the water molecules from the solvation layer/hydration shell of proteins, thus exposing the hydrophobic moieties of the protein. This causes hydrophobic interactions between the proteins causing further precipitation. Generally, the concentration of salt in TPP is varied in the range of 20–60 percent (w/v) (Yan et al., 2018). The optimized concentrations of salt for the extraction of various oleaginous materials using TPP is presented in Table 13.1.

Apart from its role in releasing oleosomes by precipitation of proteins and the phase separation, salt used in TPP also plays a crucial role in extracting oil from sources such as microalgae by exerting osmotic shock in the initial 'incubation' step (Li et al., 2015). The high salt concentration in the incubation step exerts an osmotic shock leading to the bursting of cells and release of cellular contents, exposing the oleosomes to salt. The oleosomes are globular structures made up of phospholipids and proteins embedded between them. Triacylglycerol molecules are trapped inside these globular structures. The interaction of proteins of the released oleosomes with the salt leads to their precipitation, thus releasing triacylglycerol (Vidhate and Singhal, 2013). Higher salt content favors the extraction of oleosomes as well as the complete precipitation of proteins. The role of salt in TPP is also to induce interfacial tension between the aqueous and organic (*t*-butanol) phase, leading to their phase separation and precipitation of proteins between the phases (Dutta et al., 2015). This explains the importance of selecting

Table 13.1 Optimal conditions reported for the extraction of oleaginous material using three phase partitioning (TPP).

Oilseed	TPP extraction parameter					Recovery/yield (percent)	Reference
	Salt concentration (w/v)	Solvent ratio (aqueous slurry: solvent) (v/v)	pH	Temperature (°C)	Time (h)		
Jatropha curcas L. seed kernels	30	1:1	7.0	25	1	82[a]	(Shah et al., 2004)
Mango kernel	50	1:1	4.0	37	1	63[a]	(Gaur et al., 2007)
Soybean flour	30	1:1	4.0	37	1	90[a]	
Rice bran	30	1:2	7.0	37	1	72[a]	
Turmeric fingers	30	1:1	–	–	1	7.3[b]	(Kurmudle et al., 2011)
Kokum (*Garcinia indica*) kernel	50	1:1	2.0	45	1	95[a]	(Vidhate and Singhal, 2013)
Chlorella saccharophila	30	1:0.75	–	25	1	69[a]	(Mulchandani et al., 2015)
Cassia sophera Linn	30	1:1	7.0	–	1	–	(Mondal, 2015)
Ginger (*Zingiber officinale*) rhizome	10	1:0.5	5.0	32	4	6.16[b]	(Varakumar et al., 2017)
Sesame (*Sesamum indicum* L.) seeds	40	1:1	5.0	27	1.5	78[a]	(Juvvi and Debnath, 2020)

[a]Recovery indicates oil/oleoresin extracted by TPP per unit oil/oleoresin extracted by conventional solvent extraction method.
[b]Yield indicates oil/oleoresin extracted per unit mass of the oleaginous material.

an optimum concentration of salt in TPP for extracting oil from various oil seeds. Considering the commercial aspects, the use of precise and minimal (optimal) quantities of salt for TPP also becomes crucial if scaling-up of the process is to be done, in order to achieve economic advantage.

13.5.2 Extraction solvent

The solvent almost always used for TPP is *t*-butanol. Although one of the reasons is its higher boiling point and much less flammability than solvents such as methanol, ethanol and *n*-hexane which are solvents used for the common solvent extraction process, *t*-butanol has other beneficial effects in TPP. Solvents such as methanol and ethanol are not considered kosmotropic or as crowding agents unless subjected to near-zero temperatures. Furthermore, in comparison to solvents such as *n*-propanol, isopropanol, *n*-butanol, *n*-pentanol, isopentanol, and *n*-hexane that have been tried for TPP, *t*-butanol has proven to be efficient with low deactivation and higher interfacial precipitation of proteins by hydrophobic interactions. At temperatures ranging between 20–30 °C, *t*-butanol promotes TPP by displaying substantial kosmotropic and crowding effects (Yan et al., 2018). For instance, Sharma et al. (2002) studied TPP for the extraction of soybean oil using various solvents. They observed a maximum yield of 82 percent using *t*-butanol, in comparison to the substantially low yields of 24 percent, 26 percent, and 6.4 percent, using *n*-propanol, isopropanol, and ethanol, respectively. Similarly, Vidhate and Singhal (2013) reported the extraction of fat from *kokum* (*Garcinia indica*) kernels using TPP, where maximum yield of 94 percent was achieved using *t*-butanol, as opposed to 30 percent and 20 percent using isopropanol and ethanol, respectively.

Similar to the concentration of salt, the quantity of *t*-butanol in TPP significantly alters the yield of oil. Optimum quantity of *t*-butanol is essential to achieve maximum yield of oil from the oil seed. Lower concentrations of *t*-butanol cause hindrance in its synergy with the salt, thus affecting protein precipitation as well as phase separation. Furthermore, with an increase in the proportion of *t*-butanol beyond its optima, the overall viscosity of the solution increases, causing a decrease in molecular mobility, and hence decreased molecular interactions (Yan et al., 2018). For instance, for the extraction of palmitic acid and elaidic acid from *Cassia sophera* Linn using TPP, increase in the quantity of *t*-butanol beyond the optima (1:1 v/v of water:*t*-butanol) lead to a drop in the recovery (Mondal, 2015). Since molecular interactions are crucial in TPP, setting up an optimum for a new system is essential. The proportions of *t*-butanol generally used with the crude aqueous oil seed extract range between 1:0.5 to 1:2.5 (v/v of crude extract to *t*-butanol) (Yan et al., 2018). However, as can be seen from Table 13.1, a ratio of 1:1 in most cases has proved to be optimal for TPP.

13.5.3 pH

Considering the influence of pH on the charge of proteins, pH significantly affects the precipitation in TPP, and hence the overall oil extraction. The change in pH alters the charge on the amino acid residues, and hence the charge at the protein surface. The change in the charge of the protein also influences protein-protein interactions, thus affecting their partitioning behavior. The partitioning of proteins from the aqueous phase to the interface is greatly dependent on the pI of proteins. At pH values higher than the pI, the surface-exposed amino acid residues of proteins carry negative charge, thus imparting a net negative charge on the protein, ergo leaving them in the aqueous phase. However, at pH values lower than the pI, the surface-exposed amino acid residues of proteins carry positive charge, imparting thereon a net positive charge. This leads to an enhanced interaction with sulfate ions and causes optimal precipitation at the interface. This sheds light on the significance of optimizing the pH for TPP process (Yan et al., 2018). Generally, pH of a TPP system is adjusted at the initial incubation step (for aqueous slurry of oil seed) and is typically done using 0.1 N HCl/0.1 N NaOH (Gaur et al., 2007; Kurmudle et al., 2011; Vidhate and Singhal, 2013; Dutta et al., 2015). However, a buffer system may also be used for pH adjustments. For instance, Tan et al. (2016) carried out the extraction of oil from flaxseed (*Linum usitatissimum* L.) using enzyme-assisted TPP where the pH of the aqueous slurry was adjusted between 3.0–8.0 using McIlvaine (citrate-phosphate) buffer. The optimized pH values for the extraction of various oleaginous material using TPP are presented in Table 13.1. Although various reports indicate the optimal pH to range between 4.0–7.0, in certain cases pH values lower than 4.0 has also been reported. For instance, Vidhate and Singhal (2013) reported an optimal pH of 2.0 of the aqueous slurry for the extraction of *kokum* kernel fat using TPP.

13.5.4 Temperature

TPP involves multiple effects such as changes in conformation and protein hydration, which together lead to the extraction of oil. The optimized temperatures for the extraction of oleaginous material using TPP are presented in Table 13.1. Several researchers have reported the operation of TPP at temperatures ranging between 20–40 °C. Higher temperatures may cause an improved solubility of oil in the organic phase thus giving higher yields. Higher temperature has also been reported to give better yields primarily by disruption of cells. For instance, Li et al. (2015) reported an increase in the extraction of lipids from microalgae (*Chlorella* spp.) when the temperature was raised from 25 °C to 90 °C. However, extraction temperature affects the TPP system in multiple ways, and hence is one of the crucial factors especially when ammonium sulfate and *t*-butanol are involved. This is because at temperatures above 40 °C, the volatility of *t*-butanol increases thus decreasing its proportion in the TPP system. This causes a decrease in the optimum quantity of *t*-butanol available to interact with ammonium sulfate, leading to compromised precipitation of proteins (Yan et al., 2018). Extraction temperature also influences the rate of mass transfer. The ultimate effect is on the extraction yield of oil. Thus, most of the TPP processes have been attempted and have given comparable results at ambient temperatures.

13.5.5 Extraction time

To ensure the cost-effectiveness of an extraction process, it is necessary to achieve maximum extraction in minimum time possible. Various researchers have reported optimal extraction in a time period of 1 h using TPP (Dutta et al., 2015; Li et al., 2015). Considering the role of the oil seed matrix in the extraction of oil using TPP, the extraction time may differ with various oilseeds. For instance, Gaur et al.(2007) studied the effect of extraction time on the extraction of edible oil from soybean, rice bran, and mango kernels using enzyme-assisted TPP. The authors reported an increase in the extraction of oil from mango kernels from 70 percent to 80 percent (w/w) as the extraction time increased from 1 to 5 h. However, increasing the extraction time for soybean and rice bran did not alter the yields of oil. Thus, the time required for the optimal contact between the extraction solvent and the solute (oil seed) may vary with the composition of the matrix.

Although conventional extraction techniques have shown optimal extraction at time periods of 8–24 h, the unique mechanism of extraction of TPP makes it possible to achieve optimal extraction with much lesser time. In this regard, Shah et al.(2004) reported an optimal extraction of oil (97 percent yield) from *Jatropha curcas* L. seed kernels in 2 h using a hyphenated TPP process, as opposed to Soxhlet extraction performed for 24 h. Thus, extraction time may be reduced further with the use of hyphenated TPP techniques.

13.6 Hyphenated TPP-techniques

Several techniques of cell lysis and hence extraction have been reported as precursor to TPP before extraction of oleaginous materials from natural sources. The objective is towards increasing the rate of mass transfer and hence yield of extraction. The common techniques of cell lysis such as enzymatic degradation, ultrasonic lysis, microwave assisted heating and high-pressure homogenization have been used as pretreatment steps for subsequent TPP-assisted extraction. Fig. 13.2. schematically depicts the influence of various hyphenated techniques used with TPP for extraction of oleaginous materials. The treatments can be broadly classified as mild or severe based on the energy inputs in the processing. Enzymatic, ultrasound, microwave and high-pressure homogenization are the commonly reported pretreatments for TPP. Other pre-treatments for cell disruption include osmotic shock, thermal-lysis, freeze-thawing and chemical lysis using solvents, detergents, alkali and ionic liquids (Gomeset al., 2020).

13.6.1 Enzyme assisted TPP (EATPP)

Enzyme assisted extraction of constituents from a plant, animal or microbial matrix depends on the interactions between the enzyme(s) and the components of the matrix. This enzyme-substrate interaction leads to chemical changes in the structure of the

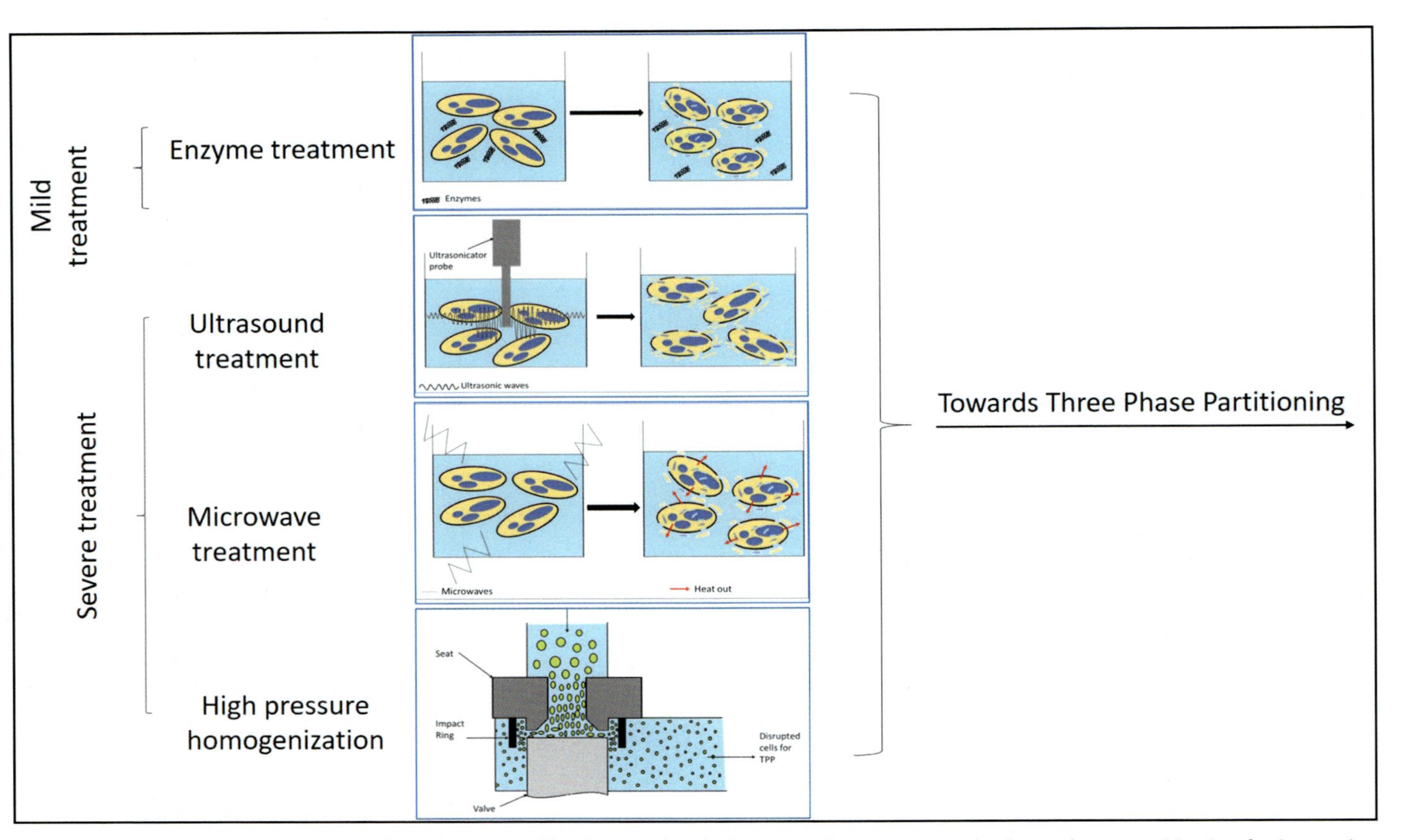

Fig. 13.2 ***Schematic representation of mechanisms of hyphenated techniques used as precursors to three phase partitioning for increasing the yield of extraction.***

cells, and eventually facilitates the extraction of the desired constituents. The enzyme substrate reactions manifest in degradation of polymers in the cells to release the bioactive embedded in the complex cell matrix. The enzymatic pretreatment before solvent-assisted extraction improves yield as well as reduces the time of extraction (Marathe et al., 2019). It produces yields equivalent to conventional solvent extraction technique (*n*-hexane) with the advantage of using a safer solvent (*t*-butanol) (Gaur et al., 2007).

The type of enzyme, its concentration, pH and temperature conditions influence the enzyme-substrate interaction and subsequently the yield. Table 13.2 shows the effect of enzymatic pretreatment on the percentage yield/recovery of extraction of oleaginous materials using EATPP. Cellulases, hemicellulases, pectinases and amylases are popularly used to degrade carbohydrates in the cell wall as well as inside the cell to increase accessibility of components to the solvent (Nde and Foncha, 2020). Protease treatment of mango kernel, soybean and rice bran powders improves yield of oils after TPP (Gaur et al., 2007). Cellulase, pectinase and protease have been studied for their effect on improving yield of flaxseed oil. Cellulase was found to be better among the three, and cocktail of the three enzymes gave the highest yield of oil (Tan et al., 2016). Cellulase was also found to be better than papain and pectinase for extraction of lipids from *Nannochloropsis* sp. biomass (Qiu et al., 2019). Enzymatic treatment with pectinase led to highest oil extraction from sesame seeds while treatment with amylase and

Table 13.2 Effect of enzymatic pretreatment on the percentage yield/recovery of oleaginous materials extracted using EATPP.

Substrate	Extract	Enzyme	Recovery/yield (percent)		Reference
			Untreated	Treated	
Mango kernel powder	Fat	Protizyme™	63[a]	79[a]	(Gaur et al., 2007)
Soybean flour	Oil	Protizyme™	90[a]	98[a]	(Gaur et al., 2007)
Rice bran powder	Oil	Protizyme™	72[a]	86[a]	(Gaur et al., 2007)
Ginger rhizome powder	Oleoresin	Stargen™	62[b]	67[b]	(Varakumar et al., 2017)
Ginger rhizome powder	Oleoresin	Accellerase	62[b]	69[b]	(Varakumar et al., 2017)
Turmeric rhizome powder	Oleoresin	α-amylase	7[b]	9[b]	(Kurmudle et al., 2011)
Sesame seeds	Oil	Pectinase	86[a]	78[a]	(Juvvi and Debnath, 2020)
Nannochloropsis sp. biomass	Total fatty acids	Cellulase	30[b]	48[b]	(Qiu et al., 2019)
Spirogyra sp. biomass	Oil	Papain	13[b]	25[b]	(Reddy and Majumder, 2014)

[a]Recovery indicates the oil/oleoresin extracted by TPP per unit oil/oleoresin extracted by conventional soxhlet technique.
[b]Yield indicates the oil/oleoresin extracted per unit mass of oleaginous material.

amyloglucosidase mixture, or protease led to lower yield (Juvvi and Debnath, 2020). Thus, depending up on the composition of matrix, various hydrolytic enzymes then increase the extraction of oleaginous material.

The use of enzymes in TPP not only increases the yield of oil but also that of proteins and/or polysaccharides (Kulkarni et al., 2017). This becomes particularly useful when TPP is planned for the extraction of two or more constituents from a single substrate. It leads to an overall increase in the yields of the desired phases.

13.6.2 Ultrasound assisted TPP (UATPP)

Ultrasonication assisted extraction of constituents of interest from plant, animal or microbial cells increases the yields due to cell disruption. This technique is considered as a severe technique of disruption, that functions by the principle of cavitation. In the cavitation phenomenon, the high frequency sonic waves lead to the formation, growth and collapse of vapor bubbles. The collapsing of bubbles generates intense elastic shock waves of the pressure of the order of 200 MPa that can disrupt the cells. A horn-type sonicator with titanium tip is typically used to generate the sonic energy for shearing action (Gomes et al., 2020). Dry biomass or aqueous suspension of cells subjected to ultrasound treatment also determine the efficiency of cell disruption (Chougle et al., 2014). Heat generated during this process needs to be controlled by ice-baths or cooling cycles to minimize the degradation of labile constituents. The oxidation quality of oils needs to be evaluated to understand the impact on the stability despite the increased yields. Comparisons among hyphenated techniques have also been evaluated. Enzyme-assisted technique *vis-à-vis* ultrasound treatment of ginger rhizome powder before TPP showed almost similar results of oleoresin yields (Varakumar et al., 2017).

The aqueous slurries of materials at suitable pH are ultrasonicated for a brief period of time ranging between 2–8 min. This is followed by addition of desired amount of ammonium sulfate and an organic solvent. The maximum oil recovery from 6 min sonicated (42 kHz, 80 W) almond, apricot and rice bran slurries post TPP are reported at 89, 79 and 88 percent, respectively. TPP alone recovered oil at 76, 67 and 72 percent, respectively. The increased yields are attributed to reduction in particle size after ultrasonication (Sharma and Gupta, 2004). Judicious application of ultrasonic power for cell disruption is important as excess may lead to degradation of oils in the extract or generate off-flavours (Chemat et al., 2004). UATPP extraction of forskolin from the roots of *Coleus forskohlii* increased extraction efficiency by 2.5-times of TPP. An added enzyme treatment increased the efficiency by 2.7-times of TPP. These treatments prior to TPP reduced the time of extraction by one-third of conventional solvent extraction (Harde and Singhal, 2012). Palmitic and elaidic acid recoveries with UATPP increased approximately 1.2 times that of TPP method (Mondal, 2015). Ultrasound pretreated (22 kHz, 120 W) flaxseed have been subjected to TPP for oil and protein extraction. The oil yield increased with power output (0–70), duty cycle (0–70 percent) and time

(0–10 min). Pertaining to the quality of oil, peroxide values of extracted flaxseed oil were in the order of commercial screw-press expeller < TPP < UATPP < soxhlet extraction. Soxhlet extraction gave highest yield and others led to 50–60 percent yields (Kulkarni et al., 2017). Algal oil extraction from *Spirogyra* sp. with TPP assisted with ultrasound treatment (40 kHz) increases the extraction efficiency by three times of TPP and ultrasound-Soxhlet extraction (Reddy and Majumder, 2014). Thus, in rigid cell matrices where conventional extraction is unable to disrupt cells and bring about mass transfer, ultrasonication facilitates the process.

13.6.3 Microwave assisted TPP (MATPP)

The implementation of microwaves in extraction of active constituents from a cell matrix is selected due to the increased rate of mass transfer. Microwave-assisted solvent extraction of oleaginous materials at higher yields has been reported due to two possible mechanisms: i) heat generated through ionic conduction and dipole moment in the sample suspended in solvent and ii) cell disruption and leaching of cellular components (Olalere et al., 2017; Gomes et al., 2020). The combination of microwave power and time influences the extraction yield and extent of degradation of bioactives in the matrix (Pasquet et al., 2011). When microwave processing is a precursor to the TPP extraction, the step involves suspension of cell biomass or plant matrix in an aqueous solution of salts such as sodium carbonate or potassium sulfate. This is followed by exposure to microwave radiation. The microwave-treated sample is cooled and then subjected to TPP by addition of hexane and 30 percent (w/w) ethanol layers. This is followed by the conventional TPP extraction steps. Highest oil recovery yield (100 percent) along with reduced extraction time from microalgae *Schizochytrium limacinum* SR21 has been reported by MATPP as against 51 percent from microwave assisted extraction, 84 percent from TPP and 95 percent from conventional extraction (Zeb et al., 2019). MATPP has been suggested as a scalable extraction technique that reduces time of extraction, and increases the yield of docosahexaenoic acid from microalgal cells (Zeb et al., 2020).

13.6.4 High-pressure homogenization assisted TPP (HPHTPP)

The disruption of plant or microbial cells can be achieved by high-pressure homogenization (HPH) prior to extraction. It is considered as a severe method of cell disruption which involves liquid-shearing action. The high efficiency and high-volume processing capacity make it easily scalable to industrial level. The process involves subjecting a suspension of cell biomass or plant material in water to pressures between 20–120 MPa or higher (250 MPa, in case of ultra-high-pressure homogenization) to result into liquid velocities as high as 400 m/s. The shear-stress, turbulence and friction promote cell disruption (Bernaerts et al., 2019; Gomes et al., 2020). HPH is particularly useful while extracting oleaginous materials from microbial cells due to their rigid cell wall and preferred over ultrasonication due to higher cell disruption efficiency (Gorte et al., 2020).

Disruption of *Chlorella* sp. to the extent of 50 percent is aided by HPH at 750 bar (Yap et al., 2014), as evidenced from the scanning electron microscopic images. Extraction of lipids from *Chlorella saccrophila* using HPHTPP has been studied. The optimized TPP conditions of 30 percent ammonium sulfate, using slurry/*t*-butanol of 1:0.75 for 60 min at 25 to 35 °C led to 69 percent (w/w) recovery of lipids. The additional step of HPH at 800 bars for 10 passes led to 90 percent (w/w) recovery of lipids. The algal cell disruption between the ring and inside valve led to increase in lipid recovery along with recovery of carotenoids (Mulchandani et al., 2015). HPH has also been used to assist disruption of yeast cells for solvent extraction of oleaginous materials (Gorte et al., 2020).

13.7 Challenges and future perspectives

Three phase partitioning as a technique for simultaneous extraction of various bioactives, proteins, enzymes, lipids, etc. is upcoming owing to its greener edge. The use of *t*-butanol is considered comparatively safer than conventional solvents such as acetone or hexane which have lower boiling points and hence safety concerns. However, in an industrial scale setup, the volumes of *t*-butanol used will also need suitable monitoring. The use of TPP at large scale leads to production of large volumes of waste that also need to be treated and discarded appropriately. In scenarios where hyphenated-TPP techniques are used at the industrial scale, the economics of the projects need evaluation. Hyphenated-TPP techniques reduce the time and increase yields but lead to addition of extra assembly lines. Enzyme-assisted TPP then seems better among the hyphenated techniques. Recovery of enzymes and immobilization become important to reduce extraction costs. Feasibility of ultrasonication of larger volumes becomes a limiting factor and needs appropriate designing of equipment. Thus, TPP which is considered as a better extraction technique than conventional solvent extraction needs proper understanding of the partitioning phenomenon for achieving maximum extraction. This technique is particularly useful when two or more constituents with economic gains can be extracted from a single substrate. This will also lead to reduction in the cost of treatment of effluent streams. With the world shifting towards the concept of integrated biorefinery by utilizing every stream from an extraction unit, economics of TPP and hyphenated TPP approaches for extraction of oleaginous materials at large scale may be worth looking into.

References

Bernaerts, T.M., Gheysen, L., Foubert, I., Hendrickx, M.E., Van Loey, A.M., 2019. Evaluating microalgal cell disruption upon ultra high pressure homogenization. Algal Res. 42, 101616.

Bhuiya, M.M.K., Rasul, M.G., Khan, M.M.K., Ashwath, N., Azad, A.K., Mofijur, M., 2015. Optimisation of oil extraction process from Australian native beauty leaf seed (*Calophyllum inophyllum*). Energy Procedia 75, 56–61.

Chemat, F., Grondin, I., Costes, P., Moutoussamy, L., Sing, A.S.C., Smadja, J., 2004. High power ultrasound effects on lipid oxidation of refined sunflower oil. Ultrason. Sonochem. 11 (5), 281–285.

Chougle, J.A., Singhal, R.S., Baik, O.D., 2014. Recovery of astaxanthin from *Paracoccus* NBRC 101723 using ultrasound-assisted three phase partitioning (UA-TPP). Sep. Sci. Technol. 49 (6), 811–818.

Dennison, C., Lovrien, R., 1997. Three phase partitioning: concentration and purification of proteins. Protein Expr. Purif. 11 (2), 149–161.

Dutta, R., Sarkar, U., Mukherjee, A., 2015. Process optimization for the extraction of oil from *Crotalaria juncea* using three phase partitioning. Ind. Crops Prod. 71, 89–96.

Gaur, R., Sharma, A., Khare, S.K., Gupta, M.N., 2007. A novel process for extraction of edible oils: enzyme assisted three phase partitioning (EATPP). Bioresour. Technol. 98 (3), 696–699.

Gomes, T.A., Zanette, C.M., Spier, M.R., 2020. An overview of cell disruption methods for intracellular biomolecules recovery. Prep. Biochem. Biotechnol. 50 (7), 1–20.

Gorte, O., Hollenbach, R., Papachristou, I., Steinweg, C., Silve, A., Frey, W., Syldatk, C., Ochsenreither, K., 2020. Evaluation of downstream processing, extraction, and quantification strategies for single cell oil produced by the oleaginous yeasts *Saitozyma podzolica* DSM 27192 and *Apiotrichum porosum* DSM 27194. Front. Bioeng. Biotechnol. 8, 355.

Guédé, S.S., Soro, Y.R., Kouamé, A.F., Brou, K., 2017. Optimization of screw press extraction of *Citrillus lanatus* seed oil and physicochemical characterization. Eur. J. Food Sci. Tech. 5 (4), 35–46.

Harde, S.M., Singhal, R.S., 2012. Extraction of forskolin from *Coleus forskohlii* roots using three phase partitioning. Sep. Purif. Technol. 96, 20–25.

Juvvi, P., Debnath, S., 2020. Enzyme-assisted three-phase partitioning: an efficient alternative for oil extraction from sesame (*Sesamum indicum* L.). Grasas Aceites 71 (1), e346.

Ketnawa, S., Rungraeng, N., Rawdkuen, S., 2017. Phase partitioning for enzyme separation: an overview and recent applications. Int. Food Res. J. 24 (1), 1–24.

Kulkarni, N.G., Kar, J.R., Singhal, R.S., 2017. Extraction of flaxseed oil: a comparative study of three-phase partitioning and supercritical carbon dioxide using response surface methodology. Food Bioproc. Tech. 10 (5), 940–948.

Kurmudle, N.N., Bankar, S.B., Bajaj, I.B., Bule, M.V., Singhal, R.S., 2011. Enzyme-assisted three phase partitioning: a novel approach for extraction of turmeric oleoresin. Process Biochem. 46 (1), 423–426.

Li, Z., Li, Y., Zhang, X., Tan, T., 2015. Lipid extraction from non-broken and high water content microalgae *Chlorella* spp. by three-phase partitioning. Algal Res 10, 218–223.

Marathe, S.J., Jadhav, S.B., Bankar, S.B., Singhal, R.S., 2017. Enzyme-assisted extraction of bioactives. In: Puri, M. (Ed.), Food Bioactives. Springer, Cham, pp. 171–201.

Marathe, S.J., Jadhav, S.B., Bankar, S.B., Dubey, K.K., Singhal, R.S., 2019. Improvements in the extraction of bioactive compounds by enzymes. Curr. Opin. Food Sci. 25, 62–72.

Mariana, I., Nicoleta, U., Sorin-Ştefan, B., Gheorghe, V., Mirela, D., 2013. Actual methods for obtaining vegetable oil from oilseeds. In: International Conference on Thermal Equipment, Renewable Energy and Rural Development. University POLITEHNICA of Bucharest.

Mondal, A., 2015. Sonication assisted three phase partitioning for the extraction of palmitic acid and elaidic acid from *Cassia sophera* Linn. Sep. Sci. Technol. 50 (13), 1999–2003.

Mulchandani, K., Kar, J.R., Singhal, R.S., 2015. Extraction of lipids from *Chlorella saccharophila* using high-pressure homogenization followed by three phase partitioning. Appl. Biochem. Biotechnol. 176 (6), 1613–1626.

Mwaurah, P.W., Kumar, S., Kumar, N., Attkan, A.K., Panghal, A., Singh, V.K., Garg, M.K., 2020. Novel oil extraction technologies: process conditions, quality parameters, and optimization. Compr. Rev. Food Sci. Food Saf. 19 (1), 3–20.

Nde, D.B., Foncha, A.C., 2020. Optimization methods for the extraction of vegetable oils: a review. Processes. 8 (2), 209.

Olalere, O.A., Abdurahman, N.H., Alara, O.R., Habeeb, O.A., 2017. Parametric optimization of microwave reflux extraction of spice oleoresin from white pepper (*Piper nigrum*). J. Anal. Sci. Technol. 8 (1), 8.

Pasquet, V., Chérouvrier, J.R., Farhat, F., Thiéry, V., Piot, J.M., Bérard, J.B., Kaas, R., Serive, B., Patrice, T., Cadoret, J.P., Picot, L., 2011. Study on the microalgal pigments extraction process: performance of microwave assisted extraction. Process Biochem. 46 (1), 59–67.

Qiu, C., He, Y., Huang, Z., Li, S., Huang, J., Wang, M., Chen, B., 2019. Lipid extraction from wet *Nannochloropsis* biomass via enzyme-assisted three phase partitioning. Bioresour. Technol. 284, 381–390.

Reddy, A., Majumder, A.B., 2014. Use of a combined technology of ultrasonication, three-phase partitioning, and aqueous enzymatic oil extraction for the extraction of oil from *Spirogyra* sp. J. Eng., 740631.

Shah, S., Sharma, A., Gupta, M.N., 2004. Extraction of oil from *Jatropha curcas* L. seed kernels by enzyme assisted three phase partitioning. Ind. Crops Prod. 20 (3), 275–279.

Sharma, A., Gupta, M.N., 2004. Oil extraction from almond, apricot and rice bran by three-phase partitioning after ultrasonication. Eur. J. Lipid Sci. Technol. 106 (3), 183–186.

Sharma, A., Khare, S.K., Gupta, M.N., 2002. Three phase partitioning for extraction of oil from soybean. Bioresour. Technol. 85 (3), 327–329.

Sharma, M., Dadhwal, K., Gat, Y., Kumar, V., Panghal, A., Prasad, R., Kaur, S., Gat, P., 2019. A review on newer techniques in extraction of oleaginous flaxseed constituents. OCL-Oilseeds and fats, Crops and Lipids 26, 14.

Tan, Z.J., Yang, Z.Z., Yi, Y.J., Wang, H.Y., Zhou, W.L., Li, F.F., Wang, C.Y., 2016. Extraction of oil from flaxseed (*Linum usitatissimum* L.) using enzyme-assisted three-phase partitioning. Appl. Biochem. Biotechnol. 179 (8), 1325–1335.

Varakumar, S., Umesh, K.V., Singhal, R.S., 2017. Enhanced extraction of oleoresin from ginger (*Zingiber officinale*) rhizome powder using enzyme-assisted three phase partitioning. Food Chem. 216, 27–36.

Vidhate, G.S., Singhal, R.S., 2013. Extraction of cocoa butter alternative from kokum (*Garcinia indica*) kernel by three phase partitioning. J. Food Eng. 117 (4), 464–466.

Yan, J.K., Wang, Y.Y., Qiu, W.Y., Ma, H., Wang, Z.B., Wu, J.Y., 2018. Three-phase partitioning as an elegant and versatile platform applied to nonchromatographic bioseparation processes. Crit. Rev. Food Sci. Nutr. 58 (14), 2416–2431.

Yap, B.H., Crawford, S.A., Dumsday, G.J., Scales, P.J., Martin, G.J., 2014. A mechanistic study of algal cell disruption and its effect on lipid recovery by solvent extraction. Algal Res. 5, 112–120.

Zeb, L., Wang, X.D., Zheng, W.L., Teng, X.N., Shafiq, M., Mu, Y., Chi, Z.Y., Xiu, Z.L., 2019. Microwave-assisted three-liquid-phase salting-out extraction of docosahexaenoic acid (DHA)-rich oil from cultivation broths of *Schizochytrium limacinium* SR21. Food Bioprod. Process. 118, 237–247.

Zeb, L., Shafiq, M., Chi, Z.Y., Xiu, Z.L., 2020. Separation of microalgal docosahexaenoic acid-rich oils using a microwave-assisted three-phase partitioning system. Sep. Purif. Technol. 252, 117441.

CHAPTER 14

Intensification of extraction of biomolecules using three-phase partitioning

Sujata S. Patil, Virendra K. Rathod
Department of Chemical Engineering, Institute of Chemical Technology, Mumbai, India

Chapter outline

14.1 Introduction

Three phase partitioning (TPP) is a fast-growing method for simultaneous extraction and purification of bioactive molecules. Various TPP applications have been explored, including improvement in purity of enzymes and their catalytic activities. Three phase separation is largely affected by salt and alcohol concentration used for the system. It

Three Phase Partitioning
DOI: https://doi.org/10.1016/B978-0-12-824418-0.00009-6

can be applied in two ways to obtain high recovery of biomolecules, either by the separation of biomolecules from aqueous extract (in the absence of solid particles) or by the simultaneous extraction and separation of biomolecules from an aqueous slurry of raw materials. In the second case, the extraction and separation process are combined, which reduces the number of unit operations and lowers the production cost. TPP can be combined with process intensification tools like ultrasound, microwave, and use of enzymes. There are modified TPP processes as well, such as two-step TPP, micro- [affinity ligand] facilitated and ionic-liquid based TPP. Till date, more than 100 reports have been available on the TPP technique for the simultaneous extraction and purification of various bioactive products. Amongst these, about 70 percent have concentrated on the extraction and purification of enzymes and proteins, while the remaining 30 percent have reported the extraction and separation of lipids, oils, carbohydrates, and low molecular weight organic components (Fig. 14.1).

14.1.1 History

In 1972, the experiments led to the TPP development were done by Tan and Lovrein (Tan and Lovrien, 1972). They discovered that large number of enzymes retained their activity in the mixture of t-butanol/ water. Further, Lovrein's research group established the method, first for the separation of constituents of the cellulose complex from

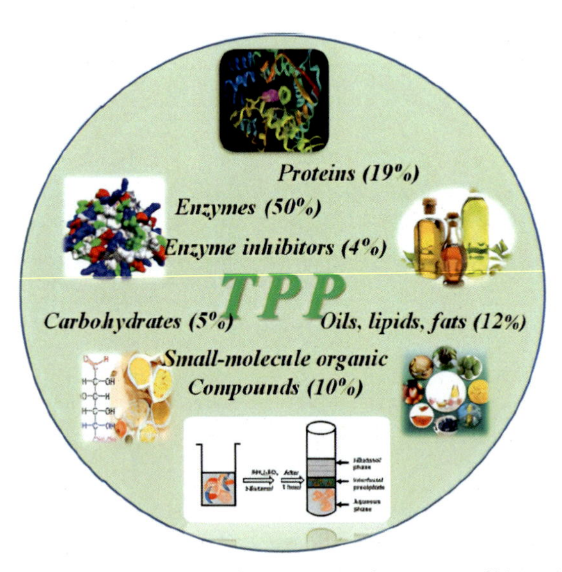

Fig. 14.1 ***TPP technique for the extraction and separation of a variety of bioactive molecules from different natural sources.***
Credit: Reproduced with permission from Yan, JK, Wang, YY, Qiu, WY, Ma, H., Wang, ZB, & Wu, J. Y. (2018). Three-phase partitioning as an elegant and versatile platform applied to nonchromatographic bioseparation processes. Critical reviews in food science and nutrition, 58(14), 2416–31.

Trichoderma reesei (Odegaard et al., 1984) and then as a common method for protein separation (Lovrein et al., 1987). This report was the first one on the upstream application of TPP in the precipitation of crude cellulases and enzymes on a small scale (Odegaard et al., 1984). The recovery of many enzymes using TPP in previous work exhibited enhancement in enzyme activity and indicated that t-butanol possibly attaches to the protein (Lovrein et al., 1987). At an optimum concentration of ammonium sulphate and t-butanol, it was found that the mixture formed three phases, with a precipitated protein phase in between aqueous and organic (t-butanol) phases. Ever since, TPP has been extensively explored for the effective extraction and purification of a wide range of biomolecules.

14.1.2 TPP process and mechanism

TPP requires the mixture of an aqueous salt solution and a water miscible aliphatic alcohol for three phase formation. Though alcohols like ethanol, methanol, 2- propanol, n-propanol, and t-butanol form a homogeneous mixture with water, they are immiscible in aqueous salt solutions prepared with an anti-chaotropic salt, ammonium sulphate ($(NH_4)_2SO_4$). Subsequently, two phases of liquid will be formed, where the top phase comprises of alcohol, and the bottom phase contains aqueous solution of the salt. In particular, ammonium sulphate and t-butanol are employed for protein precipitation from aqueous solutions. By the salt addition, t-butanol forms a two phase system (i.e., upper t-butanol and lower aqueous) (Rachana and Lyju Jose, 2014). The third phase is formed by the proteins which are present in the aqueous solution and thus forming a three phase system. The phase diagram of the three constituents [i.e. water, ammonium sulphate and t-butanol] is represented by Kiss et al. in 1998. The liquid phase is depicted at the water curve, the composition zone with two tie lines represents two immiscible phases, and above area denotes excess salt in solid form (Kiss et al., 1998). The 1 to 9 numbers symbolize different concentrations of the partitioning systems. The short tie-lines indicate lower (70–80 percent protein yield) efficacy in comparison with extended (95–100 percent protein yield) tie lines of the system. Extended tie-line systems comprise an extra quantity of t-butanol or ammonium sulphate to produce greater interfacial tension. An area of the triangle with tie-lines denotes the desired compositional range for three phase partitioning trials. The protein layer formation is mainly affected by the concentration of salt added. The upper layer or organic phase is composed of lipids, pigments, enzyme inhibitors, and small organic molecules, whereas the lower phase is concentrated with polar moieties like saccharides. The design of partitioning method should be based on its efficacy, as various configurations alongside a tie line would lead to equilibrium liquid stages by a similar configuration. This results in a similar concentration ratio within different systems. Also, the original concentration of protein does not affect the quantity of protein separated in an intermediate layer (Kiss et al., 1998).

The t-butanol addition to ammonium sulphate solution separates the proteins from the mixture, in the middle phase (precipitate) in between bottom aqueous and top organic phases. The molecules of the organic solvent bind to the hydrophobic regions of the protein that decreases the density of proteins, which is responsible for the floating of protein molecules at the intermediate layer (Choonia and Lele, 2013). The most used organic solvent for TPP is t-butanol owing to its branched structure and size, which does not easily penetrate into folded or unfolded proteins (Dennison and Lovrien, 1997; Pike and Dennison, 1989). These properties of t-butanol also inhibit the denaturation of proteins due to solvent. The graphic demonstration of TPP is given in Fig. 14.2. The precipitated proteins float above the denser salt solution because of the increase in buoyancy effect by the binding of t-butanol. The capacity of sulphate anion and t-butanol to push and pull improves the TPP's capability of protein precipitation. The protein molecules which are present in aqueous solution have a more flexible conformation than in the precipitated form as the result of TPP, and this leads to a net negative change in Gibbs free energy (Dennison and Lovrien, 1997). Thus, TPP is also a spontaneous process similar to salting out. The hydrophobic groups on the protein surface produce highly ordered hydration shells, which results in a slight decrease in enthalpy and a large decrease in entropy of the ordered water molecules compared to the bulk solution. The salt added to the aqueous extract causes movement of water molecules from the hydration shell to the bulk phase, which leads to a considerable increase in their entropy. Hence, precipitation of proteins occurs spontaneously due to negative Gibbs free energy. The mechanism of TPP offers the combined effect of alcohol precipitation and salting out methods. Besides, there are several results to show that it is unparalleled to the above-mentioned methods. First of all, TPP involves a minimum concentration of ammonium sulphate than most of the salting out processes. Then, partitioning in TPP is based on the concentration of the precipitating salt and amount of t-butanol (Pike and Dennison, 1989). In addition, TPP can be performed at mild conditions, while alcohol precipitation is generally needed to be conducted at specific conditions like low temperature (Mondal et al., 2006).

14.2 Key factors affecting the TPP method

14.2.1 Salt concentration

As discussed earlier, the selection of precipitants for protein purification is a very crucial step. The most popular inorganic salt, ammonium sulphate $(NH_4)_2SO_4$, is widely used for the precipitation of proteins based on the "salting out" due to its high ionic strength and solubility. The solubility of this salt is around 3.6 M, with the ionic strength of 1M concentration about 3 times the ionic strength of 1M NaCl. Along with this, some ions such as SO_4^{2-} and NH_4^+ are known to stabilize the protein structure. The saturation or concentration of ammonium sulphate in the solution plays a key role in the

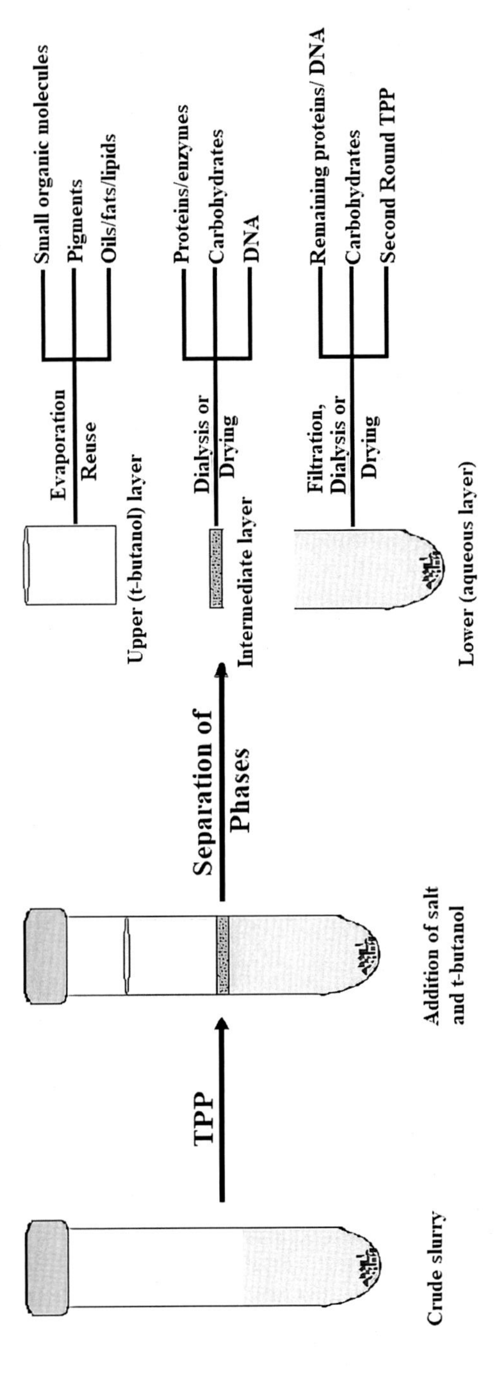

Fig. 14.2 *Schematic representation of TPP system.*

TPP system as it helps in the precipitation and protein-protein interaction. Hence, it is required to optimize the highest yield, recovery, and activity of the desired enzymes or proteins. The salting out efficacy depends on ammonium sulphate concentration and secondly on the total charge on the protein molecules. Generally, at least 20 percent (w/v) salt concentration is required as the starting concentration for the optimization of partitioning parameters during the TPP process (Dennison and Lovrien, 1997). With an increase in the concentration of ammonium sulphate, there is an increase in the water molecules leaving the solvation layers to the protein molecules. Due to this, the hydrophobic patches present on the surface of the proteins get uncovered and then interacts with the hydrophobic patches present on the surface of other protein molecules. At high salt concentration, the adsorption of water molecules takes place by salt ions; therefore, a strong interaction of protein-protein molecules leads to the precipitation of protein molecules through hydrophobic interactions (Choonia and Lele, 2013; Gagaoua et al., 2014). In TPP, the most common concentrations of ammonium sulphate vary from 20 to 60 percent (w/v). Pike and Dennison (1989) and Lovrein et al. (1987) found that 30 percent (w/v) salt concentration is best to carry out TPP process most efficiently (Lovrein et al., 1987; Pike and Dennison, 1989). Similar to the traditional kosmotropic salt, ammonium sulphate carries out the "salting out" of proteins from the aqueous solution. The "salting out" of proteins works on five different principles with SO_4^{2-} ions: 1. The effect of ionic strength 2. Kosmotropy 3. Dehydration, i.e., cavity surface tension enhancement osmotic stressors 4. Elimination of crowding mediators and 5. An interaction of SO_4^{2-} with cationic sites on proteins (Dennison, 2011). These five possibilities and possible interactions among them work at various proportions, which mainly depend on the concentration of sulphate and charge on protein molecules. Along with these, there is also a sixth factor that emerged slowly with similar importance. This is the divalent sulphate SO_4^{2-} anion, which tends to bind on few cationic sites of many protein molecules while these proteins possess a net positive charge Z_H^+. Hence, this behavior helps in conformational shrinkage and macromolecular contraction. It is the primary reason for the strong dependency on pH in the aforementioned five mechanisms or principles of sulphate "salting out" (Lovrein et al., 1987).

14.2.2 Crude extract to alcohol ratio

During TPP process, the solvent t-butanol is the preferred solvent due to its high boiling point (84 °C), and also its lower flammability in comparison with other solvents such as acetone, n-hexane, ethanol, or methanol, which are used in conventional solvent extraction (Patil et al., 2019; Vetal et al., 2014). Various other organic solvents, such as isoamyl alcohol, n-propanol, isopropanol, n-hexane, n-butanol, and n-pentanol, are also reported as selective co-solvents in the TPP process (Dennison and Lovrien, 1997). But, in comparison to t-butanol, these organic solvents carry lower interfacial precipitation and higher deactivation through the hydrophobic interactions with proteins (Choonia and Lele, 2013; Lovrein et al., 1987). Moreover, other solvents such as ethanol and methanol are not used as protein precipitation agents, except at temperatures below or near zero, and are not kosmotropic. Due to the big size and branched structure, it is difficult for t-butanol to get natural permeation inside the folded protein molecules, which prevents denaturation (Dennison et al., 2000). In addition, at 20–30 °C, t-butanol significantly shows crowding and kosmotropic effects, which enhance the protein yield in TPP (Dennison, 2011). The less t-butanol concentration may not be sufficient to synergize the crowding effect with ammonium sulphate (A. Sharma and Gupta, 2001); whereas the high t-butanol concentration may cause protein denaturation (Chaiwut et al., 2010). Furthermore, there is no significant increase in the extent of extraction of the desired biomolecule with an increase in t-butanol concentration during TPP process. The explanation is an increase in the viscosity of solution at high t-butanol concentration, which decreases the molecular mobility, thus making mass transfer difficult. Therefore, the optimization of t-butanol is necessary to facilitate the TPP separation process. The most common ratio of crude to t-butanol studied is in the range of 1.0:0.5, 1.0:1.0, 1.0:1.5, 1:0:2.0 and 1.0:2.5 (v/v) along with 30 percent (w/v) concentration of ammonium sulphate (Avhad et al., 2014; Kulkarni and Rathod, 2014; Patil et al., 2019).

14.2.3 pH

During TPP, pH plays a very crucial role and impacts the efficiency of protein separation, purification, and enrichment (Dennison and Lovrien, 1997). The effect is due to the resultant variation in the residues of amino acid at the protein surface owing to the pH changes in the extracting medium. The electrostatic interface amongst charged proteins and phases affect the partitioning behavior to a certain extent. Generally, the separation of the targeted protein at the intermediate phase or the aqueous phase in TPP mainly relies on its pI. Besides, there are four to five pH values in the range of 3.0 to 7.0 that are suitable for evaluating the distribution of proteins in the TPP process (Çalci et al., 2009; Dennison and Lovrien, 1997; Özer et al., 2010). It has been reported that sulfate anions binds to the proteins more efficiently when the pH is less than their pI; when the net charge on the protein is positive (Dennison and Lovrien, 1997; Wang et al., 2011). Hence, considering the above facts, it is required to study the impact of pH and obtain the optimum pH during the TPP process of targeted biomolecules such as enzymes and proteins.

14.2.4 Temperature

As the TPP process involves multiple effects that include the changes in protein hydration and conformational tightening, the effect of varying the process temperature during TPP is important to study. The majority of the separation processes are mainly operated at low temperatures as at low temperatures, the salt or solvent precipitation system is beneficial for the dissipation of the heat generated, which prevents the denaturation of protein (Dogan and Tari, 2008; Garg and Thorat, 2014). Even though there is no literature based on low temperature requirements during the TPP process; temperature plays a critical role in the performance of the TPP process containing ammonium sulphate and t-butanol. The temperature affects the rate of mass transfer, and almost all TPP processes are performed at room temperatures. The optimum temperature range to carry out the TPP process is between 20 and 40 ℃ (Garg and Thorat, 2014; A. Sharma and Gupta, 2001). Further, with an increase in the temperature above 40 °C, the t-butanol volatility increases, and hence there will be a lesser amount of solvent left for the extraction. The reduced quantity of t-butanol during extraction is no longer adequate to have a synergistic effect with ammonium sulphate, hence it results in lesser extraction of the targeted product. The high temperature also leads to higher consumption of energy that may make the TPP process economically unviable.

14.3 Advanced TPP processes

14.3.1 Two-step TPP

It is observed in some cases that when crude protein solution is subjected to the TPP, three phases are generated wherein the targeted protein molecules are present mainly in the aqueous layer. The target enzymes or proteins were expected to show higher partitioning to the interphase with the higher purity and yield. Hence, the formed layers are removed as soon as they are formed, and the lowest aqueous layer is subjected to a subsequent TPP by the addition of extra t-butanol and $(NH_4)_2SO_4$. Next, there is the formation of three phases, and most of the targeted proteins are separated in the interfacial region (Raghava et al., 2008). The process of TPP is carried out two times. This approach is especially useful when TPP is used for simultaneous refolding and purification of proteins. The first step consists of the solubilized inclusion body saturated with 5 percent (w/v) ammonium sulphate is vortexed and addition of 1:1 (v/v) ratio of butanol to solubilized sample is made (Raghava et al., 2008). The mixture is incubated further for 1 h at 25 °C, followed by centrifugation to get the well-formed three phases. With the solubilized inclusion bodies, the first step is important to remove proteins of the host cell as the interfacial precipitate. The aqueous layer on the lower side is subjected to the next TPP round. It is further saturated using ammonium sulphate (35 percent w/v) and t-butanol addition. The formed three phases were collected again, and activity measurement was carried out by dissolution and dialysis of formed precipitate at the interfacial layer. It has been observed that two step TPP process gives 2 to 4 times higher purity of proteins as compared to single step TPP (Jain et al., 2004; Rajeeva and Lele, 2011; Saxena et al., 2007). In one of the reports, it was observed that the aqueous phase contained maximum targeted proteins, and when a two-step TPP was used, this method resulted in the significant separation of the desired protein. Compared to traditional TPP techniques, recent literature reports that the two-step TPP processes are used more often for the recovery and purification of enzymes with considerable yields. For instance, the protease enzyme from *Calotropis procera* latex (Rawdkuen et al., 2010), alkaline phosphatase from chicken intestine (Sharma et al., 2000), phospholipase D from *Dacus carota* (S. Sharma and Gupta, 2001), and laccase from submerged cultures of *Ganoderma* sp. WR-1 (Rajeeva and Lele, 2011) were all purified using two-step TPP. In order to recover the enzyme, only one-step TPP is sufficient, and to get an enzyme in pure form two-step TPP is required.

14.3.2 Macro-affinity ligand-facilitated three-phase partitioning (MLFTPP)

Recently, MLFTPP as an extension to the TPP process has been developed by the Gupta and research group for the purification and separation of enzymes (Sharma and Gupta, 2002). It employs the introduction of polymer into the extracts for the separation of proteins. Generally, synthetic polymers such as eudragit L-100, S-100, and natural polymers such as chitosan, alginate, and k-carrageenan are used for protein precipitation in MLFTPP. Till date, the MLFTPP is explored for the direct extraction and purification of several enzymes like glucoamylase, pectinase, xylanase, cellulase, pullulanase, and α-amylases and also for the refolding of some the recombinant proteins (Gautam et al., 2012; Roy and Gupta, 2003; Mondal et al., 2003a, 2003b; Sharma et al., 2004, 2003a; Sharma and Gupta, 2002). Therefore, it is important to consider that the consecutive MLFTPP as a bio-separation process can be utilized to remove many important bio-actives such as enzymes and proteins from crude broth directly for some important usage during the processing of food. Additionally, TPP based on the metal-affinity has been established as a highly selective method for the recovery and purification of proteins with surface histidine residues by soybean trypsin inhibitor (Roy and Gupta, 2002).

14.3.3 Ionic liquid three-phase partitioning (ILTPP)

The TPP process is considered as a simple, rapid, and an easy to scale-up process for the purification of enzymes and proteins that can be directly applied to the crude mixtures. However, the critical limitation of the TPP process is the usage of t-butanol (co-solvent) as it shows volatility and flash point similar to ethanol, which limits the scaling up of the process (Przybycien et al., 2004). Hence, a new method known as ILTPP, where the combined benefits of TPP and ionic liquid-based aqueous two-phase systems (ILATPS), have been explored recently for the recovery and purification of BSA and lactoferrin (Alvarez-Guerra et al., 2014; Alvarez-Guerra and Irabien, 2015, 2014; Irabien and Irabien, 2014). The ILATPS are formed using hydrophilic ionic liquid and salt, which are mutually immiscible but miscible in water; forming two aqueous phases upon mixing, one is salt-rich, and another is ionic liquid-rich (Freire et al., 2012). Subsequently, most of the targeted protein gets collected at the liquid-liquid interface between the aqueous salt-rich phase and the aqueous ionic liquid rich phase. Presently, some three phase systems have been studied for ILTPP development using 1-butyl-3-methylimidazolium tetrafluoroborate ($BmimBF_4$)/ NaH_2PO_4/H_2O and 1-butyl-3-methylimidazolium trifluoromethanesulfonate (BmimTfO)/ (NaH_2PO_4/Na_2HPO_4)/ H_2O. Although, ILTPP is considered as a good substitute for the conventional TPP process, the reusability of the ionic liquid during the process should be evaluated in terms of the environmental and economic issues linked with these compounds. A study was carried out for the same with two alternatives to improve the recyclability of ionic liquid. These involved (a) an addition of salt to increase the ionic liquid concentration in the ionic liquid-rich phase, and (b) the use of vacuum filtration for the concentration of salt-rich phase (Irabien and Irabien, 2014). Next, estimation of the main constituents of ILTPP (ionic liquid, salt, and water) at thermodynamic equilibrium was carried out, and from the thermodynamic data and mass balance, it was found that more than 99 percent of ionic liquid used during the process could be recycled in ILTPP (Alvarez-Guerra et al., 2014).

14.3.4 Use of dimethyl carbonate (DMC) as an organic phase

t-Butanol serves as the best organic phase for the TPP process. Apart from t-butanol, solvents with low solubility in water also form a successful TPP system and prevent the protein configuration. DMC has been successfully studied for the first time by Panadare and Rathod in 2017 (Panadare and Rathod, 2017). DMC has 90.08 g/mol molecular weight, which is marginally more than t-butanol. DMC, like t-butanol has branched structure albeit not as much as t-butanol. Though, there is no definite data on whether it penetrates the protein interiors, the high recoveries of the activities in TPP using DMC indicates that probably it does not cause any significant amount of protein denaturation. Whereas, DMC has restricted solubility in the aqueous phase as it seems to work well within the narrower range of concentrations for protein precipitation during TPP

Panadare and Rathod used DMC for the purification of peroxidase from bitter guard (Momordica charantia) by TPP and reported 177 percent recovery of activities along with 4.84 fold increase in the purity with 20 percent sodium citrate as salt, 120 rpm agitation speed, pH 7, temperature 30 °C and extraction time of 3h (Panadare and Rathod, 2017). Recently, Salvi and Yadav (Salvi and Yadav, 2020) reported partial purification of epoxide hydrolase from Glycine max with 121 percent recovery of enzyme activities and 2.14 fold increase in the purity of the hydrolase with 40 percent ammonium sulphate concentration, 1:1.5 (v/v) crude extract to DMC ratio at 45 °C. Further, the process intensification using microwave (MTPP) resulted in 1.17 times increase in percent recovery of the enzyme activity. Moreover, DMC was used for TPP of exopolysaccharide from the fermentation broth of Phellinus baumii with 71.01 percent extraction yield at 0.5:1.0 (v/v) DMC to fermentation broth ratio, 19 percent sodium citrate salt concentration, 30 °C temperature and pH 4 was obtained (Wang et al., 2019). Therefore,

DMC has excellent potential as an alternative solvent for t-butanol in the TPP process of various natural biomolecules.

14.4 Process intensification of TPP

Process intensification aims towards a reduction in energy consumption and the overall processing cost by enhancing heat, mass, and momentum transfer rate (Tian et al., 2018). Stankiewicz and Moulijn (Etchells, 2005) defined process intensification as the drastic advancement in processing and manufacturing, significantly decreasing the volume of equipment and total energy consumption, leading to safer, cheaper, and sustainable technologies. Process intensification study expects the following outcomes: i. smaller equipment size; ii. higher throughput; iii. lower inventory usage and iv. the higher performance or material stroke and energy efficiency. It can also increase the throughput of the process by changes in internals in the reactor or separator. Van Gerven and Stankiewicz defined process intensification in terms of four principles: structure domain, energy domain, functional, and time domain (Van Gerven and Stankiewicz, 2009). These principles can be either implemented individually or can be combined and applied to obtain a high degree of process intensification. Process intensification can be classified into process intensifying equipment and methods. The process intensifying methods include multifunctional reactors (reactive distillation, extraction, crystallization, fuel cells, and chromatographic reactors), hybrid separations (distillation, adsorptive distillation, and membrane adsorption), alternative energy sources (solar energy, microwave, ultrasound, plasma technology, and electric fields) and other methods (supercritical fluids) (Boffito and Van Gerven, 2019; Etchells, 2005).

Process intensification has the advantage of increasing the productivity and overall rate of the process. The adoption of a new design of equipment to TPP for enhancing the mass transfer rate will help in achieving higher extraction ability. The role of intensification in TPP of biomolecules from the aqueous extract is only to improve the mass transfer rate between liquid-liquid extraction (aqueous-organic layers), thus reducing the time required for the separation. On the other hand, process intensification plays an important role in the simultaneous extraction and separation of biomolecules from an aqueous slurry of raw material. First, it improves mass transfer rate in solid-liquid (raw material-aqueous layer) by disrupting cell wall and enhancing the diffusion of solvent into plant cells; thus achieves maximum extraction of bioactive molecules from plant matrix into solvent medium. Secondly, similar to TPP of aqueous extract, it enhances mass transfer between the aqueous and organic layer and facilitates the concentration of desired biomolecules in respective phases. Therefore, the synergism between the process intensification tool and TPP is imperative in the extraction process of various biomolecules. Hence, this book chapter highlights the synergistic application of some of the recent innovative developments in the field of extraction with particular focus on the application of various intensification tools such as ultrasound, microwave, and enzyme in TPP towards the development of sustainable processes for the production of industrially relevant bioactive molecules. The intensification of TPP process is explored to increase the process efficiency and reduce time. Each of these approaches used for TPP process intensification is briefly described in the following sections.

14.4.1 Ultrasound assisted TPP (UA-TPP)

Ultrasound is well known to intensify various processes by acoustic cavitation phenomenon that reduces mass transfer resistance between solid and extracting medium and enhances the extraction yield of biomolecules in a shorter time (Dey and Rathod, 2013). Ultrasound intensifies both the modes of TPP. If TPP is applied on aqueous extract (absence of solids), then it helps to enhance mass transfer and mixing between the three phases. Whereas, when TPP is applied to aqueous slurry of raw material, ultrasound plays a role in cell bursting as well as decreasing mass transfer resistance. It must be noted that the effect of ultrasound is higher when TPP is performed directly using a natural source. Here, the ultrasound not only breaks the particles but also helps to rupture the cells due to high shear and microjets developed during the cavitation.

In the field of ultrasound, acoustic cavitation usually denotes the formation of bubbles, their growth, and collapse during the transmission of an ultrasonic wave through the extracting medium (Sandra and Ashokkumar, 2011). The bubbles produced are intact by attractive forces in the liquid medium (Suslick, 1989). The propagation of ultrasonic wave across the medium generates a sequence of compression and rarefaction stages that leads to the longitudinal displacement of constitutive molecules. During the compression phase, the liquid phase is dislocated transiently and can collide with the adjacent molecules. At the rarefaction cycle, molecules pull apart due to exerted negative pressure (Suslick, 1989). The magnitude of negative pressure is attributed to the type and purity of the extracting medium (Mason and Lorimer, 2002; Suslick, 1989). During the rarefaction stage, the force between them might be surpassed, causing a cavity in the medium at a high intensity of sound wave (Mason and Lorimer, 2002). These generated cavities into the liquid are cavitation bubbles. The dissolved gas of the liquid enters into these cavitational bubbles that result in the growth of the bubbles by coalescence (Ashokkumar, 2011; Leong et al., 2011) during the rarefaction phase and will not get completely removed at the compression cycle. Cavitation bubbles occur in two forms, namely stable and transient (Leong et al., 2011; Mason and Lorimer, 2002). Stable cavitation bubbles go through numerous compression and rarefaction phases and often non-linearly vibrate around an equilibrium size. Whereas bubbles in transient cavitation remain for one or for few acoustic cycles, through which these bubbles expand rapidly to double of its original size ahead of violent collapse into smaller bubbles (Leong et al., 2011; Mason and Lorimer, 2002). Though transient cavitation bubbles are said to be "active cavitation bubbles," Ashokkumar in 2011 (Ashokkumar, 2011) emphasized that both cavitational bubbles can collapse with high energy. As soon as the bubble size exceeds its threshold value, they burst during the compression phase, and a transient hot spot is generated (Flint and Suslick, 2007). A bubble collapse produces extreme local surroundings, i.e., measured temperature around 5000 K (Flint and Suslick, 2007) and assesses pressure around 50–1000 atm (Suslick and Price, 1999). These generated hotspots intensely accelerate the reactivity of the medium (Flint and Suslick, 2007; Suslick et al., 1999). When the collapse of these acoustic cavitational bubbles occurs close or onto the solid material surface, numerous physical effects have been seen (Suslick and Price, 1999). The collapse leads to the generation of high speed liquid jets into the surface and develops shockwave damages. These effects can result in the disintegration of materials and confined erosion. In solid-liquid mixture, the shockwaves and acoustic cavitation give rise to micro-mixing, macro-turbulence, and, consequently, collisions between particles. Eventually, collisions result in mechanical stress, and local energy density change induces the mass transfer over the different phases of TPP (Gogate and Kabadi, 2009). Fig. 14.3 depicts the process of acoustic cavitation and the effects involved, such as nucleation, diffusion, and formation of radicals, shock waves, and microjets.

Ultrasound assisted extraction is greatly affected by two main parameters, such as ultrasonic intensity and frequency. The understanding of these parameters is important

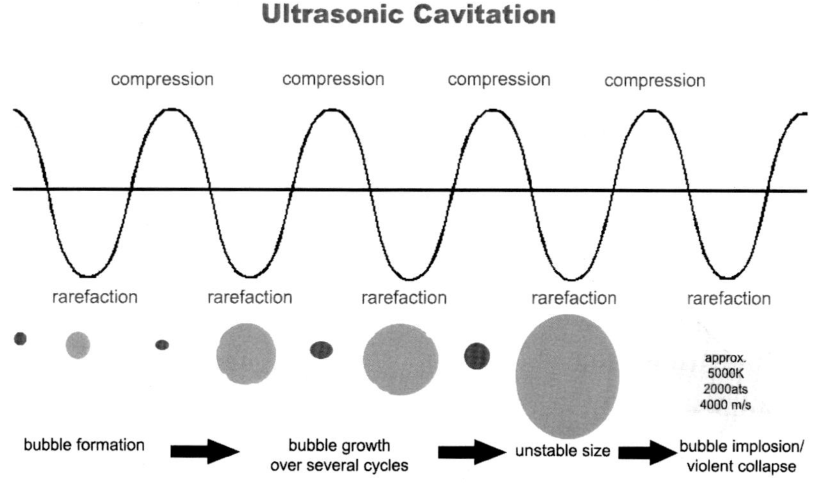

Fig. 14.3 ***The acoustic cavitaion process and it's effect.***
Credit: Reproduced with permission from Pokhrel, N., Vabbina, PK, & Pala, N. (2016). Sonochemistry: science and engineering. Ultrasonics sonochemistry, 29, 104–128.

in order to achieve high extraction efficiency by obtaining the maximum extent of extraction. However, the increase in extraction yield is not always the only objective, often use of minimum non-renewable resources and reduction of energy consumption are also the desired goals.

14.4.1.1 Ultrasonic physical parameters

Ultrasonic characteristics like power intensity, actual power input, and frequency impact the acoustic cavitation and, thus, extraction of biomolecules. The design of reactor and probe shape also affects the extraction (Chemat et al., 2017). These parameters will be discussed in this part.

14.4.1.1.1 Ultrasonic intensity. Ultrasonic intensity (UI) is given as the energy transmitted per second per square meter of the emitting surface (Tiwari, 2015). The UI is directly proportional to the amplitude of the transducer and, therefore, to the amplitude of the sound wave pressure (Santos et al., 2009). Cavitational bubbles collapse more violently with an increase in amplitude. To attain the cavitational threshold, the least value of UI is necessary. In the case of extraction, a relevant UI input is influential in achieving extraction efficiency. Ultrasonic intensity is determined using Eq. (14.1) (Tiwari, 2015).

$$Ultrasonic\ Intensity\ (UI) = \frac{P}{\pi D^2} \tag{14.1}$$

Where, P = Ultrasound power (Watt) and D = Internal diameter of ultrasonic reactor (cm)

Generally, the impact of sonochemical effects enhances with an increase in UI (Mason and Lorimer, 2002). It is observed that an increase in the amplitude of UI, there is rapid corrosion of the ultrasonic transducer, which leads to liquid agitation rather than cavitation effect and reduced diffusion of the ultrasonic wave through the extraction medium. However, while working with viscous liquids like oil, high ultrasonic intensity is required (Santos et al., 2009). In three phase partitioning, moderate use of UI is applied for uniform mixing as TPP is a heterogeneous system (Avhad et al., 2014; Kulkarni and Rathod, 2014; Panadare et al., 2020; Patil and Rathod, 2020; Rathod and Rao, 2015; Vetal et al., 2014).

14.4.1.1.2 Actual power input. The actual input power in an ultrasonic process is not always estimated, though there are some methods that allow to calculate applied power/energy. These methods evaluate the applied energy by calculating the physical/chemical changes in the medium under ultrasound exposure. The chemical methods include estimation of OH radicals produced by cavitational effect or use of chemical dosimeters (Makino et al., 1983; Suslick et al., 2011) while physical methods involve the calorimetric method, aluminium foil method, and measurement of acoustic pressure via optical microscopes and hydrophones (Chivate and Pandit, 1995; Margulis and

Margulis, 2003; Martin and Law, 1983). For example, to determine actual power input using calorimetry, it is assumed that applied power [P] by ultrasonic device is transformed into heat in the solvent medium and calculated as per Eq. (14.2) (Ratoarinoro and F. Contamine, A.M. Wilhelm*, 1995; Toma et al., 2011).

$$P = mC_p \frac{dT}{dt} \tag{14.2}$$

Where, m = amount of solvent (g), C_p = Specific heat of solvent (J g^{-1} °C^{-1}) at constant pressure, and dT/dt = temperature rise per second.

Various reports indicate that high ultrasound power induces shear force based on properties of the extracting medium, which causes key alterations in the solid material. In TPP of natural products, this parameter is generally optimized to achieve the best outcome with minimum power consumption (Daniela Bermúdez-Aguirre and Barbosa-Cánovas, 2011; Panadare et al., 2020; Patil and Rathod, 2020). Mostly, the maximum efficacy of ultrasound assisted TPP in terms of concentration and yield of the extract can be attained by optimizing applied ultrasonic power, temperature, and decreasing the moisture of raw material to augment contact between solid and solvent in minimum extraction time. Some reports indicate that the variation in power can result in evident selectivity of desired compounds, where the percentage of desired compounds is a function of the applied ultrasonic power (Chemat et al., 2004; Wei et al., 2010).

14.4.1.1.3 Ultrasonic frequency. The impact of ultrasonic frequency is also a key parameter in the ultrasound assisted extraction processes and should be optimized. Ultrasonic frequency affects the bubble size when resonance occurs. The commonly used ultrasonic frequency is between 20 and 50 kHz for extraction processes. At high ultrasound frequencies, it is difficult to induce cavitation effect as cavitational bubbles require a little delay to initiate formation during the rarefaction cycle. The length of rarefaction stage in which growth of cavitation bubbles occurs is inversely proportional to ultrasound frequency; and thus, to generate cavitation at high frequency, a large amount of intensity and amplitude is required (Mason and Lorimer, 2002). Whereas low ultrasonic frequency leads to comparatively small size transient cavitation bubbles that results in the physical changes rather than chemical (Leong et al., 2011; Mason et al., 2011). The ultrasonic frequency not only affects the size of cavitation bubble but also the mass transfer (Esclapez et al., 2011).

14.4.1.2 Ultrasonic reactor configurations

Ultrasonic cavitation for the extraction of biomolecules is achieved in two reactor configurations such as ultrasonic bath and ultrasonic horn/probe. Different approaches can be applied during UAE viz. direct extraction with ultrasonic horn/probe with or without mechanical stirring and indirect extraction using ultrasonic bath with a stirrer.

Amongst these, ultrasonic bath with stirring has been applied widely in UA-TPP of biomolecules (Avhad et al., 2014; Kulkarni and Rathod, 2014; Pakhale and Bhagwat, 2016; Rathod and Rao, 2015;Vetal et al., 2014), and very few have used direct mode of ultrasound (Panadare et al., 2020; Patil and Rathod, 2020). It is advantageous to cool the extraction solution as the ultrasonic energy may result in an increase in the temperature. The energy efficiency of ultrasonic bath is comparatively more than the horn/probe (Gogate and Pandit, 2004). Typically, a rectangular ultrasonic bath with 2–10 L volume capacity has been used, and a reactor of 50–100 mL capacity has been placed in the bath to carry out the extraction. In ultrasonic horn extraction, the horn of various tip diameter has been placed directly into the extraction medium of 50–100 mL capacity. Though ultrasonic baths are extensively applied, an ultrasound horn/probe offers a benefit of direct application of ultrasound and provides more cavitation intensity in the solution (Panadare et al., 2020; Patil and Rathod, 2020). Patil and Rathod first time reported the energy requirement by UA-TPP process of curcuminoids extraction from *C. longa* and observed that UA-TPP required significantly less energy compared to the conventional process (Patil and Rathod, 2020). However, direct sonication may impact the quality of extracted biomolecules and may degrade thermally sensitive bioactive material as they produce a high-intensity cavitation effect. Therefore, an appropriate configuration of the ultrasonic reactor should be selected based on the sensitivity of the desired biomolecule.

Although many extraction studies have been performed with ultrasonic bath and horn, the scale-up of these ultrasonic reactors are comparatively difficult. Despite the large number of research work carried out on a laboratory scale and potential application in the extraction of biomolecules, limited reports are available in the literature on industrial-scale extraction of bioactive molecules. Vinatoru published a report on the application of ultrasonic reactors at a working scale of 700–850 L for the extraction of herbs (Vinatoru, 2001). To achieve necessary intensification of the extraction process, the industrial scale reactor must achieve the constant circulation of cavitational activity above the threshold. It is important to confirm power dissipation over a broader area using multiple transducers to acquire significant cavitational yield. Higher cavitational yield is achieved with multiple transducer irradiations rather than a single transducer based reactor (Gogate and Pandit, 2004), and thus these designs exhibit good prospects for scale up. Application of multiple transducers with or without multiple frequencies can be applied in the construction of large scale ultrasonic reactors as these reactors can result in uniform and intense cavitational yield.

14.4.2 Microwave assisted TPP (MA-TPP)

Microwave irradiation technology has evolved as a green and clean source of energy. MA-TPP is a comparatively novel process that employs microwave energy to heat polar solvents in the presence of solid particles. Similar to ultrasound, microwave also intensifies both ways of TPP process by reducing the mass transfer barrier and rupturing plant cell wall. Based on solvent dielectric property and polarity, the extraction of value added components from the plant cell matrix is enhanced because of increased localized temperature and pressure produced by microwave heating. The phenomena result in more rapid partitioning of the selective bioactive products between raw material and solvent, reducing solvent and time required for extraction (Padmapriya et al., 2012; Zou et al., 2013). Microwave assisted processes have gained substantial recognition due to its incomparable heating mechanism, average cost, and high extraction efficiency. The heat loss is much lesser compared to conventional heating. Microwave heating uses the ability of an extracting medium to transform electromagnetic radiation into heat. The energy transmission is due to dielectric heating in contrast with conduction and convection, which cause conventional heating. Microwave heating enhances the diffusion of solvent into plant cell matrix and improves the extraction of desired bioactive components.

Microwave irradiation has remarkable potential and is recognized as a sustainable alternative to conventional heating. The low-temperature microwave can reduce the degradation of heat sensitive products due to various thermal effects. The microwave assisted extraction can be controlled by adjusting the microwave irradiation source, the use of cooling medium, controlling the initial temperature of the system, changing the physical properties, and the use of appropriate solvent. The advantages of microwave heating over conventional heating include rapid heat transfer, selective and uniform heating, fast on and off switching, compact reactors, and clean environment. Microwave assisted three phase partitioning process is rapid, clean, and offers better extraction yield in a short span of extraction time as compared to that of the conventional heating mechanism.

14.4.2.1 Mechanism of microwave irradiation

The microwaves are the electromagnetic waves with 0.3 to 300 GHz of frequencies. The mechanisms of the microwave involve ionic conduction and dipolar polarization (Ekezie et al., 2017; Nn, 2015). In dipolar polarization, the polar molecules try to align themselves due to the constant change of electric fields versus the intermolecular force. An adequate frequency of radiation is required to maintain sufficient oscillation between the molecules. The microwave radiation with a frequency of 2.45 GHz is sufficient enough to align polar molecules with an applied field (Veggi, PC; Martinez, J.; Meireles, 2012). In the case of ionic conduction, the interactions occur between electric components and ions in the sample medium. The oscillation of ions in the medium is generated due to the oscillating electromagnetic field. The microwave has a change in electric component at 4.9×10^4 times/second speed. There is heat generation through frictional force when polar solvent molecules try aligning themselves with respect to change in the electric field. The energy loss is much higher in ionic conduction than that in dipolar polarization. The inference from the above described mechanism is that only dielectric materials or solvents with permanent dipoles absorb microwave energy and get heated up (Tatke and Jaiswal, 2011). The value of dissipation factor (tan ∂) is an amount of the ability of extraction by which various solvents heat up under microwave (Kostas et al., 2017). The dissipation factor is depicted in Eq. (14.3) (Li et al., 2017):

$$\tan\partial = \frac{\varepsilon''}{\varepsilon'} \tag{14.3}$$

Where, ε'' is dielectric loss factor and ε' is dielectric constant

Overall, heating efficiency is higher in water than ethanol and methanol, whereas solvents with a low polarity like hexane and chloroform are transparent to microwave irradiation and do not produce heat (Lamb and Retherford, 1947; Li et al., 2017).

In microwave driven extraction, the motive in the case of dried raw material is heating the minimum amount of moisture present in a plant cell matrix. The heating of the inner moisture of plant cells with microwaves induces evaporation and enormous pressure on the cell wall. The cell wall ruptures from inside because of the pressure and ruptures. This way, the removal of bioactive components from the plant cell matrix occurs and leads to the enhanced extraction yield of bioactive constituents (Jocelyn et al., 1943; Mandal, Mohan, and Hemalatha, 2007). The extent of extraction from plant cells can be improved if the plant cell is saturated with water (Bagade and Patil, 2019). Earlier researchers reported that plant cells upon swelling with water under microwave irradiation result in the hydrolysis of the ether linkages present in the cellulosic cell wall. The glucopyranose has $-CH_2OH$ and COH polar groups which have 170–300 ps relaxation times. Also, the plant cell wall has huge molecular mass and considerable hydrogen bonding among themselves. Thus, improved heating of cellulosic cell wall leads to breakage of their order structure into different glucans due to localized oscillation under the influence of microwave (Dandekar and Gaikar, 2002; Raman and Gaikar, 2002). The maximum extraction can be attained by optimizing microwave power and temperature that promotes faster diffusion of solvent into a plant cell and also reduces diffusion path length (Azmir et al., 2013). The graphical representation is given in Fig. 14.4 for microwave driven cell rupture. In the microwave based approach, the internal water of cells gets heated, resulting in cell expansion and disruption. Owing to the cell wall opening, the extraction of value added compounds (secondary constituents or metabolites) from the plant matrix is achieved.

14.4.2.2 Important factors for microwave assisted system

14.4.2.2.1 Type of solvent. During microwave assisted extraction, the solvent choice needs to be based on the solubility of the desired compounds (L. Chen et al., 2008). The solvent has direct effects on the extraction of the plant components matrix. During the microwave process, an appropriate solvent does not get heated, but solvents with a high microwave threshold get easily heated (L. Chen et al., 2008). In order to get a higher extraction yield, the microwave solvent blends are used for the separation of

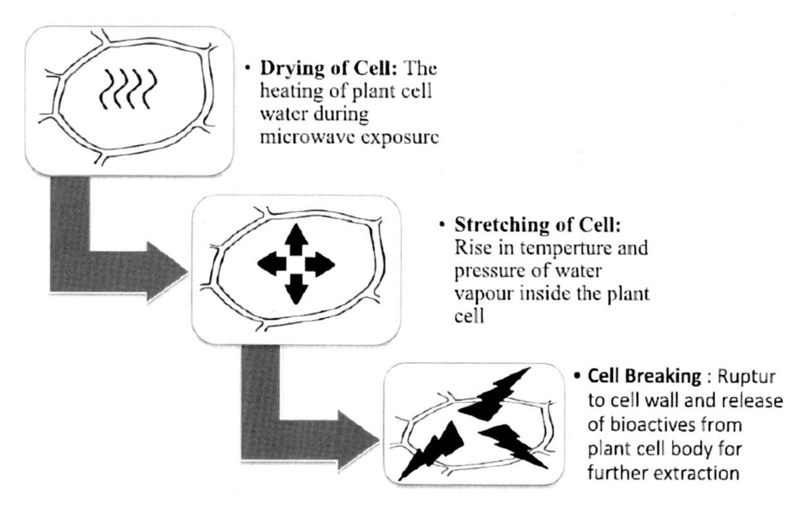

Fig. 14.4 ***The effect of microwave flux on the plant cell during MAE.***
Credit: Reproduced with permission from Bagade, SB, & Patil, M. (2019). Recent Advances in Microwave Assisted Extraction of Bioactive Compounds from Complex Herbal Samples: A Review. Critical Reviews in Analytical Chemistry, 1–12.

target molecules from plant material. For example, hexane is transparent to microwave energy, whereas ethanol is an excellent microwave absorbing solvent. Thus, the workers have also tried the mixtures of high and low absorbing microwave solvents in order to improve the extraction (Tatke and Jaiswal, 2011). For the extraction of oils from the aromatic plants, the solvent free MAE has been used. This method utilizes the moisture embedded in the plant cell walls, hence no need for additional solvents (Luque-García and Luque De Castro, 2004). The amount of plant cells also affects the solvent extraction. The higher amounts of plant cells may not result in the increase in higher extraction yield which might be attributed to the non-uniform transport of microwave flux (Tatke and Jaiswal, 2011). Recently, Panadare and Rathod reported use of microwave irradiation as a pretreatment method in the application TPP of custard apple seed oil (Panadare and Rathod, 2020).

14.4.2.2.2 Microwave irradiation power. The requirement of the needed microwave power and its irradiation time are dependent on each other. To enhance the microwave assisted process, optimization of microwave power is important. Usually, low or moderate microwave power is applied to achieve maximum extraction yield. However, at high power, there is a risk of the degradation of biomolecules or loss biomolecules activity. Some literature reported no considerable impact on the extraction yield of bioactive compounds with an increase in the microwave intensity from 500W to 1000W (Raner et al., 1993).

14.4.2.2.3 Properties of raw materials. The type of raw material has to be selected depending upon whether it contains adequate amounts of the desired value added compounds such as nicotine, volatile oils and phenols, etc. The key raw material properties that affect the extraction include the composition of raw material, types of plant cell, covalent attachment of bioactive components, particle size, matrix pre-treatment, amount of water, and size of the molecule.

14.4.2.2.4 Temperature. The temperature and microwave are strongly interconnected parameters for the design of microwave in closed vessel mode where the extraction temperature reaches above the boiling point (Letellier and Budzinski, 1999). With an increase in the temperature, the overall effectiveness of extraction using MA-TPP technique improves by the release of desired compound from the plant cell wall in the solvent medium. The increase in temperature as a key parameter for MA-TPP has been confirmed in a number of cases e.g., extraction of mangiferin, paclitaxel, and antioxidative andrographolide (Kulkarni and Rathod, 2015; Rao and Rathod, 2019; Talebi et al., 2004).

14.4.2.3 *Microwave extractors*

Researchers have explored many types of microwave extractors for extraction of biomolecules. All type of reactors have basic parts of microwave such as power source, a waveguide feed and microwave cavity. There are two modes of application viz. multi mode and single/mono mode. The important difference in the two modes is that multi vessel rotors irradiated concurrently (parallel extraction) in multimode cavities, whereas only one vessel get irradiated in mono mode at a time.

Multimode microwave extractors have been extensively reported for microwave assisted extraction. However, limited reports are available on application of microwave in TPP (Kulkarni and Rathod, 2015; Panadare and Rathod, 2020; Rao and Rathod, 2019; Salvi and Yadav, 2020). These studies have been performed in multimode open vessel extractor except the work performed by Panadare and Rathod wherein MA-TPP was carried out multimode closed vessel extractor of Anton Paar. They have studied application of microwave irradiation as pretreatment and as an assisting tool in TPP of custard apple seed oil and observed that microwave as pretreatment is more efficient for the TPP of oil than microwave assisted TPP (Panadare and Rathod, 2020). Addition of salt and t-butanol reduces the dissipation factor of the extraction mixture and thus efficiency of MA-TPP reduces compared to microwave as a pretreatment to TPP process (Panadare and Rathod, 2020).

Normally, small volume 0.2–50 mL of samples are processed in the closed vessel conditions of mono mode apparatus rather than large volume (150 mL) under reflux conditions in an open vessel (Kappe, 2004). In multimode apparatus, large volume of sample can be processed by applying closed and open vessel conditions. The extraction of material can be performed in kilograms using continuous flow reactors which exist in both mono and multimode microwave cavities. Various microwave reactors with different configurations regarding temperature and pressure monitoring, mechanization, design of vessel, data storage capacity and safety measures are provided by the manufacturing companies (Kappe, 2004).

14.4.3 Enzyme assisted TPP (EA-TPP)

Recently, the application of enzymes in pharma, food, and other industries has gained tremendous attention due to its economic and technological effectiveness. A detailed review published recently deliberates upon important aspects of the enzyme-assisted extraction to achieve a higher yield of targeted compounds and to reduce several operations and byproduct formation (Marathe et al., 2017). Till now, EA-TPP has been developed for the extraction of several biomolecules like polysaccharides, flavors, colourants, and polyphenols (Nadar et al., 2018). Enzymes can enhance the extraction when TPP can be applied to an aqueous slurry of raw materials. Here, enzymes will selectively rupture the various linkages of plant material and allow the faster release of the desired product. It is extensively used for the recovery of different plant molecules. However, to the author's best knowledge, the technique of extraction using enzymes is not yet explored for the macromolecules from microbial sources. The EA-TPP process can be seen as an alternative technique for conventional solvent extraction methods as it is gaining a lot of attention due to its efficient, eco-friendly, and sustainable nature (Nadar et al., 2018; Puri et al., 2012).

14.4.3.1 Enzyme assisted extraction [EAE] principle and factor affecting the process

The biomolecules are the natural products and are present as soluble or in insoluble conjugated forms. For example, out of 24 percent of the total phenolic content, a variety of phenolics are present in the food matrices in the bonded form. The cell wall is made up of polysaccharides such as cellulose, pectin, and hemicellulose in which the phenolic compounds are present within the cell wall patches. These polysaccharides are linked together by hydrogen bonds and hydrophobic interactions. Some phenolic acids and lignin form ether linkages by hydroxyl groups in the aromatic ring and ester linkages with proteins and structural carbohydrates through their carboxylic groups (Fig. 14.5). On the other hand, the flavonoids link covalently with sugar moieties by the glycosidic bond through carbon-carbon bonds (C-glycosides linkage) or hydroxyl group (O-glycosides linkage). The tannins and proteins show the capability to form strong interactions. Hence, enzymes such as pectinase, cellulase, protease, and hemicellulose are used to degrade the matrix of cell wall of plants, thereby increasing the release of intracellular biomolecules. The hydrolase enzymes, such as lipases, can work in water during the reaction. Additionally, it works in solvent systems such as amines, oximes, and alcohols (Hari Krishna and Karanth, 2002). While extracting flavonoids from *Ginkgo biloba*, the cellulase from *Penicillium decumbens* worked efficiently than cellulase from *Trichoderma reesei* when maltose is used as the glycosyl donor (Sharma et al., 2016). The enzymes pectinase (with pectinesterase, pectintranseliminase and

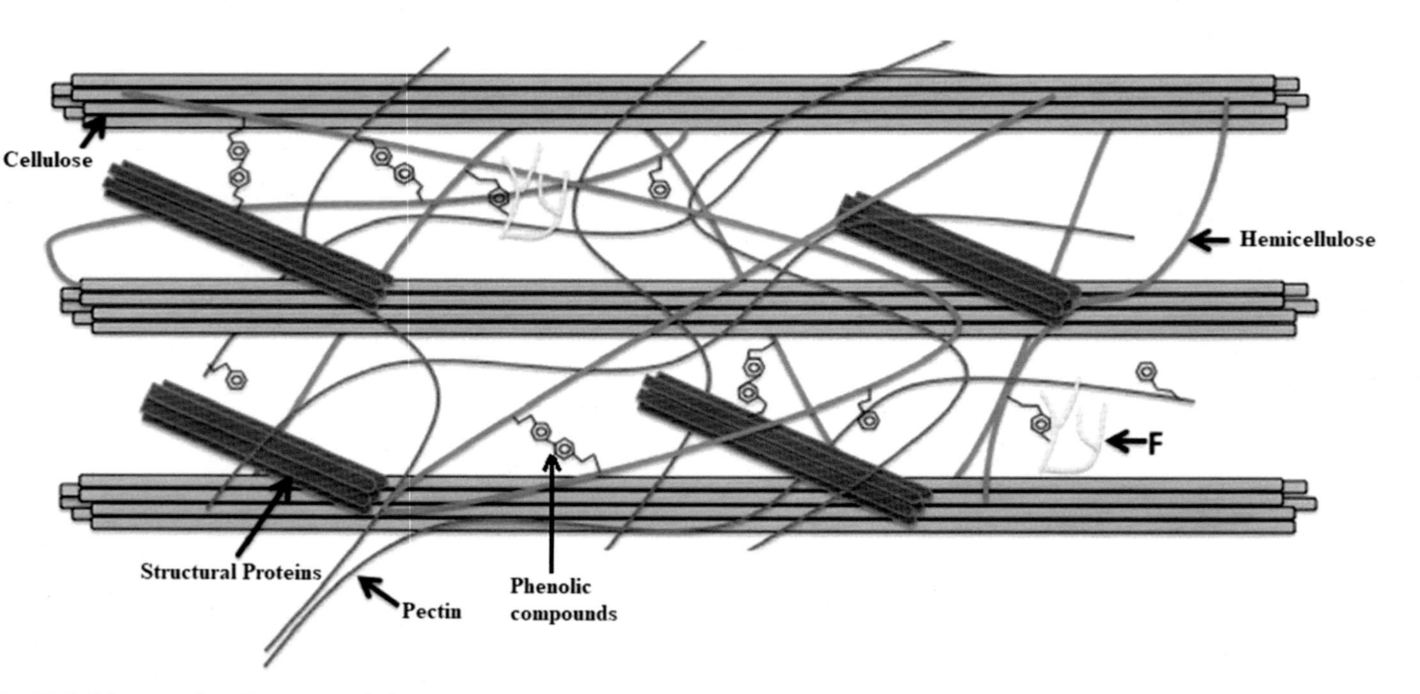

Fig. 14.5 ***Primary cell wall structure of plant material and cross-linking between structural components and phenolic compounds.***
Credit: Reproduced with permission from Acosta-Estrada, BA, Gutiérrez-Uribe, JA, & Serna-Saldívar, S. O. (2014). Bound phenolics in foods, a review. Food chemistry, 152, 46–55.

polygalacturonase as components), cellulases (with β-glucosidases, endoglucanases and cellobiohydrolases as components), and hemicellulases (with β-xylosidases and endoxylanases as components) are used for the isolation of proteins and oils. These enzymes have the capability to attack randomly on the interior sites of polysaccharide chains in the amorphous region. It results in the breakdown of polysaccharides to get small oligosaccharides with different lengths that eases the release of inside entrapped molecules (Fernandes and Carvalho, 2017). The yield and value of the extraction can be improvised using a suitable enzyme mixture and creates a synergistic effect between the enzymes. For example, the mixture of alkaline protease and cellulase was used effectively for the extraction of corn germ oil, and the yield of 80–90 percent was obtained (Moreau et al., 2009).

The efficiency of extraction depends on the time of extraction, temperature, enzyme loading, substrate availability, pH condition, and solvent system. Every enzyme exhibits maximum activity at an optimized pH (Talley and Alexov, 2010). As proteins are not soluble at the isoelectric pH range, the release of the biomolecules gets affected. Hence, it is necessary to select the pH value at which the enzyme should function properly and also not around the isoelectric point of the protein which is desired to be extracted or degraded (Ladole et al., 2018, 2017; Munde et al., 2017). Along with pH, the temperature is another important factor during the extraction process (Peterson et al., 2007). Long exposure of enzymes with extract can enhance the solubilization of different components of cell wall. While designing the large scale process, the prolonged incubation leads to energy inefficiency and reduced product quality (Babbar et al., 2016). Higher enzyme loading upto a limit increases the solubilization of cell wall (Zhang et al., 2007). Some times it also adds slight bitterness to the extract due to the extraction of some undesirable components. The enzyme loading percentage primarily depends upon the enzyme cost and the cost of biomolecules to be extracted (Jiang et al., 2010). The EA-TPP is successfully utilized for the extraction of edible oils from *Mangifera indica* using Protizyme™ (Gaur et al., 2007), and compared to TPP (82 percent), EA-TPP yielded 98 percent oil extraction at same conditions (Sharma et al., 2002). Moreover, The EA-TPP is used for the extraction of forskolin from the roots of *Coleus forskohlii* (Harde and Singhal, 2012). They have reported an increase in the extraction yield from 30.83 percent by TPP to 83.85 percent using EA-TPP. The process is also applied for the recovery of oils from *Jatropha curcas* L. seed kernels (Shah et al., 2004) and for the extraction of oleoresin from the ginger rhizome (*Zingiber officinale*) (Varakumar et al., 2017). Hence, EA-TPP is a promising method to extract oils, biosurfactants, biofuels, and other value-added products from extremophiles, including Archaea (Barnard et al., 2010; Cameotra and Makkar, 1998; Derguine-Mecheri et al., 2018; Kebbouche-Gana et al., 2009; Khemili-Talbi et al., 2015). For example, surfactin (a bacterial cyclic lipopeptide known as effective bio surfactants) is recovered from the fermentation broths using a combined TPP and membrane filtration process (H. L. Chen et al., 2008).

Moreover, recently the microalgae are gaining considerable attention due to their ability to generate a high quantity of protein and oils (Halim et al., 2011). The EA-TPP process is also applied efficiently for the extraction of oils and proteins from microalgae (Li et al., 2013; Reddy and Majumder, 2014; Waghmare et al., 2016). The high enzyme cost is the major limitation for the application of EA-TPP at a large scale. The enzyme cost limitation can be overcome by the immobilization process and thus enhancing enzyme stability and reusability. In the future, EA-TPP can be a viable tool to recover the value-added biomolecules from microbes from atypical and extreme environments.

14.5 Application of TPP for extraction and purification of biomolecules

14.5.1 Proteins

The TPP process is considered as an upstream technique for protein isolation. An optimized protocol of TPP, the target protein gets separated in the form of a concentrate as the intermediate phase, whereas unwanted components (such as lipids, oils, and pigments) are accumulated in the upper (solvent) layer. The process is well demonstrated with a noble approach for the purification of protein in some cases and can become the leading non chromatographic process for the purification of proteins in enzyme industries. In 1989, Pike and Dennison (Pike and Dennison, 1989) selected a group of some standard proteins such as hemoglobin, γ-globulin, myoglobin, ovalbumin and bovine serum albumin whose properties (like pH, solubility, concentration, molecular weight) were well known. Further, they evaluated the behavior of these proteins during the TPP method at various process parameters. Results indicated that the TPP process could be used more systematically for the concentration and fractionation of various proteins. Till date, TPP has been successfully utilized for the extraction and purification of a variety of targeted proteins using different natural sources such as recombinant green fluorescent protein (GFP) from *Escherichia coli DH5a* (Jain et al., 2004), immunoreactive-secreted proteins from *Corynebacterium pseudotuberculosis* (Paule et al., 2004), and the proteins (e.g., lipase, penicillin G acylase, GFP, alcohol dehydrogenase) from permeabilized microbial cells *viz Thermus thermophilus*, *Saccharomyces cerevisiae*, and *E. coli*. (Raghava and Gupta, 2009).

14.5.2 Enzymes

Biocatalyst or enzymes are the class of biomolecule which selectively act on the substrates to give the desired products. The advantages of a biocatalyst over other catalysts is the high reaction rate, high specificity, biodegradability, nontoxicity, and reaction at mild environmental conditions (pH, temperature, and pressure). So far, a lot of techniques are developed for the purification of enzymes from different sources such as microbial sources, plant, and animal, like dialysis, ion exchange chromatography, gel filtration chromatography, affinity chromatography, hydrophobic interaction chromatography, membrane filtration, precipitation (Azmir et al., 2013; Kula et al., 1981). These methods have certain limitations as they are costly, lengthy, and problematic to scale-up. TPP has been successfully utilized and reported as an efficient method for the recovery and purification of more than 50 industrially important enzymes (Dennison, 2011; Dennison and Lovrien, 1997; Gagaoua et al., 2016; Gagaoua and Hafid, 2016; Pol et al., 1990; Rachana and Lyju Jose, 2014), several enzyme inhibitors, including bifunctional amylase/protease inhibitors (Saxena et al., 2007; A. Sharma and Gupta, 2001), trypsin inhibitor (Roy and Gupta, 2002; Wati et al., 2009), and α-amylase inhibitor (α-AI) (Wang et al., 2011) from crude extracts.

The TPP process varies from enzyme to enzyme; hence the optimization is required for the purification of a specific enzyme. In several cases, TPP increases the activity of certain enzymes, with the reported yield of more than 100 percent (Singh et al., 2001). Thus, the purified enzyme after the TPP process is characterized for their physicochemical characteristics such as structure, stability, and catalytic activity at certain pH and temperature ranges using XRD, SDS-PAGE, and estimation of kinetic parameters such as (K_m and V_{max}). Along with this, the purification and excess concentration using TPP were found comparable with chromatographic techniques. Reports stated the application of TPP for the recovery of various enzyme activities (such as β-glucosidase, cellulase, xylanase, α-chymotrypsin, and cellobiose) from their denatured form (Roy and Gupta, 2005, 2004; Sardar et al., 2007). These findings indicated that TPP could be a valuable process for the purification and simultaneous renaturation of the several enzymes present in the protein mixture.

14.5.3 Carbohydrates

In recent years, the extraction and purification of carbohydrates using TPP has been explored. For the first time, Sharma and Gupta in 2002 reported the use of TPP for separation and purification of three commercial forms of alginates (A. Sharma and Gupta, 2002). Later, in 2003 Sharma et al. studied the application of TPP for chitosan and observed changes in certain characteristics (such as structure, biodegradability, and solubility) of this polysaccharide (Sharma et al., 2003b). It is worth noticing that TPP is a novel and simple process to obtain required variations in the properties of chitosan for certain applications. Further, TPP is applied to extract starch from tapioca and potato. Over 90 percent yield of starch and its improved forms have recovered, and the TPP process leads to significant changes in the exposure of potato starch to amylolytic hydrolysis (Mondal et al., 2004). Subsequently, the TPP method successfully precipitated and separated levan and fructooligosaccharide from aqueous solutions (Coimbra et al., 2010). As per the reports, above 90 percent polysaccharide was accumulated in the intermediate phase during the TPP process. On the other hand, Tan et al. in 2015 studied the TPP method for simultaneously purifying aloe protein and polysaccharide from the aqueous slurry of aloe powder in one step extraction (Tan et al., 2015). Aloe polysaccharide was concentrated in the bottom layer, and protein was separated in the intermediate layer with optimized parameters of ammonium sulphate 26.35 percent (w/w), t-butanol 20.82 percent (w/w), and pH 6.5 at 30 °C. In line with these findings, antioxidative polysaccharide is simultaneously extracted and purified in the lower phase using TPP from *Corbicula fluminea* (Yan et al., 2017). The maximum extraction yield of 9.32 percent was achieved in 30 min extraction time with 20 percent (w/w) salt concentration, 9.8 mL t-butanol, and 6 pH at 35.3 °C temperature. The purified polysaccharide of 86.5 percent purity exhibited in vitro antioxidant properties. Thus, TPP is a simple, green, and promising bioseparation process for simultaneous extraction and purification of active polysaccharide (PS), PS-peptide complexes from various natural resources.

14.5.4 Oil fats and lipids

The conventional solvent extraction method with n-hexane has been extensively applied to extract edible oils or fats from different plant resources. However, n-hexane is inflammable, unsustainable, and produced from oilseed extraction, which is an origin of volatile organic components. Moreover, n-hexane is commonly known as an air contaminant due to its reactivity with other pollutants to form photochemical oxidants and ozone (Gandhi et al., 2003). In the TPP process, the organic solvent (t-butanol) used has a higher boiling point (84 °C) than n-hexane (69 °C). t-Butanol also has a freezing point of 11 °C. Furthermore, it is cost-effective and enables separation by chilling instead of heating (Hari Krishna and Karanth, 2002). Hence, TPP is a promising process for the extraction of oils, fats, and lipids. In 2002, Sharma et al. initially reported the use of TPP for oil extraction from soybean, and the extraction yield of oil was found to be 82 percent by simultaneously mixing of t-butanol along with 30 percent salt concentration (1:1, v/v) in the soybean aqueous slurry in 1 h at 25 °C (Sharma et al., 2002). Generally, oils in oilseeds exist in the lipid forms, which are entrapped in a protein linkage, and hence, enzyme pretreatment prior to extraction has exhibited significant enhancement in the recovery of oil (Rosenthal et al., 1996). Proteases (Protizyme™) enzyme pretreatment with the TPP method was established to extract oils from soybean, rice bran, and mango kernel and achieved the highest oil yields of 86 percent, 79 percent, and 98 percent (w/w) respectively at optimized parameters (Gaur et al., 2007). This process is superior in comparison with conventional solvent (n-hexane) extraction, with an added advantages of room temperature operation as well as the easy solvent recovery. Thereafter, in 2013 Vidhate and Singhal studied the extraction of fat from kokam (*Garcinia indica*) kernels using TPP and achieved 95 percent (w/w) recovery of fat at optimized conditions such as 50 percent (w/v) ammonium sulphate, 1:1 (v/v) ratio of slurry to t-butanol, 2.0 pH and extraction time 2 h (Vidhate and Singhal, 2013). Recently, Panadare et al. (2020), reported the extraction of oil from custard apple seeds with ultrasound coupled with TPP (Panadare et al., 2020). They have studied two approaches for the application of ultrasound to TPP, namely ultrasound pretreatment and simultaneous ultrasound assisted TPP. The study confirmed that ultrasound pretreatment followed by TPP process is superior to simultaneous ultrasound assisted TPP that yield extraction of oil 33.6 percent (w/w) in just 150 s. Moreover, many researchers have reported TPP or TPP combined with different methods (for instance, enzyme, ultrasound, high pressure homogenization) for the recovery of oils and lipids from different natural sources as reviewed by Panadare and Rathod (Panadare and Rathod, 2017). For example, oil and lipid extraction from Jatropha curcas L. seed kernels using EA-TPP (Shah et al., 2004), almond, apricot, and rice bran using UA-TPP (Sharma and Gupta, 2004), *Crotalaria juncea* using TPP (Dutta et al., 2015), *Cassia sophera* Linn. (Caesalpiniaceae) using UA-TPP (Mondal, 2015) and *Chlorella saccharophila* using homogenization followed by TPP (Mulchandani et al., 2015). As a result, based on previous literature, TPP or TPP coupled with other methods, is a simple, inexpensive, and green method to achieve maximum yield and better quality of oil compared to conventional processes. It has major potential for both pilot and large scale extraction of oil, lipids, and fats from plant materials.

14.5.5 Small molecular weight compounds

In recent times, the separation of small molecular weight organic molecules from natural sources has gained tremendous attention as these chemicals are extensively used in pharmaceutics, food, cosmetics, perfumes, and paints. Hence, the extraction of bioactive products and identifying new natural sources for such small organic molecules have acquired industrial and scientific significance. Various conventional methods like solvent and batch extraction and novel extraction approaches such as TPP, UAE, SFE, and MAE processes have been presently applied for the isolation of small molecular weight organic compounds which act as bioactive drugs from animals, microorganisms, and plants (Wang and Weller, 2006). However, TPP is widely assessed for the simultaneous extraction and purification of enzymes and proteins from crude extracts. Along with these, small molecular weight organic compounds are recovered in the upper organic phase (t-butanol). Initially, Kurmudle et al. reported the extraction of turmeric oleoresin from *Curcuma longa* L. using EA-TPP in 4 h (Kurmudle et al., 2011). Later, TPP and TPP combined with enzyme and ultrasound were explored for extraction of forskolin and astaxanthin from *Coleus forskohlii* roots and bacterial biomass of Paracoccus NBRC 101,723, respectively (Chougle et al., 2014; Harde and Singhal, 2012). Further, Rathod and the research group extensively applied TPP method for the extraction of ursolic acid and oleanolic acid, mangiferin, antioxidative andrographolide, curcuminoids and corosolic acid from *Ocimum sanctum* leaves, *Mangifera indica* leaves, *Andrographis paniculata* leaves, *Curcuma longa* L roots and *Lagerstroemia speciosa* respectively (Kulkarni and Rathod, 2016; Patil et al., 2019; Rao and Rathod, 2015; Sonar and Rathod, 2020;

Vetal et al., 2014). Additionally, to facilitate the extraction yield of these molecules, ultrasound, and microwave as process intensification tools were utilized (Kulkarni and Rathod, 2015, 2014; Patil and Rathod, 2020; Rao and Rathod, 2019; Rathod and Rao, 2015; Vetal et al., 2014). The obtained results confirmed that the use of ultrasound and microwave as a process intensification tool has a synergistic effect on TPP system. UA-TPP and MA-TPP considerably enhanced the extraction yield and decreased the extraction time in comparison with TPP. Thus, TPP coupled with ultrasound/ microwave, not only improves both the methods but also gives an alternative process for down-streaming approaches.

14.6 Challenges in TPP

Organic solvent t-butanol has been utilized in nearly all the reports of TPP till date (Gagaoua et al., 2015; Garg and Thorat, 2014; Harde and Singhal, 2012; Pakhale and Bhagwat, 2016; Rao and Rathod, 2019; Rawdkuen et al., 2012). Though, the problem with application of TPP at a large-scale is the quantity of t-butanol used, which has similar volatility and flash point as of ethanol (Przybycien et al., 2004). t-Butanol also possesses certain hazards like acute toxicity, skin, eye irritation, dizziness, nausea, and others. Besides, TPP has a slower rate of mass transfer for the separation of biomolecules within three phases from natural sources that may not result in a maximum recovery of bioactive molecules (Patil et al., 2019; Rao and Rathod, 2015). To overcome this resistance, TPP coupled with advanced extraction methods such as enzyme, ultrasound, and microwave assisted techniques has been used but that leads to an increase in the cost of process (Acosta-Estrada et al., 2014; Bagade and Patil, 2019; Yan et al., 2018).

Moreover, the industrial growth of TPP is comparatively less than chromatographic separation or aqueous two phase separation (ATPS) based extraction (Gautam et al., 2012; Szamos et al., 1998). The cost related to the protein separation at an industrial scale, mainly in terms of required facilities and partitioning equipment has to be considered. The phase formation is dependent on the concentration limit of components present in it and can be evaluated by the relative phase diagrams (Kroner et al., 1984). The large scale process of TPP will create a problem in the exact evaluation of concentration of phase formation and biomass, and that might impact the separation of biomolecules. The large scale processes of TPP will probably result in high process losses and is a current challenge to efficiently reduce the cost of the TPP method.

The commercial application of TPP will consume a large quantity of chemicals (raw materials), but the cost of these raw materials can be acceptable by considering the high yield and purity achieved during TPP. However, this will produce a large volume of waste chemicals that is required to be controlled efficiently by reusing. Extra attention should be given to the recovery and recycling of used chemicals for further extraction processes.

14.7 Conclusions

TPP is extensively applied as a developing bioseparation method for the extraction and purification of various biomolecules, which include enzymes, proteins, carbohydrates, oils, fats, lipids, and small molecular weight organic compounds from a natural source like fundamental food materials/ nutraceuticals and drugs. These value added products can be potentially used in food, medicine, and cosmetics. Though TPP has gained tremendous attention and benefits compared to conventional extraction methods have been confirmed, there are still certain points to understand regarding TPP method. Firstly, the defined mechanism of partitioning and behavior of phases should be given more consideration to obtain detailed understanding. Then, rapid and efficient hybrid TPP methods based on conventional TPP should be more focused on the green and efficient extraction and purification of various bioactive components. Further exploration of the application of other harmless and biocompatible solvents such as ionic liquids rather than t-butanol organic solvent in TPP system is required. Additionally, the influence of TPP method on molecular structure, bioactivity and physicochemical properties of extracted compounds should be evaluated. Lastly, the industrial or large scale application of TPP for extraction of purification of natural compounds needs to be tried in the future. In summary, TPP offers an opportunity and a challenge for researchers to examine in-depth the partitioning process and to establish the application of TPP for the simultaneous extraction and purification of different biomolecules other than proteins.

References

Mondal, A., 2015. Sonication assisted three phase partitioning for the extraction of palmitic acid and elaidic acid from *Cassia sophera* Linn. Sep. Sci. Technol. 50, 1999–2003. https://doi.org/10.1080/01496395.2015.1005314.

Nn, A., 2015. A Review on the extraction methods use in medicinal plants, principle, strength and limitation. Med. Aromat. Plants. 04, 3–8. https://doi.org/10.4172/2167-0412.1000196.

Reddy, A., Majumder, A.B., 2014. Use of a combined technology of ultrasonication, three-phase partitioning, and aqueous enzymatic oil extraction for the extraction of oil from *Spirogyra sp.* J. Eng. 2014, 1–6. https://doi.org/10.1155/2014/740631.

Rosenthal, A., Pyle, D.L., Niranjan, K., 1996. Aqueous and enzymatic processes for edible oil extraction. Enzyme Microb. Technol. 19, 402–420. https://doi.org/10.1016/S0141-0229(96)80004-F.

Sharma, A., Roy, I., Gupta, M.N., 2004. Affinity precipitation and macroaffinity ligand facilitated three-phase partitioning for refolding and simultaneous purification of urea-denatured pectinase. Biotechnol. Prog. 20, 1255–1258. https://doi.org/10.1021/bp0342295.

Sharma, A., Mondal, K., Gupta, M.N., 2003a. Separation of enzymes by sequential macroaffinity ligand-facilitated three-phase partitioning. J. Chromatogr. A. 995, 127–134. https://doi.org/10.1016/S0021-9673(03)00522-3.

Sharma, A., Mondal, K., Gupta, M.N., 2003b. Some studies on characterization of three phase partitioned chitosan. Carbohydr. Polym. 52, 433–438. https://doi.org/10.1016/S0144-8617(03)00002-X.

Sharma, A., Gupta, M.N., 2002. Macroaffinity ligand-facilitated three-phase partitioning (MLFTPP) for purification of xylanase. Biotechnol. Bioeng. 80, 228–232. https://doi.org/10.1002/bit.10364.

Sharma, A., Gupta, M.N., 2004. Oil extraction from almond, apricot and rice bran by three-phase partitioning after ultrasonication. Eur. J. Lipid Sci. Technol. 106, 183–186. https://doi.org/10.1002/ejlt.200300897.

Sharma, A., Gupta, M.N., 2001. Three phase partitioning as a large-scale separation method for purification of a wheat germ bifunctional protease/amylase inhibitor. Process Biochem. 37, 193–196. https://doi.org/10.1016/S0032-9592(01)00199-6.

Sharma, A., Gupta, M.N., 2002. Three phase partitioning of carbohydrate polymers: separation and purification of alginates. Carbohydr. Polym. 48, 391–395. https://doi.org/10.1016/S0144-8617(01)00313-7.

Sharma, A., Tewari, R., Rana, S.S., Soni, R., Soni, S.K., 2016. Cellulases: classification, methods of determination and industrial applications. Appl. Biochem. Biotechnol. 179, 1346–1380. https://doi.org/10.1007/s12010-016-2070-3.

Sharma, A., Sharma, S., Gupta, M.N., 2000. Purification of alkaline phosphatase from chicken intestine by three-phase partitioning and use of phenyl-Sepharose 6B in the batch mode. Bioseparation 9, 155–161. https://doi.org/10.1023/A:1008195729472.

Sharma, A., Khare, S.K., Gupta, M.N., 2002. Three phase partitioning for extraction of oil from soybean. Bioresour. Technol. 85, 327–329. https://doi.org/10.1016/S0960-8524(02)00138-4.

Waghmare, A.G., Salve, M.K., LeBlanc, J.G., Arya, S.S., 2016. Concentration and characterization of microalgae proteins from *Chlorella pyrenoidosa*. Bioresour. Bioprocess. 3. https://doi.org/10.1186/s40643-016-0094-8.

Gandhi, A.P., Joshi, K.C., Jha, K., Parihar, V.S., Srivastav, D.C., Raghunadh, P., Kawalkar, J., Jain, S.K., Tripathi, R.N., 2003. Studies on alternative solvents for the extraction of oil-I soybean. Int. J. Food Sci. Technol. 38, 369–375.

Odegaard, B., Anderson, P., Lovrien, R., 1984. Resolution of the multienzyme cellulase complex of *Trichoderma reesei* QM 9414. J. Appl. Biochem. 6, 156–183.

Özer, B., Akardere, E., Çelem, E.B., Önal, S., 2010. Three-phase partitioning as a rapid and efficient method for purification of invertase from tomato. Biochem. Eng. J. 50, 110–115. https://doi.org/10.1016/j.bej.2010.04.002.

Acosta-Estrada, B.A., Gutiérrez-Uribe, J.A., Serna-Saldívar, S.O., 2014. Bound phenolics in foods, a review. Food Chem. 152, 46–55. https://doi.org/10.1016/j.foodchem.2013.11.093.

Paule, B.J.A., Meyer, R., Moura-Costa, L.F., Bahia, R.C., Carminati, R., Regis, L.F., Vale, V.L.C., Freire, S.M., Nascimento, I., Schaer, R., Azevedo, V., 2004. Three-phase partitioning as an efficient method for extraction/concentration of immunoreactive excreted-secreted proteins of *Corynebacterium pseudotuberculosis*. Protein Expr. Purif. 34, 311–316. https://doi.org/10.1016/j.pep.2003.12.003.

Tiwari, B.K., 2015. Ultrasound: a clean, green extraction technology. TrAC - Trends Anal. Chem. 71, 100–109. https://doi.org/10.1016/j.trac.2015.04.013.

Dennison, C., Moolman, L., Pillay, C.S., Meinesz, R.E., 2000. t-Butanol: nature gift for protein isolation. S. Afr. J. Sci. 96, 159–160.

Dennison, C., Lovrien, R., 1997. Three Phase Partitioning: concentration and Purification of Proteins. Protein Expr. Purif. 11, 149–161. https://doi.org/10.1006/prep.1997.0779.

Ed. C.Dennison, WalterGruyter, Ger.Berlin, 2011. Three-phase partitioning. In: Tschesche, H. (Ed.), Methods Protein Biochem., 1–5 Ed.

de O. Coimbra, C.G., Lopes, C.E., Calazans, G.M.T., 2010. Three-phase partitioning of hydrolyzed Levan. Bioresour. Technol. 101, 4725–4728. https://doi.org/10.1016/j.biortech.2010.01.091.

Martin, C.J., Law, A.N.R., 1983. Design of thermistor probes for measurement of ultrasound intensity distributions. Ultrasonics 21, 85–90. https://doi.org/10.1016/0041-624X(83)90008-2.

Kappe, C.O., 2004. Controlled microwave heating in modern organic synthesis. Angew. Chemie - Int. Ed. 43, 6250–6284. https://doi.org/10.1002/anie.200400655.

Rachana, C.R., Lyju Jose, V., 2014. Three phase partitioning - A novel protein purification method. Int. J. ChemTech Res. 6, 3467–3472.

Barnard, D., Casanueva, A., Tuffin, M., Cowan, D., 2010. Extremophiles in biofuel synthesis. Environ. Technol. 31, 871–888. https://doi.org/10.1080/09593331003710236.

Dandekar, D.V., Gaikar, V.G., 2002. Microwave assisted extraction of curcuminoids from *Curcuma longa*. Sep. Sci. Technol. 37, 2669–2690. https://doi.org/10.1081/SS-120004458.

Boffito, D.C., Van Gerven, T., 2019. Process Intensification and Catalysis, in: ref. Modul. Chem. Mol. Sci. Chem. Eng., 1–16. https://doi.org/10.1016/B978-0-12-409547-2.14343-4.

Panadare, D.C., Gondaliya, A., Rathod, V.K., 2020. Comparative study of ultrasonic pretreatment and ultrasound assisted three phase partitioning for extraction of custard apple seed oil. Ultrason. Sonochem. 61, 104821. https://doi.org/10.1016/j.ultsonch.2019.104821.

Panadare, D.C., Rathod, V.K., 2017. Extraction of peroxidase from bitter gourd (*Momordica charantia*) by three phase partitioning with dimethyl carbonate (DMC) as organic phase. Process Biochem. 61, 195–201. https://doi.org/10.1016/j.procbio.2017.06.028.

Panadare, D.C., Rathod, V.K., 2020. Process intensification of Three Phase Partition for extraction of custard apple seed oil using Microwave Pretreatment. Chem. Eng. Process. - Process Intensif. 157, 108095. https://doi.org/10.1016/j.cep.2020.108095.

Panadare, D.C., Rathod, V.K., 2017b. Three phase partitioning for extraction of oil: a review. Trends Food Sci. Technol. 68, 145–151. https://doi.org/10.1016/j.tifs.2017.08.004.

Avhad, D.N., Niphadkar, S.S., Rathod, V.K., 2014. Ultrasound assisted three phase partitioning of a fibrinolytic enzyme. Ultrason. Sonochem. 21, 628–633. https://doi.org/10.1016/j.ultsonch.2013.10.002.

Alvarez-Guerra, E., Irabien, A., 2014. Ionic liquid-based three phase partitioning (ILTPP) for Lactoferrin recovery. Sep. Sci. Technol. 49, 957–965. https://doi.org/10.1080/01496395.2013.878722.

Alvarez-Guerra, E., Irabien, A., 2015. Ionic liquid-based three phase partitioning (ILTPP) systems for whey protein recovery: ionic liquid selection. J. Chem. Technol. Biotechnol. 90, 939–946. https://doi.org/10.1002/jctb.4401.

Alvarez-Guerra, E., Ventura, S.P.M., Coutinho, J.A.P., Irabien, A., 2014. Ionic liquid-based three phase partitioning (ILTPP) systems: ionic liquid recovery and recycling. Fluid Phase Equilib. 371, 67–74. https://doi.org/10.1016/j.fluid.2014.03.009.

Çalci, E., Demir, T., Biçak Çelem, E., Önal, S., 2009. Purification of tomato (*Lycopersicon esculentum*) α-galactosidase by three-phase partitioning and its characterization. Sep. Purif. Technol. 70, 123–127. https://doi.org/10.1016/j.seppur.2009.09.004.

Kiss, É., Szamos, J., Tamás, B., Borbás, R., 1998. Interfacial behavior of proteins in three-phase partitioning using salt-containing water/tert-butanol systems. Colloids Surfaces A Physicochem. Eng. Asp. 142, 295–302. https://doi.org/10.1016/S0927-7757(98)00361-6.

Irabien, E.A.-.G., Irabien, A., 2014. Ionic Liquid Recovery Alternatives in Ionic Liquid-Based Three-Phase Partitioning (ILTPP) Enrique. AIChE J. 60, 3577–3586. https://doi.org/10.1002/aic.

Flint, E.B., Suslick, K.S., 2007. The Temperature of Cavitation. Science 253, 1397–1399.

Kostas, E.T., Beneroso, D., Robinson, J.P., 2017. The application of microwave heating in bioenergy: a review on the microwave pre-treatment and upgrading technologies for biomass. Renew. Sustain. Energy Rev. 77, 12–27. https://doi.org/10.1016/j.rser.2017.03.135.

Chemat, F., Rombaut, N., Sicaire, A.G., Meullemiestre, A., Fabiano-Tixier, A.S., Abert-Vian, M., 2017. Ultrasound assisted extraction of food and natural products. Mechanisms, techniques, combinations, protocols and applications. A review. Ultrason. Sonochem. 34, 540–560. https://doi.org/10.1016/j.ultsonch.2016.06.035.

Ekezie, F.G.C., Sun, D.W., Cheng, J.H., 2017. Acceleration of microwave-assisted extraction processes of food components by integrating technologies and applying emerging solvents: a review of latest developments. Trends Food Sci. Technol. 67, 160–172. https://doi.org/10.1016/j.tifs.2017.06.006.

Raman, G., Gaikar, V.G., 2002. Microwave-assisted extraction of piperine from *Piper nigrum*. Ind. Eng. Chem. Res. 41, 2521–2528. https://doi.org/10.1021/ie010359b.

Vidhate, G.S., Singhal, R.S., 2013. Extraction of cocoa butter alternative from kokum (*Garcinia indica*) kernel by three phase partitioning. J. Food Eng. 117, 464–466. https://doi.org/10.1016/j.jfoodeng.2012.10.051.

Wang, H.H., Chen, C.L., Jeng, T.L., Sung, J.M., 2011. Comparisons of α-amylase inhibitors from seeds of common bean mutants extracted through three phase partitioning. Food Chem. 128, 1066–1071. https://doi.org/10.1016/j.foodchem.2011.04.015.

Tan, H.K., Lovrien, R., 1972. Enzymology in aqueous-organic cosolvent binary mixtures*. J. Biol. Chem. 247, 3278–3285.

Chen, H.L., Chen, Y.S., Juang, R.S., 2008. Recovery of surfactin from fermentation broths by a hybrid salting-out and membrane filtration process. Sep. Purif. Technol. 59, 244–252. https://doi.org/10.1016/j.seppur.2007.06.010.

Salvi, H.M., Yadav, G.D., 2020. Extraction of epoxide hydrolase from *Glycine* max using microwave-assisted three phase partitioning with dimethyl carbonate as green solvent. Food Bioprod. Process. 124, 159–167. https://doi.org/10.1016/j.fbp.2020.08.010.

Santos, H.M., Lodeiro, C., Capelo-Martnez, J.-.L., 2009. The power of ultrasound. Ultrasound Chem, 1–16. https://doi.org/10.1002/9783527623501.ch1.

Choonia, H.S., Lele, S.S., 2013. Three phase partitioning of β-galactosidase produced by an indigenous *Lactobacillus acidophilus* isolate. Sep. Purif. Technol. 110, 44–50. https://doi.org/10.1016/j.seppur.2013.02.033.

Roy, I., Gupta, M.N., 2005. Enhancing reaction rate for transesterification reaction catalyzed by *Chromobacterium* lipase. Enzyme Microb. Technol. 36, 896–899. https://doi.org/10.1016/j.enzmictec.2005.01.022.

Roy, I., Gupta, M.N., 2002. Three-phase affinity partitioning of proteins. Anal. Biochem. 300, 11–14. https://doi.org/10.1006/abio.2001.5367.

Roy, I., Gupta, M.N., 2004. A-Chymotrypsin shows higher activity in water as well as organic solvents after three phase partitioning. Biocatal. Biotransformation. 22, 261–268. https://doi.org/10.1080/10242420400010523.

Roy, I., Gupta, M.N., 2003. Smart polymeric materials: emerging biochemical applications. Chem. Biol. 10, 1161–1171. https://doi.org/10.1016/j.

Azmir, J., Zaidul, I.S.M., Rahman, M.M., Sharif, K.M., Mohamed, A., Sahena, F., Jahurul, M.H.A., Ghafoor, K., Norulaini, N.A.N., Omar, A.K.M., 2013. Techniques for extraction of bioactive compounds from plant materials: a review. J. Food Eng. 117, 426–436. https://doi.org/10.1016/j.jfoodeng.2013.01.014.

Jocelyn, J.R., Sigouin, Michel, Lapointe, Jacques, 1943. Microwave-Assisted Natural Products Extraction.

Szamos, J., Jánosi, A., Tamás, B., Kiss, É., 1998. A novel partitioning method as a possible tool for investigating meat. I, Eur. Food Res. Technol. 206, 208–212. https://doi.org/10.1007/s002170050244.

Chougle, J.A., Singhal, R.S., Baik, O.D., 2014. Recovery of Astaxanthin from *Paracoccus* NBRC 101723 using ultrasound-assisted three phase partitioning (UA-TPP). Sep. Sci. Technol. 49, 811–818. https://doi.org/10.1080/01496395.2013.872146.

Etchells, J.C., 2005. Process Intensification. Process Saf. Environ. Prot. 83, 85–89. https://doi.org/10.1205/psep.04241.

Yan, J.-.K., Wang, Y.-.Y., Qiu, W.-.Y., Ma, H., Wang, Z.-.B., Wu, J.-.Y., 2018. Three-phase partitioning as an elegant and versatile platform applied to nonchromatographic bioseparation processes. Crit. Rev. Food Sci. Nutr. 58, 2416–2431. https://doi.org/10.1080/10408398.2017.1327418.

Yan, J.K., Wang, Y.Y., Qiu, W.Y., Shao, N., 2017. Three-phase partitioning for efficient extraction and separation of polysaccharides from *Corbicula fluminea*. Carbohydr. Polym. 163, 10–19. https://doi.org/10.1016/j.carbpol.2017.01.021.

Luque-García, J.L., Luque De Castro, M.D., 2004. Focused microwave-assisted Soxhlet extraction: devices and applications. Talanta 64, 571–577. https://doi.org/10.1016/j.talanta.2004.03.054.

Kroner, K., Hustedt, H., Kula, M.R., 1984. Extractive enzyme recovery: economic considerations. Process. Biochem. 19, 170–179.

Makino, K., Mossoba, M.M., Riesz, P., 1983. Chemical effects of ultrasound on aqueous solutions. Formation of hydroxyl radicals and hydrogen atoms. J. Phys. Chem. 87, 1369–1377. https://doi.org/10.1021/j100231a020.

Mondal, K., Sharma, A., Lata, M.N. Gupta, 2003a. Macroaffinity ligand-facilitated three-phase partitioning (MLFTPP) of α-amylases using a modified alginate. Biotechnol. Prog. 19, 493–494. https://doi.org/10.1021/bp025619e.

Mondal, K., Sharma, A., Gupta, M.N., 2003b. Macroaffinity ligand-facilitated three-phase partitioning for purification of glucoamylase and pullulanase using alginate. Protein Expr. Purif. 28, 190–195. https://doi.org/10.1016/S1046-5928(02)00673-3.

Mondal, K., Sharma, A., Gupta, M.N., 2004. Three phase partitioning of starch and its structural consequences. Carbohydr. Polym. 56, 355–359. https://doi.org/10.1016/j.carbpol.2004.03.004.

Mondal, K., Jain, S., Teotia, S., Gupta, M.N., 2006. Emerging options in protein bioseparation. Biotechnol. Annu. Rev., 1–29. https://doi.org/10.1016/S1387-2656(06)12001-3.

Mulchandani, K., Kar, J.R., Singhal, R.S., 2015. Extraction of lipids from *Chlorella saccharophila* using high-pressure homogenization followed by three phase partitioning. Appl. Biochem. Biotechnol. 176, 1613–1626. https://doi.org/10.1007/s12010-015-1665-4.

Padmapriya, K., Dutta, A., Chaudhuri, S., Dutta, D., 2012. Microwave assisted extraction of mangiferin from *Curcuma amada*. 3 Biotech 2, 27–30. https://doi.org/10.1007/s13205-011-0023-7.

Sandra, K., Ashokkumar, M., 2011. The Physical and Chemical Effects of Ultrasound. Ultrasound Technol. Food Bioprocessing, H. Feng Al., 65–105. https://doi.org/10.1007/978-1-4419-7472-3.
Talley, K., Alexov, E., 2010. On the pH-optimum of activity and stability of proteins. Proteins Struct. Funct. Bioinforma. 78, 2699–2706. https://doi.org/10.1002/prot.22786.
Raner, K.D., Strauss, C.R., Vyskoc, F., Mokbel, L., 1993. A Comparison of Reaction Kinetics Observed under Microwave Irradiation and Conventional Heating. J. Org. Chem. 58, 950–953. https://doi.org/10.1021/jo00056a031.
Suslick, K.S., Price, G.J., 1999. Applications of ultrasound to materials chemistry. Annu. Rev. Mater. Sci. 29, 295–326.
Suslick, K.S., Eddingsaas, N.C., Flannigan, D.J., Hopkins, S.D., Xu, H., 2011. Extreme conditions during multibubble cavitation: sonoluminescence as a spectroscopic probe. Ultrason. Sonochem. 18, 842–846. https://doi.org/10.1016/j.ultsonch.2010.12.012.
Suslick, K.S., 1989. The Chemical Effects of Ultrasound. Sci. Am. 260, 80–86. https://doi.org/10.1038/scientificamerican0289-80.
Suslick, K.S., Didenko, Y., Fang, M.M., Taeghwan Hyeon, K.J.K.M.M.M. and M.W. William B. McNamara III, 1999. Acoustic cavitation and its chemical consequences. Phil. Trans. R. Soc. Lond. A. 357, 335–353.
Chen, L., Song, D., Tian, Y., Ding, L., Yu, A., Zhang, H., 2008. Application of on-line microwave sample-preparation techniques. TrAC - Trends Anal. Chem. 27, 151–159. https://doi.org/10.1016/j.trac.2008.01.003.
Derguine-Mecheri, L., Kebbouche-Gana, S., Khemili-Talbi, S., Djenane, D., 2018. Screening and biosurfactant/bioemulsifier production from a high-salt-tolerant halophilic *Cryptococcus* strain YLF isolated from crude oil. J. Pet. Sci. Eng., 712–724. https://doi.org/10.1016/j.petrol.2017.10.088.
Jiang, L., Hua, D., Wang, Z., Xu, S., 2010. Aqueous enzymatic extraction of peanut oil and protein hydrolysates. Food Bioprod. Process. 88, 233–238. https://doi.org/10.1016/j.fbp.2009.08.002.
Saxena, L., Iyer, B.K., Ananthanarayan, L., 2007. Three phase partitioning as a novel method for purification of ragi (*Eleusine coracana*) bifunctional amylase/protease inhibitor. Process Biochem. 42, 491–495. https://doi.org/10.1016/j.procbio.2006.09.016.
Wang, L., Weller, C.L., 2006. Recent advances in extraction of nutraceuticals from plants. Trends Food Sci. Technol. 17, 300–312. https://doi.org/10.1016/j.tifs.2005.12.004.
Ashokkumar, M., 2011. The characterization of acoustic cavitation bubbles - An overview. Ultrason. Sonochem. 18, 864–872. https://doi.org/10.1016/j.ultsonch.2010.11.016.
Gagaoua, M., Hafid, K., Hoggas, N., 2016. Data in support of three phase partitioning of zingibain, a milk-clotting enzyme from *Zingiber officinale* Roscoe rhizomes. Data Br 6, 634–639. https://doi.org/10.1016/j.dib.2016.01.014.
Gagaoua, M., Hafid, K., 2016. Three phase partitioning system, an emerging non-chromatographic tool for proteolytic enzymes recovery and purification. Biosens. J. 5, 1–4. https://doi.org/10.4172/2090-4967.1000134.
Gagaoua, M., Boucherba, N., Bouanane-Darenfed, A., Ziane, F., Nait-Rabah, S., Hafid, K., Boudechicha, H.R., 2014. Three-phase partitioning as an efficient method for the purification and recovery of ficin from Mediterranean fig (*Ficus carica* L.) latex. Sep. Purif. Technol. 132, 461–467. https://doi.org/10.1016/j.seppur.2014.05.050.
Gagaoua, M., Hoggas, N., Hafid, K., 2015. Three phase partitioning of zingibain, a milk-clotting enzyme from *Zingiber officinale* Roscoe rhizomes. Int. J. Biol. Macromol. 73, 245–252. https://doi.org/10.1016/j.ijbiomac.2014.10.069.
Letellier, M., Budzinski, H., 1999. Microwave assisted extraction of organic compounds. Analusis 27, 259–271.
Puri, M., Sharma, D., Barrow, C.J., 2012. Enzyme-assisted extraction of bioactives from plants. Trends Biotechnol. 30, 37–44. https://doi.org/10.1016/j.tibtech.2011.06.014.
Kula, M.-.R, Kroner, K.H., Hustedt, H., Schütte, H., 1981. Technical aspects of extractive enzyme purification. Ann. N.Y. Acad. Sci. 369, 341–354. https://doi.org/10.1111/j.1749-6632.1981.tb14201.x.
Sardar, M., Sharma, A., Gupta, M.N., 2007. Refolding of a denatured α-chymotrypsin and its smart bioconjugate by three-phase partitioning. Biocatal. Biotransformation. 25, 92–97. https://doi.org/10.1080/10242420601050914.
Talebi, M., Ghassempour, A., Talebpour, Z., Rassouli, A., Dolatyari, L., 2004. Optimization of the extraction of paclitaxel from *Taxus baccata* L. by the use of microwave energy. J. Sep. Sci. 27, 1130–1136. https://doi.org/10.1002/jssc.200401754.

Toma, M., Fukutomi, S., Asakura, Y., Koda, S., 2011. A calorimetric study of energy conversion efficiency of a sonochemical reactor at 500kHz for organic solvents. Ultrason. Sonochem. 18, 197–208. https://doi.org/10.1016/j.ultsonch.2010.05.005.

Vinatoru, M., 2001. An overview of the ultrasonically assisted extraction of bioactive principles from herbs. Ultrason. Sonochem. 8, 303–313. https://doi.org/10.1016/S1350-4177(01)00071-2.

Margulis, M.A., Margulis, I.M., 2003. Calorimetric method for measurement of acoustic power absorbed in a volume of a liquid. Ultrason. Sonochem. 10, 343–345. https://doi.org/10.1016/S1350-4177(03)00100-7.

Veggi, M.A.A., Martinez, P.C., Meireles, J, 2012. Fundamentals of Microwave ExtractionMicrowave-Assisted Extr. Bioact. Compd. Bost. Springer, pp. 15–52.

Pol, M.C., Deutsch, H.F., Visser, L., 1990. Purification of soluble enzymes from erythrocyte hemolysates by three phase partitioning. Int. J. Biochem. 22, 179–185. https://doi.org/10.1016/0020-711X(90)90181-2.

Esclapez, M.D., García-Pérez, J.V., Mulet, A., Cárcel, J.A., 2011. Ultrasound-Assisted Extraction of Natural Products. Food Eng. Rev. 3, 108–120. https://doi.org/10.1007/s12393-011-9036-6.

Vetal, M.D., Shirpurkar, N.D., Rathod, V.K., 2014. Three phase partitioning coupled with ultrasoundfor the extraction of ursolic acid and oleanolic acidfrom *Ocimum sanctum*. Food Bioprod. Process. 92, 402–408. https://doi.org/10.1016/j.fbp.2013.09.002.

Peterson, M.E., Daniel, R.M., Danson, M.J., Eisenthal, R., 2007. The dependence of enzyme activity on temperature: determination and validation of parameters. Biochem. J. 402, 331–337. https://doi.org/10.1042/BJ20061143.

Freire, M.G., Cláudio, A.F.M., Araújo, J.M.M., Coutinho, J.A.P., Marrucho, I.M., Canongia Lopes, J.N., Rebelo, L.P.N., 2012. Aqueous biphasic systems: a boost brought about by using ionic liquids. Chem. Soc. Rev. 41, 4966–4995. https://doi.org/10.1039/c2cs35151j.

Chivate, M.M., Pandit, A.B., 1995. Quantification of cavitation intensity in fluid bulk. Ultrason. - Sonochemistry. 2, 19–25. https://doi.org/10.1016/1350-4177(94)00007-F.

Sonar, M.P., Rathod, V.K., 2020. Extraction of type II antidiabetic compound Corosolic acid from *Lagerstroemia speciosa* by batch extraction and three phase partitioning. Biocatal. Agric. Biotechnol. In press. https://doi.org/10.1016/j.bcab.2020.101694.

Ladole, M.R., Mevada, J.S., Pandit, A.B., 2017. Ultrasonic hyperactivation of cellulase immobilized on magnetic nanoparticles. Bioresour. Technol. 239, 117–126. https://doi.org/10.1016/j.biortech.2017.04.096.

Ladole, M.R., Nair, R.R., Bhutada, Y.D., Amritkar, V.D., Pandit, A.B., 2018. Synergistic effect of ultrasonication and co-immobilized enzymes on tomato peels for lycopene extraction. Ultrason. Sonochem. 48, 453–462. https://doi.org/10.1016/j.ultsonch.2018.06.013.

Babbar, N., Van Roy, S., Wijnants, M., Dejonghe, W., Caligiani, A., Sforza, S., Elst, K., 2016. Effect of extraction conditions on the saccharide (neutral and acidic) composition of the crude pectic extract from various agro-industrial residues. J. Agric. Food Chem. 64, 268–276. https://doi.org/10.1021/acs.jafc.5b04394.

Dogan, N., Tari, C., 2008. Characterization of three-phase partitioned exo-polygalacturonase from *Aspergillus sojae* with unique properties. Biochem. Eng. J. 39, 43–50. https://doi.org/10.1016/j.bej.2007.08.008.

Kurmudle, N.N., Bankar, S.B., Bajaj, I.B., Bule, M.V., Singhal, R.S., 2011. Enzyme-assisted three phase partitioning: a novel approach for extraction of turmeric oleoresin. Process Biochem. 46, 423–426. https://doi.org/10.1016/j.procbio.2010.09.010.

Chaiwut, P., Pintathong, P., Rawdkuen, S., 2010. Extraction and three-phase partitioning behavior of proteases from papaya peels. Process Biochem. 45, 1172–1175. https://doi.org/10.1016/j.procbio.2010.03.019.

Fernandes, P., Carvalho, F., 2017. Microbial enzymes for the food industryBiotechnol. Microb. Enzym. Prod. Biocatal. Ind. Appl. Elsevier Inc., pp. 513–544. https://doi.org/10.1016/B978-0-12-803725-6.00019-4.

Tatke, P., Jaiswal, Y., 2011. An overview of microwave assisted extraction and its applications in herbal drug research. Res. J. Med. Plant. 5, 21–31. https://doi.org/10.3923/rjmp.2011.21.31.

Munde, P.J., Muley, A.B., Ladole, M.R., Pawar, A.V., Talib, M.I., Parate, V.R., 2017. Optimization of pectinase-assisted and tri-solvent-mediated extraction and recovery of lycopene from waste tomato peels. 3 Biotech 7, 206. https://doi.org/10.1007/s13205-017-0825-3.

Gogate, P.R., Pandit, A.B., 2004. Sonochemical reactors: scale up aspects. Ultrason. Sonochem. 11, 105–117. https://doi.org/10.1016/j.ultsonch.2004.01.005.

Gogate, P.R., Kabadi, A.M., 2009. A review of applications of cavitation in biochemical engineering/biotechnology. Biochem. Eng. J. 44, 60–72. https://doi.org/10.1016/j.bej.2008.10.006.

Rao, P.R., Rathod, V.K., 2019. Microwave assisted three phase extraction of andrographolide from *Andrographis paniculata*. J. Biol. Act. Prod. from Nat. 9, 215–226. https://doi.org/10.1080/22311866.2019.1641430.

Rao, P.R., Rathod, V.K., 2015. Rapid extraction of andrographolide from *Andrographis paniculata* Nees by three phase partitioning and determination of its antioxidant activity. Biocatal. Agric. Biotechnol. 4, 586–593. https://doi.org/10.1016/j.bcab.2015.08.016.

Dutta, R., Sarkar, U., Mukherjee, A., 2015. Process optimization for the extraction of oil from *Crotalaria juncea* using three phase partitioning. Ind. Crops Prod. 71, 89–96. https://doi.org/10.1016/j.indcrop.2015.03.024.

Garg, R., Thorat, B.N., 2014. Nattokinase purification by three phase partitioning and impact of t-butanol on freeze drying. Sep. Purif. Technol. 131, 19–26. https://doi.org/10.1016/j.seppur.2014.04.011.

Gaur, R., Sharma, A., Khare, S.K., Gupta, M.N., 2007. A novel process for extraction of edible oils. Enzyme assisted three phase partitioning (EATPP). Bioresour. Technol. 98, 696–699. https://doi.org/10.1016/j.biortech.2006.01.023.

Halim, R., Gladman, B., Danquah, M.K., Webley, P.A., 2011. Oil extraction from microalgae for biodiesel production. Bioresour. Technol. 102, 178–185. https://doi.org/10.1016/j.biortech.2010.06.136.

Li, R., Zhang, S., Kou, X., Ling, B., Wang, S., 2017. Dielectric properties of almond kernels associated with radio frequency and microwave pasteurization. Sci. Rep. 7, 1–10. https://doi.org/10.1038/srep42452.

Lovrein, R., Goldensoph, C.P.C. Anderson, O. B, 1987. Three phase partitioning (TPP) via t-butanol: enzymes separations from crudes. In: Burgess, R. (Ed.), Protein Purif. Micro to Macro. A.R. Liss Inc New York, pp. 131–148.

Moreau, R.A., Dickey, L.C., Johnston, D.B., Hicks, K.B., 2009. A process for the aqueous enzymatic extraction of corn oil from dry milled corn germ and enzymatic wet milled corn germ (E-Germ). J. Am. Oil Chem. Soc. 86, 469–474. https://doi.org/10.1007/s11746-009-1363-x.

Singh, R.K., Gourinath, S., Sharma, V., Roy, I., Gupta, M.N., Betzel, C., Srinivasan, A., Singh, T.P., 2001. Enhancement of enzyme activity through three-phase partitioning: crystal structure of a modified serine proteinase at 1.5 Å resolution. Protein. Eng. 14, 307–313. https://doi.org/10.1093/protein/14.5.307.

Wati, R.K., Theppakorn, T., Benjakul, S., Rawdkuen, S., 2009. Three-phase partitioning of trypsin inhibitor from legume seeds. Process Biochem. 44, 1307–1314. https://doi.org/10.1016/j.procbio.2009.07.002.

Pike, R.N., Dennison, C., 1989. Protein fractionation by three phase partitioning (TPP) in aqueous/t-butanol mixtures. Biotechnol. Bioeng. 33, 221–228. https://doi.org/10.1002/bit.260330213.

Ratoarinoro, J.B., Contamine, H.D .F, Wilhelm*, A.M., 1995. Power measurement in sonochemistry. Ultrason. Sonochem. 2, S43–S47. https://doi.org/10.1016/0969-7012(96)00002-0.

Chemat, S., Lagha, A., AitAmar, H., Bartels, P.V., Chemat, F., 2004. Comparison of conventional and ultrasound-assissted extraction of carvone and limonene from caraway seeds. Flavour Fragr. J. 19, 188–195. https://doi.org/10.1002/ffj.1339.

Dey, S., Rathod, V.K., 2013. Ultrasound assisted extraction of β-carotene from *Spirulina platensis*. Ultrason. Sonochem. 20, 271–276. https://doi.org/10.1016/j.ultsonch.2012.05.010.

Gautam, S., Mukherjee, J., Roy, I., Gupta, M.N., 2012. Emerging trends in designing short and efficient protein purification protocols. Am. J. Biochem. Biotechnol. 8, 230–254. https://doi.org/10.3844/ajbbsp.2012.230.254.

Gautam, S., Dubey, P., Singh, P., Varadarajan, R., Gupta, M.N., 2012. Simultaneous refolding and purification of recombinant proteins by macro-(affinity ligand) facilitated three-phase partitioning. Anal. Biochem. 430, 56–64. https://doi.org/10.1016/j.ab.2012.07.028.

Krishna, S.H., Karanth, N.G., 2002. Lipases and lipase-catalyzed esterification reactions in nonaqueous media. Catal. Rev. - Sci. Eng. 44, 499–591. https://doi.org/10.1081/CR-120015481.

Jain, S., Singh, R., Gupta, M.N., 2004. Purification of recombinant green fluorescent protein by three-phase partitioning. J. Chromatogr. A. 1035, 83–86. https://doi.org/10.1016/j.chroma.2004.01.007.

Kebbouche-Gana, S., Gana, M.L., Khemili, S., Fazouane-Naimi, F., Bouanane, N.A., Penninckx, M., Hacene, H., 2009. Isolation and characterization of halophilic *Archaea* able to produce biosurfactants. J. Ind. Microbiol. Biotechnol. 36, 727–738. https://doi.org/10.1007/s10295-009-0545-8.

Khemili-Talbi, S., Kebbouche-Gana, S., Akmoussi-Toumi, S., Angar, Y., Gana, M.L., 2015. Isolation of an extremely halophilic arhaeon *Natrialba sp.* C21 able to degrade aromatic compounds and to produce stable biosurfactant at high salinity. Extremophiles 19, 1109–1120. https://doi.org/10.1007/s00792-015-0783-9.

Mandal, S., Mohan, V., Hemalatha, Y., 2007. Microwave assisted extraction - An innovative and promising extraction tool for medicinal plant research. Pharmacogn. Rev. 1, 7–18.

Raghava, S., Barua, B., Singh, P.K., Das, M., Madan, L., Bhattacharyya, S., Bajaj, K., Gopal, B., Varadarajan, R., Gupta, M.N., 2008. Refolding and simultaneous purification by three-phase partitioning of recombinant proteins from inclusion bodies. Protein Sci. 17, 1987–1997. https://doi.org/10.1110/ps.036939.108.

Raghava, S., Gupta, M.N., 2009. Tuning permeabilization of microbial cells by three-phase partitioning. Anal. Biochem. 385, 20–25. https://doi.org/10.1016/j.ab.2008.10.013.

Rajeeva, S., Lele, S.S., 2011. Three-phase partitioning for concentration and purification of laccase produced by submerged cultures of Ganoderma sp. WR-1. Biochem. Eng. J. 54, 103–110. https://doi.org/10.1016/j.bej.2011.02.006.

Rawdkuen, S., Vanabun, A., Benjakul, S., 2012. Recovery of proteases from the viscera of farmed giant catfish (*Pangasianodon gigas*) by three-phase partitioning. Process Biochem. 47, 2566–2569. https://doi.org/10.1016/j.procbio.2012.09.001.

Rawdkuen, S., Chaiwut, P., Pintathong, P., Benjakul, S., 2010. Three-phase partitioning of protease from *Calotropis procera* latex. Biochem. Eng. J. 50, 145–149. https://doi.org/10.1016/j.bej.2010.04.007.

Shah, S., Sharma, A., Gupta, M.N., 2004. Extraction of oil from *Jatropha curcas* L. seed kernels by enzyme assisted three phase partitioning. Ind. Crops Prod. 20, 275–279. https://doi.org/10.1016/j.indcrop.2003.10.010.

Sharma, S., Gupta, M.N., 2001. Purification of phospholipase D from *Dacus carota* by three-phase partitioning and its characterization. Protein Expr. Purif. 21, 310–316. https://doi.org/10.1006/prep.2000.1357.

Pakhale, S.V., Bhagwat, S.S., 2016. Purification of serratiopeptidase from *Serratia marcescens* NRRL B 23112 using ultrasound assisted three phase partitioning. Ultrason. Sonochem. 31, 532–538. https://doi.org/10.1016/j.ultsonch.2016.01.037.

Varakumar, S., Umesh, K.V., Singhal, R.S., 2017. Enhanced extraction of oleoresin from ginger (*Zingiber officinale*) rhizome powder using enzyme-assisted three phase partitioning. Food Chem. 216, 27–36. https://doi.org/10.1016/j.foodchem.2016.07.180.

Bagade, S.B., Patil, M., 2019. Recent advances in microwave assisted extraction of bioactive compounds from complex herbal samples: a review. Crit. Rev. Anal. Chem. 0, 1–12. https://doi.org/10.1080/10408347.2019.1686966.

Zhang, S.B., Wang, Z., Xu, S.Y., 2007. Optimization of the aqueous enzymatic extraction of rapeseed oil and protein hydrolysates. J. Am. Oil Chem. Soc. 84, 97–105. https://doi.org/10.1007/s11746-006-1004-6.

Marathe, S.J., Jadhav, S.B., Bankar, S.B., Singhal, R.S., 2017. Enzyme-assited extraction of bioactivesPuri M. Food Bioact. Springer, Cham, pp. 171–201. https://doi.org/10.1007/978-3-319-51639-4.

Harde, S.M., Singhal, R.S., 2012. Extraction of forskolin from *Coleus forskohlii* roots using three phase partitioning. Sep. Purif. Technol. 96, 20–25. https://doi.org/10.1016/j.seppur.2012.05.017.

Cameotra, S.S., Makkar, R.S., 1998. Synthesis of biosurfactants in extreme conditions. Appl. Microbiol. Biotechnol. 50, 520–529. https://doi.org/10.1007/s002530051329.

Nadar, S.S., Rao, P., Rathod, V.K., 2018. Enzyme assisted extraction of biomolecules as an approach to novel extraction technology: a review. Food Res. Int. 108, 309–330. https://doi.org/10.1016/j.foodres.2018.03.006.

Patil, S.S., Bhasarkar, S., Rathod, V.K., 2019. Extraction of curcuminoids from *Curcuma longa*: comparative study between batch extraction and novel three phase partitioning. Prep. Biochem. Biotechnol. 49, 407–418. https://doi.org/10.1080/10826068.2019.1575859.

Patil, S.S., Rathod, V.K., 2020. Synergistic Effect of Ultrasound and Three Phase Partitioning for the Extraction of Curcuminoids from *Curcuma longa* and its Bioactivity Profile. Process Biochem. 93, 85–93. https://doi.org/10.1016/j.procbio.2020.02.031.

Leong, T., Ashokkumar, M., Sandra, K., 2011. The fundamentals of power ultrasound - A review. Acoust. Aust. 39, 54–63.

Van Gerven, T., Stankiewicz, A., 2009. Structure, energy, synergy, time: the fundamentals of process intensification. Ind. Eng. Chem. Res. 48, 2465–2474. https://doi.org/10.1021/ie801501y.

Zou, T., Wu, H., Li, H., Jia, Q., Song, G., 2013. Comparison of microwave-assisted and conventional extraction of mangiferin from mango (*Mangifera indica* L.) leaves. J. Sep. Sci. 36, 3457–3462. https://doi.org/10.1002/jssc.201300518.

Mason, T.J., Cobley, A.J., Graves, J.E., Morgan, D., 2011. New evidence for the inverse dependence of mechanical and chemical effects on the frequency of ultrasound. Ultrason. Sonochem. 18, 226–230. https://doi.org/10.1016/j.ultsonch.2010.05.008.

Mason, T.J., Lorimer, J.P., 2002. General Principles. In: Mason, T.J., Lorimer, J.P. (Eds.), General Principles. Appl. Sonochemistry– Uses Power Ultrasound Chem. Process., 25–74. https://doi.org/10.1002/jctb.957.

Daniela Bermúdez-Aguirre, T.M., Barbosa-Cánovas, G.V., 2011. Ultrasound Technologies for Food Bioprocessing. In: Feng, H. (Ed.), Ultrasound Technologies for Food Bioprocessing. Ultrasound Technol. Food Bioprocessing, 65–105. Al. (Eds.). https://doi.org/10.1017/CBO9781107415324.004.

Przybycien, T.M., Pujar, N.S., Steele, L.M., 2004. Alternative bioseparation operations: life beyond packed-bed chromatography. Curr. Opin. Biotechnol. 15, 469–478. https://doi.org/10.1016/j.copbio.2004.08.008.

Rathod, V.K., Rao, P., 2015. Effect of Three phase extraction with ultrasound on recovery and antioxidant activity of *Andrographis paniculata*. J. Biol. Act. Prod. from Nat. 5, 264–275 http://www.tandfonline.com/doi/abs/10.1080/22311866.2015.1102081.

Kulkarni, V.M., Rathod, V.K., 2015. A novel method to augment extraction of mangiferin by application of microwave on three phase partitioning. Biotechnol. Reports. 6, 8–12. https://doi.org/10.1016/j.btre.2014.12.009.

Kulkarni, V.M., Rathod, V.K., 2014. Extraction of mangiferin from *Mangifera indica* leaves using three phase partitioning coupled with ultrasound. Ind. Crops Prod. 52, 292–297. https://doi.org/10.1016/j.indcrop.2013.10.032.

Kulkarni, V.M., Rathod, V.K., 2016. Utilization of waste dried *Mangifera indica* leaves for extraction of mangiferin by conventional batch extraction and advance three-phase partitioning. Green Process. Synth. 5, 79–85. https://doi.org/10.1515/gps-2015-0090.

Lamb, W.E., Retherford, R.C., 1947. Fine structure of the hydrogen atom by a microwave method. Phys. Rev. 72, 241–243. https://doi.org/10.1103/PhysRev.72.241.

Wei, X., Chen, M., Xiao, J., Liu, Y., Yu, L., Zhang, H., Wang, Y., 2010. Composition and bioactivity of tea flower polysaccharides obtained by different methods. Carbohydr. Polym. 79, 418–422. https://doi.org/10.1016/j.carbpol.2009.08.030.

Tian, Y., Demirel, S.E., Hasan, M.M.F., Pistikopoulos, E.N., 2018. An overview of process systems engineering approaches for process intensification: state of the art. Chem. Eng. Process. - Process Intensif. 133, 160–210. https://doi.org/10.1016/j.cep.2018.07.014.

Wang, Y.Y., Ma, H., Yan, J.K., Di Wang, K., Yang, Y., Wang, W.H., Zhang, H.N., 2019. Three-phase partitioning system with dimethyl carbonate as organic phase for partitioning of exopolysaccharides from *Phellinus baumii*. Int. J. Biol. Macromol. 131, 941–948. https://doi.org/10.1016/j.ijbiomac.2019.03.149.

Li, Z., Jiang, F., Li, Y., Zhang, X., Vidhate, T., 2013. Simultaneously concentrating and pretreating of microalgae *Chlorella spp.* by three-phase partitioning. Bioresour. Technol. 149, 286–291. https://doi.org/10.1016/j.biortech.2013.08.156.

Tan, Z.J., Wang, C.Y., Yi, Y.J., Wang, H.Y., Zhou, W.L., Tan, S.Y., Li, F.F., 2015. Three phase partitioning for simultaneous purification of aloe polysaccharide and protein using a single-step extraction. Process Biochem. 50, 482–486. https://doi.org/10.1016/j.procbio.2015.01.004.

Index

Page numbers followed by "*f* ", "*t*" and "*b*" indicate, figures, tables and boxes respectively.